T0094284

Replayed

Replayed

Essential Writings on Software Preservation and Game Histories

Henry Lowood

Edited by Raiford Guins
With a Foreword by Matthew G. Kirschenbaum
and an Interview by T. L. Taylor

JOHNS HOPKINS UNIVERSITY PRESS BALTIMORE

Printed in the United States of America on acid-free paper
9 8 7 6 5 4 3 2 1

Johns Hopkins University Press
2715 North Charles Street
Baltimore, Maryland 21218
www.press.jhu.edu

Library of Congress Cataloging-in-Publication Data

Names: Lowood, Henry, author. | Guins, Raiford, editor.
Title: Replayed : essential writings on software preservation and game histories / Henry E. Lowood ; Edited by Raiford Guins With a Foreword by Matthew G. Kirschenbaum and an Interview by T. L. Taylor.
Description: Baltimore : Johns Hopkins University Press, [2023] | Includes bibliographical references and index.
Identifiers: LCCN 2022031405 | ISBN 9781421445946 (hardcover) | ISBN 9781421445953 (ebook)
Subjects: LCSH: Video games—Design—History. | Computer games—Design—History. | Digital preservation. | Computer software—History. | BISAC: TECHNOLOGY & ENGINEERING / History | HISTORY / Essays
Classification: LCC GV1469.3 .L73 2023 | DDC 794.8—dc23/eng/20230214
LC record available at https://lccn.loc.gov/2022031405

A catalog record for this book is available from the British Library.

Special discounts are available for bulk purchases of this book. For more information, please contact Special Sales at specialsales@jh.edu.

To Heidi, Karl, and Paul

Contents

Foreword
The Hard Work of Henry Lowood

Let's take a peek into the Google Drive folder Raiford Guins set up for the small group of us involved in putting together this book. We needed a place to collect and share all of Henry's writing that would be going into the volume, harvested as it was from different sources. At a glance, what's on the screen is twenty or so thumbnails tiled in orderly rows and columns, some of them accented with brightly colored images (thumbnails within the thumbnails), while others reveal grayish squiggles of text, line after line in a microfont that's actually readable if you get your nose close enough to the screen. Some of the thumbnails are wrapped in the red livery of an Adobe PDF document; the other half sport the bright blue of Microsoft Word. (Coincidence, but a neat reminder of Henry's writing on the *Red vs. Blue* machinima.) More to the point, it is immediately apparent that these documents represent different stages in the life cycle of publications. Some are self-evidently proofs: facsimiles, complete with formatting and typesetting, of how a piece would eventually appear in print. Some, by contrast, are what still look like drafts, the thumbnail exposing ragged right margins and workaday double-spacing. At least one document is a download from the Web, its thumbnail trailing spaghetti entrails of dropped CSS callouts to style sheets it will never be able to locate again.

This little scene of document work is, of course, entirely unremarkable. There are countless other Google Drive folders just like it, used every day for equally mundane purposes. Everything about what I have just described is so ordinary that it can be hard to really see it—we pause only long enough to select the file we're looking for and then double-click to get to the text to either read on the screen or dump to a printer. All the more remarkable, then, that Henry Lowood did see this—a scene of documentary work very much like it—some twenty years ago. In an essay entitled "The Hard Work of Software History" (chapter 1 in this volume), Lowood describes an encounter at a computer expo with a journalist who tells him he has just written and sold an article about the event to a magazine. The article was composed on a laptop, illustrated with photos from a digital camera, emailed to an editor, and then published (exclusively) online. Nothing remarkable about any of that either, except that it all happened before the turn of the millennium. "Every trace of these activities," Lowood noted in his own essay, "has been born digital."

This was the first piece of writing by Henry Lowood I ever encountered. I was doing research for my book *Mechanisms*, dutifully scanning for what archivists and curators were saying about so-called born-digital media—word processing files to be sure, but also more complex kinds of digital objects like games or interactive fiction. Not just documents, in other words, but software, the software that is in the title of the essay.

But software wasn't always what we thought it was—this is something Lowood clarified for me early on. Software is as much a legalistic construct as it is anything else, specifically the offshoot of a 1969 decision when IBM, then under threat of antitrust legislation, agreed that it would no longer "bundle" computer programs with the hardware the company sold. As a result, software came into its own, first as a legal entity, then as a branch of computer engineering, then as a saleable object—a product, like the stuff in shrink-wrapped boxes that eventually found its way into my own adolescent hands when I got my first computer in the 1980s—and finally, as an object for curation, a cultural and historical artifact.

But even that is too easy. Software is strange stuff, and it can seem like no one really knows exactly what it is or, even more to the point, where it is. (Where, exactly, is "your" "copy" of Microsoft Office 365? The scare quotes expose the vagaries bedeviling any potential answer.) Lowood opens his essay by quoting the German media theorist Friedrich Kittler, who famously proclaimed, "There is no software." What Kittler meant is that software was ultimately an abstraction, code on top of code, layers of signals and signifiers n-dimensions deep until finally, inevitably, we crash on the cold, hard circuitry of the machine. If so, what is there left for the historian to preserve, other than maybe a hunk of metal?

Lowood had some answers, which I leave for you to discover in the pages that follow. The essay stayed in my source repertoire for a number of years after. But I also wanted to meet this guy: the software archivist who could sling a citation from Kittler.

And now a non sequitur, but bear with me. I play tabletop wargames. Wargames are painstakingly crafted models and simulations of battles and campaigns, typically fought on maps overlain with a hexagonal grid that mirrors the real-world terrain. Units take the form of small cardboard chits (or "counters": you will also read more about them in these pages) that, depending on the topic and scale of the game, can represent anything from a single squad or vehicle to a full armored division.

Anyway, one day not so long after reading Henry on software history for the first time, I was leafing through an old issue of one of the hobby's review organs, a magazine called *Paper Wars*. There was a long, measured piece on a game entitled *Enemy at the Gates*, which was a study of the 1942–43 winter operations in and around a city called Stalingrad (like the Jude Law movie, the game takes its title from William Craig's classic work of military history). As befitting its subject, *Enemy at the Gates* was huge: four maps providing a roughly 4 × 6-foot playing area, two 36-page rulebooks, multiple booklets of charts and tables, and over two thousand of the aforementioned counters. "This is a difficult game to review," the author observed. He meant that not because of the game's physical complexity alone, but because the evident seriousness of the simulation demanded an equally serious response from any prospective reviewer: historical accuracy, the workings of the game's internal systems, the components, and yes, even fun all had to weigh in the balance. The author of the piece, who was respectful but sharply critical? Henry Lowood.

Now I really wanted to meet this guy.

The review was followed by an equally vigorous response from the game's designer, Dean Essig. Reading the two in tandem was its own titanic clash—if not quite the Eastern Front, then at least two fierce intellects coming to grips over an object of play (made out of paper and cardboard, no less) with the utmost earnestness and conviction. The editor, Bill Gibbs, would later remark that this exchange (it can be found in the December 1994 issue) helped establish the reputation of the magazine.

From this I learned several things. First, Henry Lowood takes his games seriously, both as a player and as a scholar. In ways that seem obvious to us now, he understood games as a major form of cultural expression, worthy of real analysis and critique. Second, he understood the deep entanglement of paper (or tabletop) and the screen, something game studies people are only now coming to fully appreciate as scholars like Mary Flanagan and Jon Peterson are revising our understanding of the history of computer games.

Above all, though, Henry's interest in the game struck me as wholly compatible with his interests in software. Wargames like *Enemy at the Gates* are some of the most complex ludic entities ever devised—certainly in the scope and scale of the components, but also in the interactions of all the different systems and subsystems. Moving a unit across the map is not just a function of lifting the cardboard chit—one must take into account the terrain it is crossing, the presence of other units (both friendly and enemy), its supply status (does it have enough gas?), its formation (is it deployed for battle or strung out in road column?), the potential for aerial interdiction, and more. The point is these games function very much like software, or paper computers. Yet unlike a computer program where the code is often proprietary or black-boxed, here all of the source code must be visible and available. Thus the massive rulebooks: the player in essence "runs" the software every step of the way, unit after unit, move after move, combat after combat, each interaction altering the overall state of the game system. All tabletop games—but wargames especially in my view—promote the kind of mindset needed to understand how software, with its nests of procedure and dependencies, really works.

Lowood, of course, appreciated all this from the get-go, as pieces like "War Engines" (chapter 12) in this volume make plain. (Improbably at first glance, he finds the roots of today's computer game engines in just such bellicose analog systems as I have been describing.) And in a few years' time we would indeed get to meet and get to know one another, chiefly through our collaboration on the Preserving Virtual Worlds project funded first by the Library of Congress and then the Institute of Museum and Library Services. Since then, we have collaborated in other ways and, yes, even matched forces in a wargame (where I got crushed).

The very first thing you learn about Henry when you work with him is also right there in the title of that first essay: you *work* with him, and he works very hard. I don't just mean that in the vaguely paternalistic sense of his being a hard worker. I mean that he understands that preservation, like writing, is labor. Sometimes this is manual labor, as when you have to load the boxes yourself, or root around in a warehouse for just the right object (not for nothing do job descriptions for archivists routinely call on them to be able to lift forty pounds overhead). But there are other kinds of labor: the endless cycle of grants and reporting, for example; the networking and outreach; the research into arcane hardware or devices; the carefully composed finding aid; the ever-present financial constraints; and so on. With such labor also comes an awareness and appreciation of institutions, and of the people who inhabit them—always evident in everything Henry does.

This volume is distinctive in that it gathers and showcases one particular form of Lowood's work, the intellectual labor of writing, thinking, and historicizing games and other kinds of software. It is hard work in its own right, as those rough-edged drafts I mentioned at the outset suggest. The pages that follow will be deceptively clean, but make no mistake—each one has been dipped in sweat of the brow, been held in the palm of the practitioner's hand, born the weight of the archivist's appraising eye, and been the beneficiary of the scholar's churning mind. Henry Lowood does nothing by half measure. As these essays show, the hard work of software history means labors of love are necessary but not sufficient.

Matthew G. Kirschenbaum
University of Maryland

Acknowledgments

Henry Lowood

It has been a pleasure to work once again with my friend and colleague Ray Guins. Not only has it been fun, but I am also deeply honored by his commitment to this book project. My appreciation and thanks also go to T. L. Taylor and Matt Kirschenbaum for their contributions. It would be difficult for me to imagine a trio of scholars for whom I have more respect, and here they all are!

My love and thanks go also to my sons Karl and Paul. I expect they will want to read this volume from cover to cover . . . many times over. And as always, my thanks for everything to Heidi.

Raiford Guins

I wish to thank Matt Kirschenbaum and T. L. Taylor for their amazing contributions to this book. Having colleagues who have worked closely with Henry over the years share their insights on his scholarship and professional practice makes this book all the richer.

This project was originally conceived with Henry while sharing a meal at a (fittingly) German biergarten in San Mateo, California, in 2019. Actually, it was a setup: I invited Henry, his wife, Heidi, and his son Paul to dinner to "drop the bomb" that I would be adding to Henry's workload somewhat substantially over the next few years should he agree to pencil in this project on his ever-daunting to-do list. Fortunately for us, he did! His willingness to revisit his own intellectual history in the form of crafting three original section introductions plus being the subject of a lengthy interview with T.L. along with his enduring patience with my constant barrage of questions is something all of us will greatly benefit from when reading these pages.

Henry took a "hands-off" approach by opting out of reviewing my editor's introduction until the production stage. I can only hope that I've been able to convey the importance of his writings on software preservation and game history in a manner that is both accessible and of interest to readers while living up to Lowood's high standards.

I wish to also thank journal editors and publishers for their assistance in compiling this volume. Information about original publication of each essay (with the exception of chapters 2 and 13, which appear in print for the first time) is given in a footnote on the first page of each chapter.

Replayed

Editor's Introduction

Henry's Many Hats

Raiford Guins

We pace on the edge, wondering where to cut in.

—Michael S. Mahoney, "The History of Computing in the History of Technology" (1988)

I will shift my historian's hat to one side of my head now and slide up my curator's cap.

—Henry Lowood, "It Is What It Is, Not What It Was" (2016)

"What is the history of games a history of?" That line may strike a familiar chord for historians of technology, resonating with Michael S. Mahoney's landmark article, "The History of Computing in the History of Technology," in which he considers how the history of computing has been written and by whom, what the object of study and its history entails, and "its parent discipline," the history of technology.[1] When Mahoney published his article in the *Annals of the History of Computing* in 1988, he addressed a readership largely consisting of "computer experts rather than trained historians," according to Thomas Haigh in his collection of Mahoney's works, *Histories of Computing*.[2] The article in question and historiographic works in its wake nevertheless made sizeable waves in the history of technology with Mahoney's postulation that "despite the pervasive presence of computing in modern science and technology, not to mention modern society itself, the history of computing has yet to establish a significant presence in the history of science and technology."[3]

Mahoney's perturbation was adapted to the historical study of games when Henry Lowood, then Curator for History of Science and Technology

Collections and Film and Media Studies at Stanford University Libraries, gave his keynote address at the first International Conference on the History of Games (Montreal, June 22–23, 2013). His brief, though prudent, conclusion to a talk devoted to the game engine and its history bore the heading, "A Future for Game History." Lowood, whose scholarly and professional work speaks to both game history and software preservation, summarized the works that Mahoney identified as constitutive of the history of computing in the late 1980s: insider histories establishing a series of firsts and factual events, journalist accounts with a penchant for the unusual and spectacular, social impact statements and social criticism coupled with a small body of professionally written historical studies. At this point, Lowood stopped pacing to cut in with his own imperative: "See if you recognize them in game history."[4] The historical study of games has its own versions of the works Mahoney once identified as prominent: writers who in the first decade of the 21st century produced many insider, enthusiast, generalist, descriptive histories of games, plotting firsts and chronicling events, with scholarly labor representing an even smaller concentration compared to Mahoney's survey of histories of computing.[5] Lowood's overarching intention by introducing Mahoney's argument into a forum of game historians—many of whom, at the time, were writing their dissertations, or inaugural articles and books—was to propose to those who would do the actual work of history writing that the "history of games is in a similar situation to the history of computing, only twenty-five years later and with a less clear notion of its natural parent discipline."[6] Add the time since Lowood's keynote and we find ourselves adjusting that number to nearly thirty-five years. Hence, we can fittingly modify Mahoney's charge accordingly: despite the pervasive presence of *gaming* in modern industry, education, and technology, not to mention modern society itself, the history of *games* has yet to establish a significant presence in the history of ________ (like *Mad Libs*, choose your noun).

Lowood advises that the history of games will remain a toddler until it starts to shed itself from the prevalent forms of history writing that have defined the subject. That is not to fault celebratory tomes or journalistic accounts of game history, but to call for an expansion of the types of histories being written to include more critical, interpretative, and contextual methods. Additionally, Lowood underscores the importance of Mahoney's claim that the history of computing will need to address the "big questions" that have occupied the thought of historians of technology: "How do new technologies establish themselves in society, and how does society adapt to them? To what extent and in what ways do societies engender new technologies? What are the patterns by which technology is transferred from one culture to another?"[7] Taking on his own "big question," Lowood asks: "What are the patterns by which games are transferred from one culture to another?" He presses the history of games to establish connections with "other areas of historical research and cultural studies, which in turn will invigorate our own work with fresh perspectives and interpretive frameworks."[8] In no way is Lowood advocating to collapse the history of games *into* the history of technology. He turns to the discipline—its methods, debates, and approaches—to consider how game historians might benefit from this perspective.

This is not the first time that Lowood, who long served as the bibliographer for the Society for the History of Technology's journal, *Technology and Culture*, has turned to his parent discipline to tackle challenges confronting the historical study of games. In another forum, the launch of the journal *Games and Culture* in 2006, Lowood asked readers of the anticipated premier issue to explore what the emerging field of game studies could learn from the established discipline and profession of the history of science (citing postwar historians of science at length).[9] Lowood would soon speak to the community that Mahoney once addressed by editing a themed issue of the *IEEE Annals of the History of Computing* twenty years later. Devoted to the history of computer games, Lowood's editorial insists that the "serious study of computing used for recreational or entertainment purposes such as games or virtual worlds" provides new opportunities "for critically engaging the history of computing."[10] It is worth stressing that Lowood's words are not employed to reunite a lost child (history of games) with its proper parent (in this instance, the history of computing). He seeks partnership with shared benefits instead. Members of the Society for the History of Technology would witness this fruitful enterprise firsthand when he co-organized "Videogame History as History of Technology?," a panel at the professional organization's 2015 annual conference in Albuquerque, New Mexico.

If Lowood has an agenda for the historical study of games, it is one of generosity, awareness, commitment to research, and intellectual inclusivity—asking fellow travelers to learn from other disciplines that have experienced similar crises of identity when ascertaining their own objects of study, methods, histories, disciplinary and professional status. Those invested in the history of games need not "go it alone"; precedents exist and ought to be learned from. The history of games' strength resides in the connections its proponents make with other fields of study. To help realize the importance of disciplinary connections in yet another foundational moment, Lowood asks in the launch title for our Game Histories book series, *Debugging Game History: A Critical Lexicon*: "How do we encourage new historical work on games to stop digging and lean on its shovel by attending to related areas of scholarly research, from game studies generally to histories of media, technology, materiality, design, culture, and society?"[11] The call to "lean on its shovel" is an opportunity to take stock and plot one's next move cautiously, wisely. Lowood's decision to import Mahoney's voice into a conference on game history was not to argue that the ripening field needs to become more "disciplined" or to grant the history of technology custodial rights. It was an instance of Lowood leaning on his shovel, surveying the landscape from his keynote lectern. We now find ourselves bolstered by a different shovel, leaning on it to take in an expansive view of Lowood's presence as an immensely influential contributor to and advocate for the history of games and software preservation for over thirty years. Lowood is to the history of games what Mahoney was for the history of technology and more.

The Hat Rack

At the conference and via the publications referenced above, Lowood donned only one of his many hats, namely historian. Harold C. Hohbach Curator, History of Science and Technology Collections, and Curator, Film and Media Collections is Henry Lowood's official title at the Stanford University Libraries as of 2019. He began his tenure at Stanford University as a rare books specialist in the Department of Special Collections in 1980, to then become a Bibliographic Assistant to the Director for Collection Development until 1983, when he took up the position of Head Librarian of the Physics Library into the early 1990s. Lowood's first decade at Stanford University overlapped with his doctoral studies across the Bay in History (area: history of science and technology) and his earning of a master's in library and information science, both at the University of California, Berkeley. Over the years Lowood has served as Curator for Germanic, Film and Media, and History of Science and Technology collections and codirected the Stanford Humanities Laboratory. In December 2018, Stanford University Libraries received a $25 million gift from the Harold C. and Marilyn A. Hohbach Foundation to renovate the Cecil H. Green Library's East Wing, renamed Hohbach

Hall, and endow the Silicon Valley Archives program.[12] Lowood was bestowed the Harold C. Hohbach curatorship. He has devoted his entire career to a single institution.[13]

Lowood's intellectual headdress, to squeeze the metaphor for all its worth, is appreciated best when adjudged via the complement of his many professional roles and the areas of research to which he diligently and prolifically dedicates his energy. This complement expresses Lowood's influence and impact in the field of game studies—the historical study of games in particular—as well as across the worlds of libraries, archives, and museums. Lowood possesses an acute awareness of the need to take games (whether electronic or paper based) seriously while working to ensure that their history is documented and preserved.

Whereas the verb "curating," as Hans Ulrich Obrist keenly observes, "is being used in a greater variety of contexts than ever before, in reference to everything from an exhibition of prints by Old Masters to the contents of a boutique wine shop,"[14] it is necessary to construe what Lowood's title of "curator" encompasses if we are to appreciate his enduring involvement in and facilitation of preservation work. When entertaining the question, "What is a curator?" George Adrian's *The Curator's Handbook* embraces the broad-ranging nature of the noun in its contemporary usage to declare that the role of curator "incorporates those of producer, commissioner, exhibition planner, educator, manager and organizer."[15] In the context of art, Adrian adds, a curator may be responsible for "writing wall labels, catalogue essays and other supporting content for an exhibition," while the curator writ large may also be tasked with interacting with "the press and the public giving interviews and talks."[16] Further afield, a curator might become involved in fundraising or other forms of patron-related events and provide "lectures, seminars, internships, or work placement opportunities."[17] The curator's role is multifaceted, requiring one to work across duties of subject specialist; collection development; department, library, archive, or gallery management; and educator while facing outward to the public as representative of a collection or institution.

Per Obrist, the chief action of a curator is that of "connecting cultures," that is "bringing their elements into proximity with each other—the task of curating is to make junctions, to allow different elements to touch."[18] The term, especially when embodied by Lowood, bleeds across neighboring roles of archivist, librarian, and educator while fulfilling many of the attributes of a curator detailed by Adrian. Obrist's words remind that selecting, collecting, caretaking, and stewarding are generative. The curator is a producer of knowledge, an authorial actor through the collections they locate, acquire, organize, store, and make accessible. Lowood incubates and activates materials for us to think with, to both spark and answer questions. His advocacy, leadership, subject area expertise, deep investment in archives, archival research, and preservation demonstrate curating as a significant part of the critical production of meaning—curating as a mode of critical coordination to enable the "different elements to touch."

Lowood played a leading role in establishing the Stephen M. Cabrinety Collection in the History of Microcomputing at Stanford University in the late 1990s.[19] The Cabrinety Collection is most likely the earliest and largest institutional archival collection of historical microcomputing software in existence.[20] The collection boasts close to twenty thousand software titles of multiple formats spanning the early 1970s to early 1990s, the majority acquired unused and in their original packaging (shrink wrap removed for preservation reasons), along with seventy different platforms, computer and game magazines, books, and other materials related to game and other kinds of software. It not only supports researchers by providing access to software and hardware artifacts but has become an invaluable resource to major initiatives devoted to digital preservation and standards sponsored by the Library of Congress's National Digital Information Infrastructure and Preservation Program, the National Institute of Standards and Technology, the National Endowment for the Humanities, and the Institute of Museum and Library Services.

The Cabrinety Collection became a lightning rod of sorts. For a major research institution like Stanford University to acquire a collection largely comprising game software signaled to the world that games are important artifacts to collect, that they are indicative of software and microcomputing history, and that Stanford University would be a leader in the field and practice of software preservation. It is also worth noting that the collection was obtained *before* the scholarly interest in games formed into the field of game studies, when only a handful of academic books devoted their pages to the subject and no journal dedicated exclusively to the subject existed.[21] The Cabrinety Collection was a forerunner to an academic field while exhibiting the type of archival and preservationist resources the field requires. Housing a collection of its enormity and significance given the popularity of games also sent out a beacon to prospective donors that their personal materials could live on in perpetuity through institutional care at Stanford University—an important signal given the university's prime location within Silicon Valley. Lowood remains involved in procuring materials from former Atari employees such as Allan Alcorn, Steve Bristow, David Cook, and Mike Jang as well as from other collections documenting game development and game history. The founding of the Cabrinety Collection in 1998 confirmed that a world-renowned research institution such as Stanford University recognizes the importance of collecting and preserving games materials. Many other gaming initiatives, collections, and archives at universities and museums the world over followed.[22]

The founding of the Cabrinety Collection is not, however, the catalyst for Lowood's active participation in software preservation. His involvement in the conversation commenced nearly a decade prior when still in the Physics Library: in November 1989, Lowood was invited to take part in the "Planning a Center for the Preservation of Microcomputing Software" symposium organized by Columbia University's School of Library Service, which was held at the university's Arden House Conference Center in the Catskill Mountains from March 23–25, 1990. The symposium's general aim was to "chart a responsible and practical course for the preservation of microcomputer software, an 'intellectual audit trail' for future generations."[23] Lowood's own handwritten notes from the event highlight the following objectives as key to establishing a national center to promote and coordinate research in the history of software: the "establishment of a consortium of centers that collect and preserve computer software and archival records on the history of software," the "promotion of standards and techniques for the preservation of machine-readable information," the need to "collect resources" (e.g., relevant manuscripts and supporting records, acquiring and preserving selected software in appropriate formats), and to "provide technology platforms" (to execute software for historical purposes).[24] Some of these objectives would inform Lowood's understanding of game and software preservation in his earliest works on the subject when entertaining the question, "What can be done?"[25]

Lowood was also in conversation with the Library of Congress regarding its Machine-Readable Collections Reading Room (MRCRR), gathering information on its establishment in 1988, its aims, goals, and services offered along with general press coverage and institutional reports.[26] Lowood's rush to the century's end was spent devouring publications on preservation and archival software collections while attending meetings (e.g., the first National Software Archives Conference in Seattle, WA, October 15–16, 1993) and speaking at symposia (e.g., the Software Publishers Association in San Francisco, CA, March 13–16, 1994), not to mention being involved in numerous conversations across institutions and organizations.

Lowood would revisit this period in software preservation work in his piece "Software Archives and Software Libraries" (chapter 6 in this volume) to illustrate that the Cabrinety Collection "aligns neatly with the patterns of collection building that have for centuries brought private collections from the idiosyncratic to the monumental into museums and libraries."[27] A few years after David Bearman's landmark study on software collections, *Collecting Software: A New Challenge for Archives and Museums* (1987), which asserted that no major entity—universities, professional associations, or companies—was currently forming software collections or actively collecting software documentation, Lowood notes that at the time of planning the Arden symposium in 1989 Stephen Cabrinety had already "set up the Computer History Institute for the Preservation of Software (CHIPS)—at the ripe age of 23."[28] The goal of CHIPS, Lowood shares by way of correspondence with Patricia Cabrinety, mother of Stephen, was to preserve software and hardware for future generations to better understand how the industry evolved. Stephen Cabrinety's massive collection was doing the work that Bearman could not locate and that those in attendance at the Arden symposium would propound in their discussions of collecting strategies and centers, standards for preservations of software, and the challenges confronting executing historical software. And this collection would soon come into the curating hands of Lowood.

Building on a decade's worth of work on software preservation and the increased momentum gained by establishing the Cabrinety Collection, Lowood—with Tim Lenoir—spearheaded the research project How They Got Game: History and Culture of Interactive Simulations and Video Games (HTGG). It proves challenging to settle on a single descriptor to capture the Swiss Army knife multifunctional quality of HTGG. Launched two years after the founding of the Cabrinety Collection in 2000, the HTGG project was created to "work on the history and preservation of digital games and interactive simulations and continue an established project in software history and archives carried out in the Silicon Valley Archives at Stanford."[29] HTGG's website remains an invaluable resource, supporting and sharing research and news on preservation, blog posts (the last being in 2016) on the utilization of the Cabrinety Collection, and other posts devoted to game history and preservation. The project hosts links to articles on and from the Game Developers Conference (GDC) and Electronic Entertainment Expo (E3), scholarly conferences (e.g., Play Machinima Law Conference, April 24–25, 2009), exhibitions (e.g., *Fictional Worlds, Virtual Experiences: Storytelling and Computer Games* at the Cantor Center for Visual Arts, November 12, 2003–March 28, 2004), and workshops (one being an early engagement with competitive gameplay and e-sports, "E-Sports and Cyberathleticism," May 8, 2009). HTGG moreover served as the research/public-facing front for the Machinima Archive, the result of a collaboration between HTGG, the Internet Archive, the Academy of Machinima Artists and Science, and Machinima.com.

Lowood represented Stanford University and its software preservation mission as one of the principal investigators for the Preserving Virtual Worlds project (2007–2010). The grant was awarded to the University of Illinois Urbana-Champaign, the University of Maryland, the Rochester Institute of Technology, Linden Lab, and Stanford University in conjunction with Preserving Creative America, an initiative of the National Digital Information Infrastructure and Preservation Program at the Library of Congress. According to its executive summary, the project sought to "investigate issues surrounding the preservation of video games and interactive fiction through a series of case studies of games and literature from various periods in computing history, and to develop basic standards for metadata and content representation of these digital artifacts for long-term archival storage."[30] Nearing two hundred pages, the *Preserving Virtual Worlds Final Report* and its forerunner, "Before It's Too Late: A Digital Game Preservation White Paper,"[31] make a compelling case for the need to preserve electronic games while raising awareness of the challenges facing software preservation (e.g., software and hardware obsolescence; intellectual property rights). Going into the Preserving Virtual Worlds project, Lowood was already garnering public attention in the *New York Times* for his proposal for a national game canon, modeled after the National Film Preservation Board, which he submitted to the Library of Congress in 2006.[32] Many of the titles on Lowood's list, which he presented at the 2007 Game Developers Conference, became the case studies for Preserving Virtual Worlds (e.g., *DOOM*, *Star Raiders*, and *Spacewar!*).

Preserving Virtual Worlds (PVW) was followed by Preserving Virtual Worlds 2 (2010–2013), resulting from a grant by the Institute of Museum and

Library Studies. PVW2 had a mission to investigate the "significant properties" of digital artifacts that "must remain intact over time"; it also "analyzed the likelihood that various preservation strategies such as virtualization and data migration can preserve these properties."[33] Beyond PVW and PVW2, Lowood would receive grant support from the National Software Research Lab of the National Institute of Standards and Technology for imaging the Cabrinety Collection (2012–2014), and collaborate with the University of California, Santa Cruz, on a National Endowment for the Humanities grant for "A Unified Approach to Preserving Cultural Software Objects and Their Development Histories" (2013–2014). This collaboration between UC Santa Cruz and Stanford University Libraries would continue in the form of another grant for "From Descriptive Metadata to Citation: Building a Framework for Search and Communication in Game Studies" from the Institute of Museum and Library Studies (2013–2016).

Lowood's curator's cap with its ability to "connect cultures" is neither limited to his active presence in preservation-related projects like those mentioned above nor his ongoing advocacy for and stewardship of the Cabrinety Collection. Lowood has also served as curator for the Machinima Archive (since 2003) and the Archiving Virtual Worlds project (since 2008). He collaborated with René de Guzman to curate the exhibition *Bang the Machine: Computer Gaming Art and Artifacts* at the Yerba Buena Center for the Arts and soon found himself working on another exhibition, *Fictional Worlds, Virtual Experiences: Storytelling and Computer Games*, for the Cantor Arts Center at Stanford University. The band of Lowood's curator's cap expands further: he has also taught courses on game history and digital preservation at Stanford University, San Jose State University, and the University of California, Santa Cruz.[34] Lowood has reviewed games and served as a judge for e-sports events while maintaining the following active service roles: advisory committee member to the Lemelson Center for the Study of Invention and Innovation, judge for the Strong's World Video Game Hall of Fame, and member of numerous editorial boards.[35] In all of his spare time, Lowood continues to play games as well as write about them.

When adjusting his curator's cap to slip up his historian's hat, Lowood prioritizes the following research areas: military simulation, wargames and wargaming, machinima, virtual worlds, game engines, game designers and developers (across the platforms of coin-op machines via Allan Alcorn, computer games with John Carmack and John Romero, and tabletop wargames focusing on Charles S. Roberts and James Dunnigan), the history of software preservation, game documentation, performance capture and replays, game art,[36] game historiography, and sports while also lending his hand to general histories of games.[37] These subjects have been disseminated across conference panels, keynote addresses, interviews, book chapters, articles, and in collections that Lowood himself has edited or coedited. I intentionally forgo the anticipated summaries of the above areas. Lowood lends his own voice directly to these areas when contextualizing his research in the form of introductions written specifically for each of the book's first three parts. Instead, I offer a few observations on Lowood's research.

The first observation should be fairly obvious: preservation and history are intertwined in Lowood's thinking. A few examples will suffice. When writing on the history of "play by mail" (PBM) wargames commencing in the mid-1960s to explain what the seemingly anachronistic act can tell us about player communities, game design, and the organization of play as a creative action, Lowood foregrounds the role that documentation plays to better understand this historical mode of gameplay. He shares Stanford University's acquisition of an eBay auction listed as "Rare big box Avalon Hill Stalingrad + PBM History & early letters from clubs!" What piqued Lowood's interest in acquiring the collection was the inclusion of correspondence (letters) between players; PBM forms used for gameplay; the Avalon Hill PBM kit from 1964; documents from wargamer clubs; and the premier issue of *The Informer* (May 1965), a publication of the Avalon Hill Modern Wargamers PBM League, among other materials. For Lowood, this acquisition was an archive. He elaborates:

> In short, these documents provide a relatively complete picture of the kinds of materials that a first-generation PBM wargamer would have accumulated. They include record-keeping for turns, club communications, corrections of moves, comments on results, questions about game rules, and more. The correspondence, in particular, documents the intermingling of social correspondence and game moves, as well as implications of the pace of postal play for these interactions.[38]

Lowood writes about the history of PBMs while explaining the value of the paper records acquired with Avalon Hill's *Stalingrad* as primary sources for documenting the experience of wargame play, identifying such idiosyncratic materials as indispensable for historical game studies. He does so via the inclusion of a personal note that explains why both areas—history writing and preservation—are collapsed: the collection, Lowood writes, "reminded me that my engagement with this subject has been shaped by my life with games in ways other than being a historian: specifically, as a curator of library collections and as a player."[39] Such happy collision is neither isolated nor reserved for PBMs.

His study of the *World of Warcraft* community "Nogg-aholic Collaboration" illustrates how a group of player exploits reveals hidden artifacts (e.g., models and maps), new areas under construction, and Blizzard Entertainment archives in the massively multiplayer game world.[40] Working "inside the game," Lowood expresses how game players archived and shared their discoveries of unfinished or inaccessible areas in *World of Warcraft*—how their exploration efforts documented and preserved evidence of hidden locations or inaccessible content in a virtual world (via sharing screenshots and machinima projects, e.g., "Last Wallwalk: The Movie"). Lowood also turns his attention to figures other than the player when considering the work of preservation and documentation. In "It Is What It Is, Not What It Was," from which I've lifted my epigraph for this book's introduction, he identifies the figures of the "historian," "media archeologist," and "re-enactor" as use cases to gauge where preservation efforts should aim to establish and maintain software archives or software libraries devoted to access, installation, and executable software. Each figure demonstrates different needs, relationships, understandings, and concerns over software preservation summarized by Lowood in the form of three "lures": privileging surface properties of the screen as the prime source of value for preservation (lure of the screen); the problem of offering an authentic experience with historical software (lure of the authentic experience); and a software collection's ability to reliably provide executable historical software to researchers (lure of the executable).[41]

Preservation problems are highlighted through consideration of end users' different senses of historical enactment. In doing so, Lowood keeps the act of researching history front and center to the question of software preservation. His unique ability to easily shift hats allows us to consider the different use scenarios of historical research while thinking cautiously about the practical (and challenging) programs of access to software collections needed for such research. We can also observe a similar take in Lowood's interests in sports. Parsing viable documentation strategies for interactive digital games, he champions the example of sports—as opposed to other media like cinema or television—to evince the need for documentation efforts to consider not only game-related artifacts or textual information, but also records of performance to build a more robust account of gameplay.[42] He maintains this reliance on sport when investigating the history of televised sport's "instant replay" to draw analogies to performance capture for digital games.[43]

A final point on this observation: Lowood's own historiographic writings are reliant on materials collected in archives and museums, resultant from the efforts of preservation.[44] Of course, this is not an outstanding quality for a trained historian or the disciplinary and professional expectations for historical work. I draw attention to the de rigueur practice of a historian because of its absence in the bulk of writings on game history that were given over to chronicling events and providing descriptive surveys with no regards to archival research, argumentation, historiographic methods, or contextualization. Professionally trained historians lug an entirely different set of tools to a problem. Whether working with original artifacts (e.g., the *Pong* prototype at the Computer History Museum), conducting or drawing from oral histories, accessing legal documents, doing interviews, or using documents held in archival collections, Lowood's research exemplifies the critical historical study of games.

My second observation further evidences Lowood's historian and curator interests in player activities and communities while emphasizing his penchant for revisiting specific phenomena to diverse ends. Machinima proved a malleable object for Lowood, another area where he would take a formidable lead. He published widely on the subject beginning in 2005—when many were still attempting to grasp the neologism for filmmaking within real-time, three-dimensional virtual environments—with interest brought to full fruition in the form of a collection coedited with Michael Nitsche, *The Machinima Reader*, and a special themed issue of the *Journal of Visual Culture*, both of which appeared in 2011.[45] Machinima fit in well with the archival and preservation work already being carried out with the Cabrinety Collection and HTGG. It offered new ways to think about preservation, curation, and forms of documentation (e.g., replay, game performance capture, screen capture, demo recording), not to mention the opportunity to consider the legal ramifications swarming around the production and distribution of machinima, challenges to hosting machinima at media archives, and the chance to learn more about the creative player-maker community involved.

One game proved a linchpin in Lowood's writings: *DOOM*, developed by John Carmack and John Romero of id Software in 1993. More keystrokes have been devoted to *DOOM* than any other game across Lowood's writings; it has become a protagonist of sorts.[46] General histories typically delimit *DOOM* as controversial (its association with the Columbine High School massacre due to the student shooters—Eric Harris and Dylan Klebold—having played the game), or celebrate its innovation in gameplay and genre (first-person shooter, or FPS), or highlight its utilization of networked play (e.g., via modem or local area networks, LANs) for peer-to-peer multiplayer competitions. Lowood's interest in *DOOM* far exceeds these usual suspects to assess the game's significance for both preservation and historical study. *DOOM*'s architecture allowed players to record real-time footage of gameplay. These demos of *DOOM* gameplay, or "demo movies," were distributed in the form of discreet files to other players who could then watch or replay them via a copy of their own game. "*DOOM*," Lowood exclaims, "linked unprecedented multiplayer competition, reproduction of gameplay as demo movies, and a context for spectatorship through the creation of a player community that would distribute and replay these movies."[47] *DOOM* redefined the relation between the player and the game: the player became equal parts performer, producer, spectator, and distributor of their own gameplay. But according to Lowood, that's not all.

Enter: NoSkill. "NoSkill" was a player tag for Chris Crosby, a highly skilled player of *DOOM* who earned the title of "Doomgod" in the game's community in the mid-1990s. Crosby died in a car crash in 2001. On a memorial site erected in his honor, visitors can download demo files of Crosby's gameplay of *DOOM* and run these files via a vintage version of the game to witness, as Lowood observes frequently and somewhat cryptically, "a now-obsolete game through the eyes of a dead player."[48] The repercussions of NoSkill's demo files speak beyond their memorial function to prove vital for matters of software preservation and historical enactment. Lowood stresses this importance when stating:

> Because we are using an essentially "dead" game to produce this replay, we are also engaging in an act of software preservation and resurrection. The result is that for this FPS, it is possible to see a historical game as played—as seen—through NoSkill's eyes. The player is dead, but his avatar in some sense lives on through this act of perfect reproduction, accessible to any future historians of the game.[49]

Lowood cautions that although we can observe perfect capture of events played out in NoSkill's gameplay of *DOOM*, we know little about why he played or what his actual broader experience of gameplay entailed. In other words, the perfect capture of gameplay in the form of a *DOOM* demo tells us little about the "documentation of the events and activities—the history—that have occurred in these worlds."[50]

The reader can follow Lowood's line of argument in more detail in chapter 3, "Video Capture: Machinima, Documentation, and the History of Virtual Worlds," republished in this volume in its entirety from *The Machinima Reader*. My observation shared here is designed to highlight Lowood's willingness to pierce the surface so to speak, to look beyond the taken-for-granted ways of framing a game's significance to reveal "bigger questions" that pertain to understandings of gameplay performance, preservation, and issues of documentation. His persistent working with *DOOM* digs deeper than sensationalism, generic conventions, and gameplay to examine how performance capture, replay technologies, and in-game assets can be reimagined and reused, how machinima in general can be a tool for historical documentation to capture gameplay and, if possible, a broader experience of activities, practices, and community. Lowood values players as indicators of creative engagements with the medium of games: learning from their practices, asking how reproductions of digital data serve as possible methods to assist curators and archivists in their work on preservation and documentation of virtual world histories and player experiences.

My third observation rubs against Lowood's prolonged and multifaceted engagement with machinima as both creative and documentation/preservation practice while further attesting to his ability to work a particular game like *DOOM* over a period of time. Here I am referring to his treatment of the game engine, described by Lowood as a "particular way of organizing the structure of computer game software. This structure separates execution of core functionality by the game engine from the creative assets that define the play space or 'content' of a specific game title."[51] The term most likely made its debut in Lowood's writing during his delivery of "Shall We Play a Game: Thoughts on the Computer Game Archive of the Future," presented at the conference Bits of Culture: New Projects Linking the Preservation and Study of Interactive Media, hosted by Stanford University in 2002. The term only appears a handful of times in reference to modifications to games (modding) and for production of "movie-like game experiences," better known as machinima. Even in this early work on game preservation and library collections, Lowood is already motioning to the role that the game engine affords to play capture and replays while calling for the organization of working groups to focus on "collecting and securing rights to collect game performance, then work with a digital repository to build a small demonstration collection within one year."[52]

Jumping ahead, it was with his writings on *DOOM* that he really revved up his interest in the game engine to better understand how game assets were being used in the production of *DOOM* demos and *Quake* movies. With his keynote at the first International Conference on the History of Games and continued more recently in his entry in our collection *Debugging Game History*, Lowood turned his attention to the history of the game engine itself: seeking its etymological roots, asking where the term came from and why it was coined, considering the historical context for its employment, learning more about the chain of events leading to Carmack and Romero's adoption of the term and its enabling of the phenomenon known as "open games." What drove Lowood to a history of the game engine was his work in preservation. Consider the following passage:

> Thinking about how game engines interact with assets as input has even informed my work on game software preservation in the second Preserving Virtual Worlds project. The separation of game engine from assets in *DOOM* suggested a possible solution to the problem of auditing software in digital repositories. It turns out that the version-specific ability of a game engine to play back demo files constitutes what media preservationists call a significant property of computer games. Put another way, we can check the integrity of game software stored in a digital repository by seeing if it can run a historical data file such as a *DOOM* demo. The game asset provides a key for verifying the game software, and the game engine pays back the service by playing back historical documentation such as replays.[53]

DOOM was a case study for PVW and PVW2. It compelled Lowood to consider the *DOOM* engine as a historical moment and to fathom how its history ought to be written. *DOOM* and the game engine represent Lowood's most sustained investigations in his research.

This observation furthers the intertwined relationship between preservation and historical research, not to mention Lowood's ability to squeeze a lot of mileage out of carefully selected case studies that allow him to revisit them productively and critically. The game engine revisitation becomes a generative act: not satisfied with just producing a history of the term, Lowood deems it elastic, undetermined by computational architecture so that it demarcates other forms of games, namely wargames executed not by electricity but on tabletops. As Lowood notes, Romero's explanation for the term "engine" stems from a car enthusiast's passion: the engine "is the heart of the car, this is the heart of the game; it's the thing that powers it . . . and it kind of feels like it's the engine, and all the art and stuff is the body of the car."[54] Lowood's passion is wargames and wargaming.

Applying Romero's frame to the history of wargames as models for game systems, Lowood reassigns his research in the game engine to the "war engine" to articulate the combination of wargame systems and game design/play scenarios. Games like Avalon Hill's *Gettysburg* or *Waterloo* from the mid-1960s are deemed monographic in that they are subject specific, "covering a single conflict situation with a bespoke system, components and rules. They were fixed on a single topic."[55] By 1970 with Avalon Hill's *PanzerBlitz*, a different game system emerged based on "modularity and reusability of core components to produce a variety of scenarios." *PanzerBlitz*, in contrast to monographic games, Lowood continues, "introduced the game system as a generator for multiple mini-games," referred to as "scenarios."[56] This observation motions to Lowood's work on the *DOOM* engine that separated the game's core software components from its "art assets, game worlds, and rules of play that comprised the player's gaming experience."[57] Noting *PanzerBlitz*'s rules, Lowood highlights the inclusion of situation cards, "each describing a 'scenario' in terms of specified arrangements of three 'geomorphic' maps, unit counters used by both sides, victory conditions, the number of turns to be played, and the historical situation that would be simulated."[58] The rules suggested that players could design their own scenarios, thus encouraging the publication of games based on the *PanzerBlitz* war engine while revising other games published by Avalon Hill.

By adopting and applying the "engine" concept to the history of wargame design, Lowood is less interested in trying to assert analog influences on digital game design than trying to tease out how designers/developers in different historical moments and through different technological architecture (computer, paper) were speaking to related issues of production, innovation in design, and new roles for players as designers. In making these connections, in one of his longest works, Lowood regards this research as a means to bring "board game and computer game designers into the same conversation about wargame systems and player-created scenarios" to better grasp the "persistent challenge of designing games systems and the relevance of 'manual' wargame design for digital games."[59] We might venture that preservation questions brought Lowood to machinima/*DOOM* while historical questions propelled him to expound on these interests; in doing so, this research has generated unforeseen connections between analog and digital game design and play. Sustained engagements worked carefully, meticulously, methodically; asking different questions; seeking to push and prod historical research into unexpected directions; and demonstrating the need for meaningful, committed, and rigorous research all point to the innovative lessons that Lowood offers to the history of games. Lowood's research exemplifies the value and importance of opening up conversations, offering connections between seemingly disjointed phenomena and showing us the need for critical historical scholarship. It is not hard to hear the faint echo of Mahoney's closing words to his famous essay in Lowood's writing on game history: "What is truly revolutionary about the computer will become clear only when computing acquires a proper history, one that ties it to other technologies and thus uncovers the precedents that make its innovations significant."[60]

A last cap on Lowood's rack, hanging alongside his curator and historian hats, is that of editor. We witness this practice in both journal and book works. The former demonstrates Lowood's ability to introduce subjects into scholarly fields where little to no attention has been afforded. This was the case when editing a themed issue of the *IEEE Annals* on the history of computer games and when introducing the phenomenon of machinima into the field of visual culture studies—in doing so having the themed issue of the *Journal of Visual Culture*'s title ask, "Is Game-Based Moviemaking a New Form of Visual Culture?" while also offering a questionnaire on "Machinima as Visual Culture."[61] The latter demonstrates Lowood's emphasis on a much more concentrated and thorough investigation of a particular subject in the form of edited book collections that encourage multiple "takes," diverse research lenses on a certain phenomenon whether *The Machinima Reader*, *Debugging Game History*, or the video game sports series *FIFA*, taking the form of *EA Sports FIFA: Feeling The Game*.[62] Lowood's editorship also includes servicing the field in the form of a journal (*ROMchip: A Journal of Game Histories*, 2019–present) and book series (MIT Press's Game Histories, 2016–present). The act of editing, like that of curating, is a creative, authorial practice inhabited at the junction of assisting research, cultivating intellectual community, and representing and supporting a field of study. Editors empower possibility. Lowood's hat rack carries a lot of weight.

Why This Book

I first encountered Henry Lowood on a bus in late August 2006. It was chartered to escort over a hundred academics participating in the John D. and Catherine T. MacArthur Foundation's Digital Media and Learning initiative to dinner at a Portuguese restaurant somewhere outside of Newark, New Jersey. We were the first ones on the bus that evening. Both of us, I've learned over the years, share a penchant for punctuality, often managed by being early. In Lowood's case, arriving early isn't only a matter of personal virtue but, as shared above, indicative of an intellectual exercise: expressive of his influence and impact on the historical study of games as well as across the worlds of libraries, archives, and museums. To disturb the silence, we immediately introduced ourselves to each other. We hadn't met directly. Participants workshopped only within their specific research groups throughout most of the two-day session. This research would take the form of the MIT Press open-access series. Lowood contributed to *Digital Youth, Innovation, and the Unexpected*, edited by Tara McPherson,[63] while I provided a chapter to Anna Everett's *Learning Race and Ethnicity: Youth and Digital Media*. Most of our conversation was devoted to the differences between our groups and the fact that neither of us, with our interests in games, were in Katie Salen Tekinbas's collection, *The Ecology of Games*. That evening we sat with others interested in chatting about games over an assortment of seafood dishes. I stayed in contact with Lowood after the MacArthur event, eventually inviting him to contribute a short essay on the "game counter" for *The Object Reader*, which I was coediting with Fiona Candlin.[64] In 2011 he coedited "The Machinima Issue" for the *Journal of Visual Culture*, for which I served as founding editor from 2002 until 2018. Little did I know that this chance meeting on a bus ride in New Jersey would kick off a series of collaborations and partnerships, many of which persist today (including our shared enthusiasm for EA Sports *FIFA* and the beautiful game).

One reason why I championed the need to compile Lowood's writings on software preservation and game history into a single volume is that his works appear in notably distributed forums. As a reminder, his "day job" is that of curator, not professor. Without the monograph and peer-reviewed article pressures that weigh on tenure-track faculty, Lowood writes what he wants and chooses where he publishes his writing. While it is difficult to synthesize his work, in no way does such freedom detract from the acumen of Lowood's voluminous output, regardless of whether he is writing for games scholars who heavily cite his historical writing on *Pong*, librarians interested in metadata,[65] or curators preserving the history of software.[66] Lowood's audiences are mixed and numerous, like his publications. If anything, this freedom produces a lucid, highly readable, jargon-free style that many a tenure-track scholar might aspire to. His oeuvre is unbounded, disseminated across magazines (e.g., *Garage Magazine*), contributions to edited collections,[67] editorship of books and journal issue guest editor,[68] encyclopedia entries,[69] popular culture publications and fan sites (e.g., *Journey Planet*), scholarly journals transcending academic fields (e.g., *Games and Culture*, *Journal of Visual Culture*, *Journal of Library Metadata*, *Mediascape: Journal of Cinema and Media Studies*), white paper reports, a vast amount of interviews, contribution to works produced by the US Army and Modeling Virtual Environments and Simulations Institute,[70] and lest I forget, a long-standing tradition of having his research appear in Italian publications.[71] This list does not even attempt to account for Lowood's early writings on subjects including the history of science, Germanic studies, and forestry management.

When one scans Lowood's full CV or attempts to follow all of the talks he presents annually, stay current on the research that his editing roles enable, or keep abreast of the Silicon Valley Archives' many activities (including the recently launched project, Histories of African Americans in Silicon Valley), it is obvious that the medium of the book is simply too slow for Lowood—as it proved to be for Walter Benjamin and Stuart Hall, two other writers known for their prolific scholarly output that took forms other than traditional monographs while still proving eminently influential to a variety of disciplines.[72] Consolidating a selection of his writing provides the opportunity to read, teach, appreciate, and enjoy his scholarly output as a single corpus of thought committed to the historical study of games and indicative of the need to take seriously their preservation for purposes of research and historical posterity. It is hoped that those who come to the volume to learn more about game history will quickly realize the importance of software preservation and documentation for ensuring that they will have materials to work with. Conversely, the inquirer into the many challenges facing preservation will appreciate the need to consider the historical context of digital and analog artifacts as well as the complex actions of user experience and its value for documentation. Lowood's writing on preservation is always framed by historical concerns. His preservation work helps historians do theirs. "Take off one hat, and the other one is already on," as Lowood is prone to say.[73]

Another rationale for this book is to mark the coming of age of the critical study of games. As many within the field of game studies are already aware, Espen Aarseth announced that 2001 was "year one" of "computer game studies as an emerging, viable, international, academic field."[74] That bold assertion corresponded to the release of his own journal, *Game Studies*, and was intended to announce that the field of study was beginning its own professional accoutrement in the form of a peer-reviewed international journal, soon to be followed by the formation of the Digital Games Research Association (DiGRA) in 2003; the emergence of another academic journal, *Games and Culture* (2006); and a scattering of critical scholarship in the form of games-specific monographs indicative of the field,[75] along with puissant collections.[76] Twenty years on from Aarseth's asseveration, we find a field of study supported by diverse book series (e.g., MIT Press's Playful Thinking, Platform Studies, Game Histories), major academic presses boasting ample game studies titles in their lists (e.g., Bloomsbury, University of Minnesota Press, Duke University Press, New York University Press), even more academic journals (e.g., *Eludamos, Press Start, Analog Game Studies*), conferences (e.g., the Society for Cinema and Media Studies' Video Game Studies special interest group, games panels at the Society for the History of Technology along with its SIGCIS—the Special Interest Group for Computing, Information and Society—and the History of Games Conferences), and a new generation of researchers taking the field in important and much-needed directions.[77]

Despite such growth into its third decade, the field of game studies—and to an even lesser degree, game history—has not produced the sort of retrospective scholarship indicative of other fields' growing maturity. We observe nothing akin to Thomas Haigh's collection of Mahoney's papers for the history of technology or *Bringing It Back Home: Essays on Cultural Studies*,[78] a book compiling founding figure of US cultural studies Lawrence Grossberg's articles. I could also motion to works collected by media studies scholar Henry Jenkins in *Fans, Bloggers, Gamers: Exploring Participatory Culture* as another indicator of absence.[79] The closest game studies has come to adding such books to its shelves are Alexander Galloway's *Gaming: Essays on Algorithmic Culture*,[80] with its previously published works along with original pieces on broad subjects pertinent to games, gaming, and digital media in general;

Ian Bogost's *Things to Do with Videogames* and *How to Talk about Videogames*,[81] both of which consist of the author's miscellaneous observations on video games; and finally, the more recent *The Infinite Playground: A Player's Guide to Imagination*, a Festschrift-like collection completed during the remaining months of Bernard De Koven's life with the help of colleagues, friends, and admirers.[82] Thematic readers in game studies abound, yet no single book shares the essential writings from one of the most prominent, astute, and prolific voices who has helped shape it.

This book signals a milestone within a field, the moment when it produces its first major corpus of critical works on history and preservation. The time certainly seems fitting to produce such a book when one considers the attention that game history has received via the sheer number of documentaries available on streaming services like Amazon Prime and Netflix; academic conferences like the 2020 History of Games Conference in Krakow, Poland, devoted exclusively to the subject of game history in conjunction with panels on the subject at annual professional society conferences; academic journals running themed issues on game history;[83] game historiography receiving attention in academic journals;[84] the establishment of a journal devoted exclusively to the historical study of games (*ROMchip: A Journal of Game Histories*); and an increase in scholarly publications on the subject from across the globe along with recognition of the importance of the subject (e.g., Jaroslav Svelch's *Gaming the Iron Curtain*, winner of the Computer History Museum Prize, International Committee for History of Technology 2019).[85] It is my hope that this book, twenty years on from Aarseth's flag planting, bespeaks a much more permanent presence with new developments already underway.

Beyond flagging a milestone for game studies, I believe that this book proves something of a landmark publication for the history of technology. To my knowledge, the only book to consider electronic games written by a leading historian of technology is Carroll Pursell's *From Playgrounds to PlayStation: The Interaction of Technology and Play*.[86] The subject appears as a final chapter and offers few new insights into the history of electronic games other than what is already offered by well-known chronicles and descriptive accounts. Lowood's formal training in the history of science and technology at UC Berkeley as well as his hands-on preservation and curatorial work in game preservation gifts the study of games with one who deftly understands how to write about technology historically—something rare within the field of game studies. Publishing with Johns Hopkins Studies in the History of Technology series announces that the history of games as well as their preservation are subjects of importance to historians of technology, that Mahoney's influence isn't limited to the history of technology but, thanks to Lowood's efforts, resonates in other fields.

I have gathered Lowood's writings into two primary categories: "archives, documentation, and the preservation of historical software" and "game histories and historiography." These broad themes are designed to help structure some of Lowood's most perspicacious observations on the subjects of preservation and history. The two sections are not disconnected from one another. Readers will encounter ideas shared across both. Likewise, by no means are the works gathered here meant to be exhaustive of Lowood's twenty-year body of publications or his even longer commitment and lead in software preservation. A book's spine can only bear so much! As shared previously, Lowood has written prolifically about machinima. Here I've only privileged its relation to preservation, documentation, and player creativity (as opposed to his interests in aesthetics and game art). A similar choice was made for Lowood's excellent essay coauthored with Tim Lenoir, "Theaters of War: The Military-Entertainment Complex," which in its entirety exceeds this collection's more immediate interests in the history of wargames and wargaming to a more general history of military simulations and training scenarios (e.g., the SIMNET system).[87] Luckily for us, Lowood's contributions to that particular essay on the history of wargames find themselves elaborated on in more detail in a few of the chapters presented within this volume. The previously published essays have been lightly edited for consistency of spelling and punctuation in the volume but otherwise appear as originally written.

Two unpublished works are also included within this volume. "Shall We Play a Game: Thoughts on the Computer Game Archive of the Future" (chapter 2) marks Lowood's first direct and public treatment of game preservation outside of his more general interests in software history and preservation shared the year before as "The Hard Work of Software History," published in *RBM: A Journal of Rare Books, Manuscripts, and Cultural Heritage*. Readers of "Shall We Play a Game" will observe that Lowood does not just diagnose the many problems confronting the historical preservation of game software and hardware but prescribes solutions.

The second unpublished work is situated outside of parts I and II of the book into a third part designated "Further Directions: Sports Games and e-Sports." Chapter 13, "'Beyond the Game': The Olympic Ideal and Competitive e-Sports," was originally written in 2007 for a collection that, unfortunately, failed to materialize. This chapter, examining the annual World Cyber Games and initiatives to promote and encourage competitive gameplay, attests to a perennial research interest gaining increased momentum in recent years: e-sports and sports games. Lowood's interests have been expressed in the form of a publication in the first collection on sports and games and,[88] prior, via his organization of the HTGG workshop on e-sports. With his short pieces on EA Sports *FIFA* and *John Madden Football* for Van Burnham's *Supercade 2: A Visual History of the Videogame Age, 1985–2001* and the publication of his *EA Sports FIFA: Feeling the Game* coupled with Lowood's strong desire to produce more work on the subject of sports games, it should come as no surprise that this subject is a significant area of research at present and well into the future.

The book's foreword is written by Matthew G. Kirschenbaum, professor of English and digital studies at the University of Maryland.[89] Kirschenbaum coedited (with Pat Harrigan) *Zones of Control: Perspectives on Wargames* within which Lowood's chapter on the war engine originally appeared. Kirschenbaum also occupies the dual worlds of preservation and historical studies with an emphasis on the history of writing, serious games, and military technology. Kirschenbaum and Lowood have a history of collaborations, partnering together on preservation projects (e.g., Preserving Virtual Worlds), along with sharing a fascination in the history and play of wargames. Who better to provide a prologue than a colleague who has long observed Lowood hard at work. If Kirschenbaum launches into Lowood's scholarship and professional labor, then T. L. Taylor draws the curtain in the form of an extensive interview with Lowood. Taylor, who is a professor of comparative media studies at the Massachusetts Institute of Technology, studies online gameplay and player communities.[90] Taylor's collaborations with Lowood are numerous as well. Taylor was one of the organizers and participants at HTGG's "E-Sports and Cyberathtleticism" workshop in 2009, and their publications often share the same venues. They've collaborated more recently at a 2018 Lemelson Center workshop at the Smithsonian National Museum of American History, and Lowood interviewed Taylor for *ROMchip: A Journal of Game Histories*. Now they swap seats. Taylor's interview provides the all-important, rare glimpse into Lowood's personal and professional background.

When I originally reached out to both Kirschenbaum and Taylor, their reaction to my invitation was immediate: Kirschenbaum described the collection as a "lovely idea" and shared that he would be "honored" to write the foreword. Taylor responded, "What a fantastic idea!" Their sentiments speak vividly to Lowood's Gemütlichkeit, enduring influence, and leadership in the world of software preservation and game studies. I cannot help but agree with the conviction expressed. This book is well deserved, an opportunity for many to reread Lowood's essential works as constitutive of a project much bigger than its scattered parts, writings that remain timely, skillfully executed, rigorously researched, thought through in a manner most erudite with a brilliant knack for conciseness and clarity. I continue to learn from Lowood. Those readers coming to Lowood's works for the first time will encounter a voice that has proven central in articulating the challenges the future of games faces while being made aware of the pressing need to write about their past meaningfully. We've leaned on our shovel long enough. Time to dig in.

NOTES

1. Michael S. Mahoney, "The History of Computing in the History of Technology," *Annals of the History of Computing* 10, no. 2 (1988): 114.

2. Thomas Haigh, ed., *Histories of Computing* (Cambridge, MA: Harvard University Press, 2011), 2.

3. Mahoney, "The History of Computing in the History of Technology," 113.

4. Henry Lowood, "Game Engines and Game History," *Kinephanos* (January 2014).

5. Cultural historian Erkki Huhtamo brought the state and status of writings on game history to the fore when defining efforts as constitutive of a "chronicle era." His point was that writers of this era were mainly concerned with "organizing data" around hardware developments and landmark games, with little to no "critical and analytic attitude toward its subjects." Erkki Huhtamo, "Slots of Fun, Slots of Trouble: An Archaeology of Arcade Gaming," in *Computer Game Studies Handbook*, ed. Joost Raessens and Jeffrey Goldstein (Cambridge, MA: MIT Press, 2005), 5.

6. Lowood, "Game Engines and Game History."

7. Lowood, "Game Engines and Game History."

8. Lowood, "Game Engines and Game History."

9. Henry Lowood, "Game Studies Now, History of Science Then," *Games and Culture* 1 (January 2006): 78–82.

10. Henry Lowood, "Perspectives on the History of Computer Games," *IEEE Annals of the History of Computing* 31, no. 3 (2009): 4. In his editorial to the themed issue that Lowood edited for *IEEE*, Jeffrey R. Yost, editor in chief, echoes Huhtamo's concern from a few years earlier when writing that "little critical historical analysis has been written on computer games to date. Much of the existing literature is blindly celebratory, or merely descriptive rather than scholarly and analytical." See Jeffery R. Yost, "From the Editor's Desk," *IEEE*

Annals of the History of Computing 31, no. 3 (2009): 2. I put the question to Lowood directly about why he thinks it took until 2009 for the history of computing to consider the history of games. "In the early days of game studies (c. 2000 and earlier)," Lowood explains, "I suppose games were generally understood as an entertainment medium primarily for kids and teenagers. Historians of science and technology probably viewed that as territory for media, literature, or film studies. One area that might have offered a closer fit was history of computing, but that field was only just beginning to focus on the history of software and microcomputing c. 2000, let alone games. Rather than the obvious fact that games involve technologies, my sense of the early 2000s is that the angle that brought games more into focus for historians of science and technology was the realization that the 'users' were not just male boys and teenagers, which opened a bigger set of issues. Just speculating, but perhaps the visible rise and popularity of massively networked games and virtual worlds prompted this perception that there was a lot more to work with in game studies. Not that this was particularly new in game studies—e.g., the writing of Julian Dibbell during the 1990s, among others—but my recollection is that games as a problem area of interest to historians of science and technology did not pick up until later." Henry Lowood, email message to author, August 25, 2021.

11. Henry Lowood and Raiford Guins, *Debugging Game History: A Critical Lexicon* (Cambridge, MA: MIT Press, 2016), xiii.

12. See Gabrielle Karampelas, "Stanford Libraries' Transformative Gift Creates Hub Highlighting Silicon Valley History," *Stanford News*, January 31, 2019, https://news.stanford.edu/2019/01/31/stanford-libraries-transformative-gift-creates-hub-highlighting-silicon-valley-history/.

13. Lowood celebrated his fortieth anniversary at Stanford University in November 2020.

14. Hans Ulrich Obrist, *Ways of Curating* (New York: Farrar, Straus and Giroux, 2014), 23.

15. George Adrian, *The Curator's Handbook: Museums, Commercial Galleries, Independent Spaces* (New York: Thames and Hudson, 2015), 2.

16. Adrian, *The Curator's Handbook*, 2.

17. Adrian, *The Curator's Handbook*, 2.

18. Obrist, *Ways of Curating*, 1.

19. Stephen M. Cabrinety (1966–1995) was a former student at Stanford University who began collecting hardware and software during his teens. His life was cut short due to his death from Hodgkin's lymphoma in 1995. Cabrinety's sister reached out to Stanford University soon after his death to find a suitable home for his collection. Given Lowood's role as Curator of History of Science and Technology Collections and Film and Media Collections as well as his stewardship with the Silicon Valley Archives, Lowood negotiated the transfer of the collection as a gift to Stanford University in 1998 with further acquisitions in 1999 and 2000.

20. I have qualified this claim with the word "institutional" in order to not preclude other efforts outside of institutional officialdom. Institutional collections that I have studied personally either came into being after the Cabrinety family's gift to Stanford University (e.g., the Strong's International Center for the History of Electronic Games in Rochester, New York) or exist outside of the domain of research (e.g., American Classic Arcade Museum at Fun Spot in Lanconia, New Hampshire). See Raiford Guins. *Game After: A Cultural Study of Video Game*

Afterlife (Cambridge, MA: MIT Press, 2014). Lowood is also sensitive to staking claims of "firsts" in his discussion of both the Cabrinety Collection and the Library of Congress's Machine-Readable Collections Reading Room. He writes: "Both collections were substantial and consisted primarily of software in the categories of microcomputer software intended for consumer use, including games and various forms of entertainment and educational software. They were, as far as I know, the first major collections of historical software that researchers used in controlled reading rooms, that is, within the walls of traditional research repositories." Henry Lowood, "Software Archives and Software Libraries," in *Challenging Collections: Approaches to the Heritage of Recent Science and Technology*, ed. Alison Boyle and Johanes-Geert Hagmann. (Washington, DC: Smithsonian Institution Scholarly Press, 2017), 72.

21. The following are examples of scholarly works that address video games in the late 20th century: Geoffrey R. Loftus and Elizabeth F. Loftus, *Mind at Play: The Psychology of Video Games* (New York: Basic Books, 1983); David Sudnow, *Pilgrim in the Micro-World: Eye, Mind, and the Essence of Video Skill* (New York: Warner Books, 1983); Brian Sutton-Smith, *Toys as Culture* (Mattituck, NY: Gardner Press, 1986); Marsha Kinder, *Playing with Power in Movies, Television, and Video Games: From Muppet Babies to Teenage Mutant Ninja Turtles* (Berkeley: University of California Press, 1991); Justine Cassell and Henry Jenkins, eds. *From Barbie to Mortal Kombat: Gender and Computer Games* (Cambridge, MA: MIT Press, 1998).

22. Examples of US university collections include the UT Videogame Archive at the Dolph Briscoe Center for American History, University of Texas at Austin; the Computer and Video Game Archive, University of Michigan Library; and the now defunct William A. Higinbotham Game Studies Collection at Stony Brook University.

23. Hans Rütimann, letter of invitation to Henry Lowood, November 19, 1989.

24. Henry Lowood, personal note, March 23, 1990.

25. See "Shall We Play a Game: Thoughts on the Computer Game Archive of the Future." Originally presented as a paper at Bits of Culture: New Projects Linking the Preservation and Study of Interactive Media, Stanford University, 2002.

26. Lowood shared personal files with me containing documents on the Library of Congress's Machine-Readable Collections Reading Room Pilot dating back to 1983.

27. Lowood, "Software Archives and Software Libraries," 75. He also writes on software preservation across the 1980s in "It Is What It Is, Not What It Was," *Refractory: A Journal of Entertainment Media* 27 (2016) and in this essay's predecessor, "The Lures of Software Preservation," *Preserving.exe: Toward a National Strategy for Software Preservation* (Washington, DC: Library of Congress, October 2013).

28. Lowood, "Software Archives and Software Libraries," 75. Stephen Cabrinety was also practical: CHIPS had developed a business plan by May 1989 with concrete ideas for staffing the institute, fundraising, a projected annual collecting budget, as well as preference for where to locate CHIPS (Silicon Valley was among his top choices).

29. Henry Lowood, "Memento Mundi: Are Virtual Worlds History?" in *Digital Media: Technological and Social Challenges of the Interactive World*, ed. Megan Winget and William Aspray (Lanham, MD: Scarecrow Press, 2011), 16.

30. Library of Congress, "Preserving Virtual Worlds Final Report," August 31, 2010, https://www.ideals.illinois.edu/handle/2142/17097.

31. Lowood served as the general editor for "Before It's Too Late: A Digital Game Preservation White Paper." This white paper was coauthored by members of the International

Game Developers Association (IGDA) Game Preservation Special Interest Group (founded in 2004) and was supported by the work being conducted by the Preserving Virtual Worlds project. The white paper was first made available by Lulu Press and then published as an article: "Before It's Too Late: A Digital Game Preservation White Paper," *American Journal of Play* 2, no. 2 (2009): 139–166.

32. See: Heather Chaplin, "Is That Just Some Game? No, It's a Cultural Artifact," *New York Times*, March 12, 2007.

33. "Preserving Virtual World II: Methods for Evaluating and Preserving Significant Properties of Educational Games (2010–2013)," Stanford Libraries, https://library.stanford.edu/projects/preserving-virtual-worlds.

34. At Stanford University, Lowood has taught the following undergraduate courses: Media and Message for the Thinking Matters program and the Film and Media Studies program; The Consumer as Creator in Contemporary Media, also for the Film and Media Studies program; and co-teaches with Scott Bukatman Humans and Machines as part of Stanford's Introduction to the Humanities curriculum. At the University of California, Santa Cruz, he has, on numerous occasions, taught the History of Digital Games for the Art Department as well as various courses on games and curation for the School of Library and Information Studies at San Jose State University.

35. Lowood has written game reviews for *Grenadier*, the *Wargamer*, *Paper Wars*, *Berg's Review of Games*, *Zone of Control*, and other game publications. He has been nominated for the Charles S. Roberts Award for Best Game Review or Analysis Article for "Hitler's Last Gamble," *Wargamer* 2, no. 17 (1989) and for his review of "Ukraine '43," *Paper Wars* no. 40 (2001).

36. See Henry Lowood, "Jon Haddock, *Screenshots:* Isometric Memories," in *GameScenes: Art in the Age of Videogames*, ed. Matteo Bittanti and Domenico Quaranta (Milan: Johan and Levi editore, 2006), 15–39.

37. For general writings on the subject of game history, see "A Brief Biography of Computer Games," in *Playing Computer Games: Motives, Responses, and Consequences*, ed. Peter Vorderer and Jennings Bryant (Lawrence Erlbaum Associates, 2006), 25–41; and "Game History," in *The Johns Hopkins Guide to the Digital Media*, ed. Marie-Laure Ryan, Lori Emerson, and Benjamin J. Robertson (Baltimore: Johns Hopkins University Press, 2014), 206–212.

38. Henry Lowood, "Putting a Stamp on Games: Wargames, Players, and PBM," *Journey Planet* 26 (2015): 21.

39. Lowood, "Putting a Stamp on Games," 20.

40. Henry Lowood, "Forbidden Areas: The Hidden Archive of a Virtual World," in *Rough Cuts: Media and Design in Process. A MediaCommons Project*, ed. Kari Kraus (2012), http://mediacommons.org/tne/pieces/forbidden-areas-hidden-archive-virtual-world.

41. See Henry Lowood, "It Is What It Is, Not What It Was."

42. Lowood, "Shall We Play A Game," 7.

43. Henry Lowood, Eric Kaltman, and Joseph Osborn, "Screen Capture and Replay: Documenting Gameplay as Performance," in *Histories of Performance Documentation: Museum, Artistic and Scholarly Practices*, ed. Gabriella Giannacchi and Jonah Westerman (New York: Routledge, 2017), 149–164.

44. For an example of his utilization of archival and museum materials, please refer to Henry Lowood, "Video Games in Computer Space: The Complex History of *Pong*," *IEEE Annals in the History of Computing* 31 (2009): 5–19.

45. Lowood's interest in machinima continues to trickle: Henry Lowood, "*Quake*: Movies," in *How To Play Video Games*, ed. Matthew Thomas Paine and Nina B. Huntemann (New York: New York University Press, 2019), 285–292.

46. When I asked about his personal relation to *DOOM*, Lowood offered the following: "I was very interested in playing it not long after its release at the end of 1993. Maybe 1994. However, it was the first game (and so far the only game) that gave me motion sickness! So despite the initial interest, I stepped away for a while, but did come back to it. I am generally more likely to play simulation, strategy or real-time strategy games than shooters for recreational play anyway. I came back to *DOOM* via research, starting with my work on the history of machinima. That topic took me to *Quake*, of course. Clearly *DOOM* and *Quake* are closely related, but the topic that really pushed me towards *DOOM* in this context was wanting to know more about the development at id Software of their game engine technology as a platform for user-/player-generated projects such as game-based movies. Both games occupied a historical sweet spot for just this sort of activity, as the technologies that produced the game (both hardware and software) were both robustly capable of supporting such projects and relatively easy to comprehend with a sufficient investment of obsessive attention and time." Henry Lowood, email message to author, August 25, 2021.

47. Henry Lowood, "High-Performance Play: The Making of Machinima," in *Videogames and Art*, ed. Andy Clarke and Grethe Mitchell (Bristol, UK: Intellect Books, 2007), 64.

48. Henry Lowood, "Video Capture: Machinima, Documentation, and the History of Virtual Worlds," in *The Machinima Reader*, ed. Henry Lowood and Michael Nitsche (Cambridge, MA: MIT Press, 2011), 8.

49. Lowood, "Video Capture," 8.

50. Lowood, "Video Capture," 8.

51. Henry Lowood, "Game Engine," in *Debugging Game History: A Critical Lexicon*, ed. Henry Lowood and Raiford Guins (Cambridge, MA: MIT Press, 2016), 204.

52. Lowood, "Shall We Play A Game," 16.

53. Lowood, "Game Engines and Game History."

54. Lowood, "Game Engine," 206.

55. Henry Lowood, "War Engines: Wargames as Systems from the Tabletop to the Computer," in *Zones of Control: Perspectives on Wargaming*, ed. Matthew G. Kirschenbaum and Patrick Harrigan (Cambridge, MA: MIT Press, 2016), 89.

56. Lowood, "War Engines," 93.

57. Jason Gregory, *Game Engine Architecture* (Wellesley, MA: A. K. Peters, 2009), 11.

58. Lowood, "War Engines," 94.

59. Lowood, "War Engines," 103.

60. Mahoney, "The History of Computing in the History of Technology," 123.

61. Henry Lowood, "A 'Different Technical Approach'? Introduction to the Special Issue on Machinima," *Journal of Visual Culture* 10, no. 1 (2011): 3–5.

62. Raiford Guins, Henry Lowood, and Carlin Wing, *EA Sports FIFA: Feeling the Game* (New York: Bloomsbury Press, 2022).

63. Henry Lowood, "Found Technology: Players as Innovators in the Making of Machinima," in *Digital Youth, Innovation, and the Unexpected*, ed. Tara McPherson (Cambridge, MA: MIT Press, 2008), 165–198.

64. Henry Lowood, "Game Counter," in *The Object Reader*, ed. Fiona Candlin and Raiford Guins (Abingdon, England: Routledge, 2009), 466–469.

65. Eric Kaltman, Noah Wardrip-Fruin, Mitch Mastroni, Henry Lowood, Glynn Edwards, Marcia Barrett, Greta de Groat, and Christine Caldwell, "Implementing Controlled Vocabularies for Computer Game Platforms and Media Formats in SKOS," *Journal of Library Metadata* 16, no. 1 (2016): 1–22.

66. Henry Lowood, "The Hard Work of Software History," *RBM: A Journal of Rare Books, Manuscripts, and Cultural Heritage* 2, no. 2 (2001): 141–161; and Lowood, "Software Archives and Software Libraries."

67. Examples not cited elsewhere include Henry Lowood, "Impotence and Agency: Computer Games as a Post-9/11 Battlefield," in *Games without Frontiers—War without Tears. Computer Games as a Sociocultural Phenomenon*, ed. Andreas Jahn-Sudmann and Ralf Stockmann (London: Palgrave Macmillan, 2008), 78–86; and Henry Lowood, "*Warcraft* Adventures: Texts, Replay, and Machinima in a Game-Based Storyworld," in *Third Person: Authoring and Exploring Vast Narratives*, ed. Pat Harrington and Noah Wardrip-Fruin (Cambridge, MA: MIT Press, 2009), 407–428.

68. In addition to those volumes edited by Lowood mentioned in the chapter, I will include here the *World of Warcraft*–themed issue: Tanya Krzywinska and Henry Lowood, *Games and Culture* 1, no. 4, 2006.

69. Henry Lowood, "Computer and Video Games," *Encyclopedia of 20th-Century Technology*, ed. Colin Hempstead (Milton Park, England: Routledge, 2004), 180–181; Henry Lowood, "Animation Technology and Computer Graphics in the United States," in *Oxford Encyclopedia of the History of Science, Medicine, and Technology in America*, ed. Hugh Slotten (Oxford: Oxford University Press, 2014), 63–65.

70. Henry Lowood, "The Obstacle Course: Documenting the History of Military Simulation," in *America's Army PC Game: Vision and Realization*, ed. Margaret Davis (Monterey, CA: US Army and MOVES Institute, 2004), 18.

71. Many of Lowood's works that appear translated into Italian are due to his longstanding collaboration with Matteo Bittanti, associate professor at IULM University, Milan, Italy. Lowood's "Jon Haddock, *Screenshots*: Isometric Memories," which appeared in *GameScenes: Art in the Age of Videogames*, and selections from the Machinima Issue of the *Journal of Visual Culture* have been translated into Italian.

72. I should point out that Lowood's dissertation was published as a monograph. See *Patriotism, Profit, and the Promotion of Science in the German Enlightenment: The Economic and Scientific Societies, 1760–1815* (New York: Garland, 1991).

73. Henry Lowood, email message to author, August 25, 2021.

74. Espen Aarseth, "Computer Game Studies, Year One," *Game Studies* 1, no. 1 (2001), http://www.gamestudies.org/0101/editorial.html.

75. Nick Montfort, *Twisty Little Passages: An Approach to Interactive Fiction* (Cambridge, MA: MIT Press, 2003) and Jesper Juul, *Half-Life: Video Games between Real Rules and Fictional Worlds* (Cambridge, MA: MIT Press, 2005).

76. Joost Raessens and Jeffrey Goldstein, eds. *Computer Game Studies Handbook* (Cambridge, MA: MIT Press, 2005); and Katie Salen Tekinbas and Eric Zimmerman, *The Game Design Reader* (Cambridge, MA: MIT Press, 2005).

77. Two examples of important and new directions for game studies are Alenda Y. Chang, *Playing Nature: The Ecology of Games* (Minneapolis: University of Minnesota Press, 2019) and Bonnie Ruberg, *The Queer Games Avant-Garde* (Durham, NC: Duke University Press, 2020).

78. See Lawrence Grossberg, *Bringing It Back Home: Essays on Cultural Studies* (Durham, NC: Duke University Press, 1997).

79. See Henry Jenkins, *Fans, Bloggers, Gamers: Exploring Participatory Culture* (New York: New York University Press, 2006).

80. See Alexander Galloway, *Gaming: Essays in Algorithmic Culture* (Minneapolis: University of Minnesota Press, 2006).

81. See Ian Bogost, *Things to Do with Videogames* (Minneapolis: University of Minnesota Press, 2011); and Ian Bogost, *How to Talk about Videogames* (Minneapolis: University of Minnesota Press, 2015).

82. Bernard De Koven with Holly Gramazio, *The Infinite Playground: A Player's Guide to Imagination*, ed. Celia Pearce and Eric Zimmerman (Cambridge, MA: MIT Press, 2020).

83. See "Game History: A Special Issue," *Game Studies* 13, no. 2, 2013; "New Game History," *American Journal of Play* 10, no. 1 (2017); and "Feminist Game History," *Feminist Media History* 6, no. 1 (2020).

84. See Jaakko Suominen, "How to Present the History of Digital Games: Enthusiast, Emancipatory, Genealogical, and Pathological Approaches," *Games and Culture*, 12, no. 6 (2016): 544–562.

85. Examples of historical game research devoted to regions beyond North America include Melanie Swalwell, *Homebrew Gaming and the Beginnings of Vernacular Digitality* (Cambridge, MA: MIT Press, 2021); Melanie Swalwell, ed., *Game History and the Local* (London: Palgrave Macmillan, 2021); Alison Gazzard, *Now the Chips Are Down: The BBC Micro* (Cambridge, MA: MIT Press, 2016).

86. Carroll Pursell, *From Playgrounds to PlayStation: The Interaction of Technology and Play* (Baltimore: John Hopkins University Press, 2015).

87. Henry Lowood and Tim Lenoir, "Theaters of War: The Military-Entertainment Complex," in *Collection, Laboratory, Theater: Scenes of Knowledge in the 17th Century*, ed. Helmar Schramm, Ludger Schwarte, and Jan Lazardzig (Berlin: Walter de Gruyter, 2005), 427–456.

88. See Henry Lowood, "Joga Bonito: Beautiful Play, Sports and Digital Games," in *Sports Videogames*, ed. Mia Consalvo, Konstantin Mitgutsch, and Abe Stein (London: Routledge, 2013), 67–86.

89. See: Matthew G. Kirschenbaum, *Mechanisms: New Media and the Forensic Imagination* (Cambridge, MA: MIT Press, 2008); Matthew G. Kirschenbaum, *Track Changes: A Literary History of Word Processing* (Cambridge, MA: Belknap Press, 2016); Matthew G. Kirschenbaum, *Bitstreams: The Future of Digital Literary Heritage* (Philadelphia: University of Pennsylvania Press, 2021).

90. See: T. L. Taylor, *Play between Worlds: Exploring Online Game Culture* (Cambridge, MA: MIT Press, 2006); T. L. Taylor, *Raising the Stakes: E-Sports and the Professionalization of Computer Gaming* (Cambridge, MA: MIT Press, 2012); and T. L. Taylor, *Watch Me Play: Twitch and the Rise of Game Live Streaming* (Princeton, NJ: Princeton University Press, 2018).

I ARCHIVES, DOCUMENTATION, AND THE PRESERVATION OF HISTORICAL SOFTWARE

Author's Introduction

In November 1989, I received a letter out of the blue from Hans Rütimann of the School of Library Service at Columbia University. It was an invitation to attend a symposium called "Planning a Center for the Preservation of Microcomputer Software." Rütimann wrote that initial responses to the agenda of this symposium suggested that the topic of software preservation was "necessary and urgent. Some years hence it may even be viewed as a milestone event."[1] I accepted the invitation and from March 23 to 25 of the following year found myself at Columbia's Arden House Conference Center in the Catskills. The packed agenda for this conference covered a range of topics from the objectives of such a center—with prompts such as "What exists?" and "Are there models?"—to practical matters such as the scope and building of collections, use policies, and funding. Five papers prepared ahead of the conference by Paul Banks, Mary Bowling, Bernard Galler, C. Lee Jones, and Paul Evan Peters focused attention on topics such as "Preservation and Format Issues" (Banks), "Software Issues" (Galler), and "Machine Aspects" (Peters). The second day of hard conferencing concluded with discussion of an "Agenda for Action."

The symposium succeeded in introducing the important general topic of software preservation to the world, reflecting on its importance for cultural history, and defining a series of specific problems that would need to be solved. However, it did not produce a project that would define workflows or an archival center where collections of operable software could be preserved for the future. With the benefits of hindsight, we can say today that the technologies and institutions available during the 1990s were not ready for such projects. It strikes me that two observations can be made about the general lack of forward motion in software preservation during that decade. First, the agenda and discussion papers for the Arden House conference clearly document a general malaise about software as a category of preservable cultural objects circa 1990. Banks observed in his conference discussion paper that the "subject of this conference is interesting and significant beyond the preservation of microcomputer software itself." He meant that issues around the problem of sustaining access to historical software were a "microcosm of issues that society is facing on a broad scale for preservation of knowledge embodied in high-technology formats in the 'information age' that in certain respects resembles the oral age of evanescence more than the print age of relative permanence of information to which our preserving institutions are geared."[2] This statement connects the prediction of a "digital dark age"—an idea that became more popular in the mid- to late 1990s—to the problem of preparing an agenda for "preserving institutions," meaning libraries, archives, and museums, to deal with software.[3]

Second, despite the magnitude of the problem and the consequences of inaction, conference participants could give no examples of an institution that had set up a working program for software preservation. As we know now, significant progress would not be made for another decade or more. The failure was not in ignoring the problems around collecting and preserving historical software, but in not having practical options for addressing the problems. In "Software Archives and Software Libraries" (chapter 6 in this volume), I began with the important writing of David Bearman and Margaret Hedstrom that began to be widely read in the archival community during the late 1980s. They introduced the concept that software could be considered as an archival object. Already in 1987, Bearman produced a vision of how heritage institutions might begin to build and preserve software collections in *Collecting Software: A New Challenge for Archives and Museums*. Despite the enthusiasm and optimism that must have accompanied his ambitious vision for software collecting at that time, Bearman admitted that "no one seems to have formed a software archive nor is anyone collecting software documentation, except as an inadvertent biproduct of institutional archives."[4] Over the course of the 1990s, institutions such as the Smithsonian Institution (David Allison's National Software Archives and Awards project, circa 1993–1994), the Computer History Museum (for which Bearman consulted during the late 1980s), and the Charles Babbage Institute (CBI Software Task Force, 1998–1999) sought to gain purchase on the ways and means for creating and maintaining software archives.[5] Even if these projects failed to produce archival collections of software, they certainly stimulated my thinking (and others, I am sure) to move forward in the following decade.

I not only tracked but was also active in some of these initiatives during the 1990s. For example, I participated in the National Software Archives planning group and was one of the coauthors of the Charles Babbage Foundation report. For the most part, however, my curatorial engagement with software history focused on documentation during this time. With the launch in 1985 of the archival documentation project that became the Silicon Valley Archives, collections such as the papers of Douglas Engelbart and Edward Feigenbaum came to Stanford during the ensuing years. We also created resources, such as oral histories or SiliconBase, an online collection of scanned documents and digital objects. These collections documented the history of software, rather than providing collections of historical software. In other words, they were archives, not libraries, a distinction that I drew out in "Software Archives and Software Libraries." This was the norm for the few libraries, archives, and museums that took on the challenges of the history of software as a collecting area. Following Bearman's report, a wave of significant digital preservation[6] projects and reports[7] through the 1990s contributed more practical accomplishments than had the proposals for software archives. This was the solid foundation for larger-scale projects such as the Internet Archive (founded 1996) for Web archiving, the public launch of its Wayback Machine five years later, or the CAMiLEON Project (Leeds University and University of Michigan, launched 1999), which produced the complex and much-discussed BBC Domesday Project, launched in 2002.[8]

A sports analogy comes to mind here, which might be saying more about my personal interests than my perspective on collecting and preserving software, but it goes something like this: the activities described thus far strike me today as something like pregame preparation for a sports event, dressing in the locker room and preparing mind and body for the game. Following this analogy, two relatively gigantic acquisitions in 1998 and 1999 constituted my moment of first dashing onto the court and then hearing the opening whistle signaling the beginning of the game. These were the acquisitions by the Stanford Libraries of the Apple Computer records, over 600 linear feet of archival material and objects covering the period from 1977 to 1998 and acquired in 1998, and the Stephen M. Cabrinety Collection in the History of Microcomputing, nearly 1,400 linear feet covering the period from 1975 to

1995, acquired in 1997 and received in two tranches in 1998 and 1999. In total, these two collections would stretch the long way across nearly seven football fields of the US variety. The Apple collection was to the best of my recollection the first collection acquired by Stanford to include much in the way of hardware and software. Software was only a small part of this collection. The finding aid lists fourteen record storage boxes under "digital media," including computer diskettes, CD-ROM titles, laserdiscs, and computer tapes, a little less than 10 percent of the collection. Still, this acquisition pushed the library out of the locker room and onto the court. It established the reality that the game of software preservation would be a long game involving a complex series of problems and projects. For example, one title in the Apple collection, a courseware program developed at Stanford in 1985 as part of the Faculty Author Development (FAD) project called "The Would-Be Gentleman," became the first case study of the Emulation as a Service Infrastructure Project in the Born Digital Preservation Lab at Stanford in 2021, twenty-three years after the acquisition of this collection![9]

The Cabrinety Collection is covered in "Software Archives and Software Libraries." One point I would like to underline here is that it was the first acquisition by any cultural repository—library, museum, or archive—of a large, curated collection of historical software. But that is not the important point. As the "Would-Be Gentleman" example suggests, such an acquisition is just the opening whistle of a game with many plays. The Cabrinety Collection has since its acquisition been the focal point for a series of 21st-century projects that run through the gauntlet of defining workflows around curation, selection, preservation, description, and access for such a collection in a library setting, including Preserving Virtual Worlds, the National Institute of Standards and Technology's National Software Reference Library/Stanford software capture project, the Game Metadata and Citation Project, and, currently, the Emulation as a Service Infrastructure Project.[10] The diversity of these problems is illustrated best perhaps by the US institutions that have supported these projects: the Library of Congress, the Institute of Museum and Library Services (IMLS), the National Institute of Standards and Technology (NIST), and the Alfred P. Sloan and Andrew W. Mellon Foundations.

Acquisition of the Cabrinety Collection also stimulated the foundation of the How They Got Game project in 2000. It began as an initiative funded by the then brand-new Stanford Humanities Lab to support work in the history and preservation of digital games and simulations. Timothy Lenoir, then a professor in the History Department, and I collaborated with graduate and undergraduate students drawn to this new area of research and teaching. The result was an exciting variety of projects, from military simulation through game technology and on to art games. It also led to collaborations with other institutions, notably the Preserving Virtual Worlds project, led by the late Jerome McDonough at the School of Information Sciences of the University of Illinois. The mix of engagements (preservation and history, games and simulations, archives and exhibits) probably explains my own tendency in talks and essays such as "Video Capture" (chapter 3) and "Screen Capture and Replay" (chapter 5) to play with themes cutting across documentation and play, digital archives, and questions of historical authenticity, or the history of game-based filmmaking (machinima) and the history of performance capture.

NOTES

1. Hans Rütimann to Henry Lowood, November 19, 1989, Henry Lowood Papers, Stanford University Libraries, Box 3, folder 35 (hereafter cited as Lowood Papers).

2. Paul N. Banks, "Preservation and Format Issues," unpublished paper, Lowood Papers, 17.

3. E.g., Steven Brand, "Escaping the Digital Dark Age," *Library Journal* 124, no. 2 (February 1999): 46–49, which leads with, "Due to the relentless obsolescence of digital formats and platforms, along with the ten-year life spans of digital storage media such as magnetic tape and CD-ROMs, there has never been a time of such drastic and irretrievable information loss as right now."

4. David Bearman, *Collecting Software: A New Challenge for Archives and Museums*, Archival Informatics Technical Report 2 (Toronto: Archives and Museum Informatics, 1987), 24.

5. Charles Babbage Foundation Software Task Force, *Final Report*, November 7, 1998. One of the objectives the task force identified was "revival of selected aspects of the National Software Archives project, which would lead to preservation of major collections," p. 5.

6. A broader category of work than software preservation that includes digital documents, Web sites, media files, etc.

7. Some examples that remain on my office bookshelf to this day: Margaret Johnson et al., *Computer Files and the Research Library* (Mountain View, CA: Research Libraries Group, 1990); Commission on Preservation and Access and the Research Libraries Group, *Preserving Digital Information: Report of the Task Force on Archiving of Digital Information* (Washington, DC: Commission on Preservation and Access, 1996); Jeff Rothenberg, *Avoiding Technological Quicksand: Finding a Viable Technical Foundation for Digital Preservation* (Washington DC: Council on Library and Information Resources, 1999). Many readers with an interest in such matters would probably cite Jeff Rothenberg's "Ensuring the Longevity of Digital Documents," *Scientific American* 272 (January 1995): 42–47, as the most influential take on digital preservation of the 1990s; I see its value as being one of the few works that specifically and significantly addressed software preservation.

8. Paul Wheatley, "Digital Preservation and BBC Domesday," paper presented at the Electronic Media Group, Annual Meeting of the American Institute for Conservation of Historic and Artistic Works, Portland, Oregon, June 14, 2004, http://citeseerx.ist.psu.edu/viewdoc/download?doi=10.1.1.84.7455&rep=rep1&type=pdf. I cite this paper here because I attended this conference. Wheatley's presentation introduced me to the combination of videodisc data extraction, digital preservation, emulation, and more that this project entailed. Several other pathbreaking projects of the early 2000s were represented at this conference.

9. Annie Schweikert and Ethan Gates, "EaaSI Case Study #1: The Would-Be Gentleman," Software Preservation Network, January 2021, https://www.softwarepreservationnetwork.org/eaasi-case-study-1-the-would-be-gentleman/.

10. Jerome P. McDonough, Robert Olendorf, Matthew Kirschenbaum, Kari Kraus, Doug Reside, Rachel Donahue, Andrew Phelps, Christopher Egert, Henry Lowood, and Susan Rojo, *Preserving Virtual Worlds Final Report*, August 31, 2010, https://www.ideals.illinois.edu/handle/2142/17097; Douglas White, "NIST National Software Reference Library (NSRL) Efforts in Preserving Software in the Stanford University Libraries (SUL) Cabrinety Collection," September 11, 2013, video, https://www.loc.gov/preservation/outreach/tops/white/white.html; GAMECIP: The Game Metadata and Citation Project, https://gamecip.soe.ucsc.edu/; EaaSI: Emulation-as-a-Service Infrastructure, https://www.softwarepreservationnetwork.org/emulation-as-a-service-infrastructure/resources/.

1

The Hard Work of Software History

A few years ago, the literary and media historian Friedrich Kittler opened an essay called "There Is No Software" with, in his own words, a "rather sad statement." In his view, "the bulk of written texts—including this text—do not exist anymore in perceivable time and space but in a computer memory's transistor cells." From a scholar who, until then, had situated the cultural meaning of literary texts in discourse networks dependent on technologies of inscription (writing, gramophone, typewriter, computer) and the materiality of communication, this remark captures the essence of a significant cultural shift. At the end of the 20th century, according to Kittler, texts—and even software itself—have vanished. Our text-producing gestures merely correspond to codes built on silicon and electrical impulses; the texts themselves no longer exist materially and, indeed, we have ceased to write them: "All code operations . . . come down to absolutely local string manipulations and that is, I am afraid, to signifiers of voltage differences."[1] Following Kittler, libraries and archives have yet to determine which traces have replaced the acts and artifacts of writing, such as books and manuscripts, that were part of Goethe's or Einstein's worlds at the conclusions of the previous two centuries.

About ten years ago, the computer scientist Mark Weiser contributed an article titled "The Computer for the 21st Century" to the *Scientific American*. Weiser's essay introduced his own research program to the magazine's technologically literate readership, a program he called "ubiquitous computing." The most startling observation in this article turned his vision of the future

Originally published as "The Hard Work of Software History," *RBM: A Journal of Rare Books, Manuscripts, and Cultural Heritage* 2, no. 2 (Fall 2001): 141–161.

in an unexpected direction. According to Weiser, "the most profound technologies are those that disappear. They weave themselves into the fabric of everyday life until they are indistinguishable from it."[2] In the first phase of computing, as he saw it, many people shared one large computer, such as a time-shared IBM mainframe, the Big Iron of the Information Age. Then with the advent of the microprocessor, computers became personal—one person, one machine. In his work at Xerox's Palo Alto Research Center until his premature death two years ago, Weiser—whose papers are now at Stanford—created small, portable, and networked devices for times in which computers would far outnumber people, the third age of ubiquitous computing. He believed it significant not that computers would outnumber us but, rather, that they would become invisible in order to be useful. As he put it a few years later, the "highest ideal is to make a computer so imbedded, so fitting, so natural, that we use it without even thinking about it."[3] Indeed, he frequently referred to this third age as "the age of calm technology," meaning that in becoming omnipresent, computers also would become unremarkable.

The insights of Kittler and Weiser offer two versions of the disappearance of software in a world of computers. In this sense, they define the vanguard of writing about the cultural impact of software anno 2000. The notion that computers have taken over our lives has become fairly commonplace, but at a deeper level, Kittler and Weiser touch upon a source of malaise to some and exhilaration to others that future historians may regard as a fundamental aspect of our civilization—the notion that what is central is no longer the material, whether text or technology, but, rather, the imperceptible, the virtual, the invisible.

My topic today is the challenge these historians will face in documenting a cultural transformation that, by its very nature, has transformed the substance of historical documentation and radically altered the conditions of its preservation. For the most part, I will concentrate on the cultural medium of this transformation—software—and its history. By "software," let it be understood that I am using the term loosely to include not just code and executable programs, but also digital media dependent on software and, at times, computing generally. I also will comment here and there on the changes that efforts to preserve the history of software may impose on institutions such as libraries, archives, and museums.

My Stanford colleague Tim Lenoir has written that he is "intrigued by the notion that we are on the verge of a new renaissance, that, like the Renaissance of the fourteenth and fifteenth centuries, is deeply connected with a revolution in information technology." He describes the transformation of our times as "heralding a posthuman era in which the human being becomes seamlessly articulated with the intelligent machine."[4] Some of you in this audience might be more comfortable with the "printing revolution in early modern Europe," the title of Elizabeth Eisenstein's now-famous book, than you are with the notion that texts, technology, and even humanity have become dependent on, even integrated with, computer-based information technologies. On the other hand, who could be better situated to confront intellectual issues raised by profound transformations of media than historians of print and manuscript culture?

In my brief tour through some of the pitfalls and possibilities in building software history collections, I will begin with a short introduction on the history of software as a medium and describe a few of the characteristics of software that are likely to be important in historical perspective. From this wobbly ledge, I will dive into the turbulent problems that make it difficult to collect software. After drying off with a few examples of what has been accomplished despite the obstacles, I will finish this talk off with a cold shower by considering how providing access to these collections will raise organizational issues for archives, libraries, and museums.

The History and Historiography of Software

In light of the dependence of software on hardware, we should not be surprised that most histories of the software industry begin with their separation.[5] The short version of this story takes off from the announcement by IBM in June 1969 that it would unbundle the provision of software from the sale or lease of its computer systems. In other words, until 1969, most software came bundled with computer hardware systems, the very industry dominated by IBM. Not that independently developed and marketed software was completely unknown, but it was limited largely to special-purpose applications or academic projects.

During the 1970s, the business, culture, and technology of software production changed dramatically. The industry grew rapidly after 1969. According to Martin Campbell-Kelly, sales of software in 1970 represented less than 4 percent of the entire computer industry. The volume of sales increased from this base more than twentyfold by 1982, fiftyfold by 1985.[6] At about the same time, the term "software engineering" took hold to describe systems of software production based on theories and methods of computer science, stimulated by the first NATO Conference on Software Engineering in 1968. The proponents of software engineering applauded the establishment of computer science as a legitimate scientific field. The first academic departments in this new discipline were founded at institutions such as Purdue and Stanford in the early to mid-1960s; these new departments shifted the weight of attention to the study of software techniques, as opposed to the hardware engineering already sufficiently represented in electrical engineering and applied physics. Finally, also at the end of the 1960s, fundamental innovations in interface design and electrical engineering provided new platforms for changing the very meaning of computing in ways that redefined work on software. Douglas Engelbart's work at the Stanford Research Institute, for example, liberated the computer from its primary role as a calculating engine and headed it toward a future centered on information management and networked communications. The system designed at his Augmentation Research Center debuted spectacularly at the Fall Joint Computer Conference held in 1968 at the San Francisco Convention Center, just a few blocks from here, and inspired a generation of computer scientists to dream of new systems replete with mice, windows, icons, and desktops. Only a few months later, Stanley Mazor, Ted Hoff, and Federico Faggin designed the first single-chip central processing unit—a computer for all intents and purposes—which Intel introduced to the world as the 4004 microprocessor in 1971. A few years later, microprocessors made microcomputers possible, and soon after that, the personal computer took advantage of the capabilities of the new generation of software and computer interfaces.

Until these developments, the creation of software was inextricably tied to the relatively closed world of computer engineering. In time, the corporate world of Big Blue (IBM) gave way to the computers "for the rest of us," a change immortalized in the famous Macintosh Super Bowl television advertisement of 1984. Writing on the history of software production lagged behind these changes, focusing, until recently, on the period from roughly 1945 to 1970 and thus for the most part following early hardware development. Paul Edwards, one of a new generation of historians of computing, sites the older historiography in what he calls the tradition of "machine calculation," while also locating a distinct set of writings and historical actors in the tradition of "machine logic" (software). The ancestry of this latter tradition, he argues, "lies in mathematics and formal logic."[7] In his book on the "closed world" of Cold War computing, Edwards noted that the prevailing accounts within these internalist historiographies have rarely ventured beyond the perspectives of the scientists and engineers whose technical achievements defined them. Further, according to Edwards, "There is little place in such accounts for the influence of ideologies, intersections with popular culture, or political power."

Change in the History of Software

The changing industry, technology, and culture of computing from the 1970s to the present redefined the aspirations of software designers and programmers. As historians have begun to come to grips with these changes, limitations in the historiography of "machine calculation" and "machine logic" have become more apparent. Wider, then widespread, access to computer technology has intensified interest in the social, cultural, and business history of computing, topics of no little importance for a historical social construction of software. The PC Revolution of the late 1970s and early 1980s revealed intersections among the contributions of computer scientists, software engineers, hobbyists, and entrepreneurs, connections implicit in the founding of influential organizations such as the Homebrew Computer Club and the People's Computer Company, or the title of Ted Nelson's *Computer Lib/Dream Machines*, first published in 1974. The work of Douglas Engelbart, Ted Nelson, Alan Kay, and others active in the 1960s and 1970s provided nourishing soil for the rapid development of software technology in the 1980s and beyond; examples include the role played by Nelson's hypertext in the creation of the World Wide Web, the influence of Engelbart's SRI lab and Kay's work at Xerox PARC on the development of graphical user interfaces such as those embedded in the Macintosh and Windows, and the many spin-offs of Cold War research in artificial intelligence and other problems for information technologies, such as library catalogs. An authoritative history of software since the late 1960s has not yet been written, but when it is, its author will face the task of synthesizing a rich and variegated history extending beyond the internal development of code, languages, and protocols.

The Difficulties of Collecting Software

These brief remarks on the history of software merely set the stage for considering the difficulties—some would say the impossibility—of collecting software. Archivists, librarians, curators, and historians face the daunting task of documenting these strands of software history and providing the source materials for studying them. In the first instance, this means considering how historians ten or a hundred years from now will reflect upon a ubiquitous, but invisible, technology. The emancipation of software production from the closed and bundled world of computer engineering since the 1970s has rapidly accelerated our dependence on software, increasing not only our interest in its historical development, but also our awareness of the evolving nature of information resources and storage. The new developments of the past quarter century, including personal computer technology, graphical interfaces, networking, productivity software, electronic entertainment, the Internet, and the Web have expanded the use of software and shifted the discourse of software history, while at the same time providing an astonishing potential wealth of electronic data for historical analysis. Yet, this potential has not come without a cost.

One of the most important changes future historians of the past twenty or twenty-five years will study is the dizzying rate of change in the uses of software for supporting new media of communication, entertainment, and information management. Recall that the research of Engelbart, Nelson, and others established the computer as a communication, rather than a calculation engine. This first-order expansion of the nature of software has been followed by convergences of media and software technology that will push software historians into nearly every medium of entertainment, art, storytelling, and information management. Software has become many things to many people, occupying the work, leisure, and creative time of millions of nonprogrammers as well as software designers.

The broader social and cultural impact of computing will revolutionize (if it has not already) all cultural and scholarly production. It follows that historians (not just of software and computing) will need to consider the implications of this change, and they will not be able to do it without access to our software technology and what we did with it. Software and digital information have begun to rival printed materials, visual media, and manuscripts as primary sources in many fields of inquiry, while writers, artists, musicians, game designers, and even historians work productively in the media of computing.

Often every trace of these activities, save our memory of them, has been born digital. Consider this simple example: I recently attended the conference of the Electronic Entertainment Exposition in Los Angeles as a session panelist. E3, as it is known, is the Mecca of computer-based entertainment, and it draws the digital generation, such as journalists, to a free lunch (which, in fact, is one of the attractions). Justin Hall, the organizer of my panel on the "Computer and Console Games—A Cultural Legacy?," told me that he already sold an article written the night before our panel to the highest bidder. He had been tapping away at his laptop and downloading images from his digital camera, but only after returning home did I discover that everything about this transaction, as well as his text, was now embedded in silicon, as Kittler might say. Hall has been keeping an online, Web-based diary continuously since 1994, "Justin's Links from the Underground," and after checking this site, I found this link, as we say, in his comments about the E3 article "The Auction." A moment's further browsing, and I found the article at salon.com. In other words, this article was written, photographed, sold, and published without a single "written" or paper trace. Of course, there are millions of similar examples of commerce, entertainment, authorship, artistic creation, journalism, science, and even software engineering carried out without paper. Each one adds to the urgency of software preservation, digital archiving, and accessible electronic libraries on a front far broader than the history of computing.

The relentless advance of computer technology on an ever-expanding set of fronts is redefining the nature and scope of computing itself. It could be argued, at least from the vantage of the present, that human beings interact directly with computers more than with any other technology. In many contemporary families, computers have partly replaced television sets, radios, and telephones. In *The Road Ahead*, published in 1995, Bill Gates provided a vision of the near future of computing that explicitly includes *all* "mediated experiences," whether of commerce or culture.[8] Historians of software clearly will have to venture into every niche, nook, and cranny of society in ways that will separate their work from the work of other historians of science and technology. It has become far more difficult to locate the edges of computing as a discipline and to map the boundaries of its impact on society than for most other technical and scientific fields. The open-ended nature of computing challenges archivists, librarians, and curators, and it complicates matters for researchers looking for disparate materials in a variety of media and repositories.

So what do we do in the face of the growing volume, diversity, and importance of software? Part of the difficulty in defining next steps is that the very cat we are trying to put in the bag is ripping our heirloom luggage to shreds. This is perhaps where the history of software least resembles the history of print culture. This is not so much in the impermanence of its media—an issue upon which the dust has not yet settled—but, rather, in the flexibility of its use, with the capacity for converging previously separable realms concerned with what we now call "content": texts, stories, audiovisual experiences, interactive simulations, data processing, records management, and metadata applications such as indexing, among them. Traditional institutions and professional identities provide uncertain guidance in deciding who is responsible for the custodial care of software, given this diverse range of applications and associated knowledge. As Doron Swade points out from the perspective of a museum curator:

> Some software is already bespoke: archivists and librarians have "owned" certain categories of electronic "document": Digitised source material, catalogues, indexes, and dictionaries, for example. But what are the responsibilities of a museum curator? Unless existing custodial protection can be extended to include software, the first step towards systematic acquisition will have faltered, and a justification for special provision will need to be articulated ab initio in much the same way as film and sound archives emerged as distinct organisational entities outside the object-centred museum.[9]

Swade considers the problem as one of "preserving information in an object-centred culture," the title of his essay; that is, he ponders the relevance of artifact collections of software and the various methods of "bit-perfect" replication of their content. Libraries, and within libraries rare books and manuscript librarians, are coming to grips with related issues that might be described as "preserving information in a text-centred culture." In saying this, I realize that rare book and manuscript librarians are quite often the chief protectors of artifact-centered culture in American libraries. Nonetheless, their raison d'être is the preservation of special categories of original source materials—primarily texts—for programs of academic research and teaching. This is one of the rubs in formulating institutional approaches to the preservation of software and related digital media, for software defines a new relationship between media objects and their content, one that calls into question notions of content preservation that privilege the original object. Current debates about the best methods for preserving software, which I have no intention of rehearsing here, are partly stuck on different institutional and professional allegiances to the preservation of objects, data migration, archival functions, evidentiary value, and information content. I fear that these issues are not likely to be sorted out before it is necessary to make serious commitments at least to the stabilization, if not the long-term preservation, of digital content and software.

What Can Be Done? Some Projects and Programs

Preservation of the records of software history has benefited from archival and historical work in other areas of recent science and technology. By the late 1970s, archival organizations, historical repositories, and professional societies had begun to pay systematic attention to their history. Disciplinary history centers such as the AIP History Center, the IEEE History Center, and the Charles Babbage Institute were established, in part, to coordinate and support the preservation of historical documentation and to work with existing repositories to address issues of archival appraisal, preservation, and access. In the early 1980s, the Society of American Archivists, History of Science Society, Society for the History of Technology, and the Association of Records Managers and Administrators cosponsored a Joint Committee on Archives of Science and Technology, known as JCAST. Its report, *Understanding Progress as Process: Documentation of the History of Post-War Science and Technology in the United States*, represented an important milestone when published in 1983, especially by raising awareness among American archivists of their need to understand better the records of postwar science and technology.

A loosely knit group of archival repositories and, just as important, an evolving set of principles and practices emerged out of archival research and projects such as the JCAST report. Guidelines for appraisal of records and documentation strategies set the stage for projects. By the late 1980s, the first published guides to collections in the history of computing appeared in print: *Resources for the History of Computing*, edited by Bruce Bruemmer, and *The High-Technology Company: A Historical Research and Appraisal Guide* by Bruce Bruemmer and Sheldon Hochheiser, both published by the Babbage Institute; and *Archives of Data-Processing History: A Guide to Major U.S. Collections*, edited by James Cortada and published by Greenwood Press. Together, they effectively document the strategies and programs that guided the growth of archival resources in the history of computing up to about 1990. Yet, it was clear that the work had only begun. Cortada noted that:

> The first group of individuals to recognize a new subject area consists usually of participants followed closely after by students of the field and finally, if belatedly, by librarians and archivists. It is very frustrating to historians of a new subject, because it takes time for libraries to build collections or to amass documentary evidence to support significant historical research. This situation is clearly the case with the history of information processing.[10]

During these initial stages, the documentation of the history of computing was largely paper based. The founding of the archives of the Charles Babbage Institute at the University of Minnesota in 1979 (it had been founded at Stanford a few years earlier) was a signal event in this phase, as was the publication of a brochure on behalf of the History of Computing Committee of the American Federation of Information Processing Societies (AFIPS), called "Preserving Computer-Related Source Materials" and distributed at the National Computer Conference that year. The information in this brochure was based on earlier documentation efforts at the Center for the History of Physics of AIP. The brochure recommended that:

> If we are to fully understand the *process* of computer and computing development as well as the end results, it is imperative that the following material be preserved: correspondence; working papers; unpublished reports; obsolete manuals; key program listings used to debug and improve important software; hardware and componentry engineering drawings; financial records; and associated documents and artifacts.[11]

It focused almost entirely on the preservation of paper records, even printouts, manuals, and text listings of programs, but nowhere mentioned the preservation of data files, merely noting with a nod to the museum value of hardware artifacts that "Actual computer componentry is also of great interest. The esthetic and sentimental value of such apparatus is great, but aside from this, the apparatus provides a true picture of the mind of the past, in the same way as the furnishings of a preserved or restored house provide a picture of past society."[12]

Even in the absence of a mandate to save software, libraries, archival repositories, and museums have mobilized resources to document the history of computing. Historians of software will draw on a variety of historical documentation that includes many formats, both digital and paper based. Because of the widening realm of software applications, hundreds, if not thousands, of repositories have saved collections touching on the history of computers and computing. Consider topics such as the history of hospital information management, library database technology, scientific computation, digital typography, or computer graphics in the film industry, topics for which documentation may be found in repositories ranging from government record centers and university archives to closed private collections and corporate records centers. The spectrum ranges from the Library of Congress to the Disney Archives, and as we shall see in a few moments, to virtual collections such as the Internet Archive.

Archives of Data-Processing History provided a good overview of the major repositories in the field circa 1990, and this circle has not widened considerably since that time, even though many collections have been added since then. The core group of brick-and-mortar collections consists of the Charles Babbage Institute, the Computer Museum (now the Computer History Center), the Hagley Museum and Library, the Library of Congress, the National Archives and Records Administration, the Smithsonian Institution, and the Stanford University Libraries, plus several corporate archives (IBM, AT&T, Texas Instruments, and so on). Smaller, but nonetheless significant, collections can be found in university libraries and archives at Dartmouth, Harvard, MIT, Carnegie Mellon, Illinois, and Pennsylvania as a consequence of the historical role of these institutions, rather than active collecting programs. In short, certainly fewer than ten institutions in the United States actively collect research materials in traditional formats for the history of computing. Growth since the early 1990s in available documentation has occurred largely as a result of independent, largely Web-based initiatives such as the RFC (Request for Comment) Index of key documents on the development of the Internet, private initiatives such as the Internet Archive, and numerous collections of digitized and born-digital materials assembled and accessible via online archives, home pages, and corporate Web sites. In a sense, a second generation of software archives has emerged in its own medium, creating a recursive problem concerned with the long-term preservation of these digital archives.

The Stanford University Libraries, where I have been curator of the history of science and technology collections since 1983, maintains an active archival program in the history of computing. Let me take a few minutes now to use our program as an example of how institutions go about acquiring collections of historical records relating to software. The library's program in the history of computing grew on two legs: first, an archival orientation in the narrow sense, focused on records of activities that took place at Stanford; and second, a collecting program founded in 1984 and called the Stanford and the Silicon Valley Project, today known as the Silicon Valley Archives. The idea behind the Silicon Valley Project was straightforward: Compile documentation tracing relationships connecting Stanford faculty and graduates to emerging high-technology industries in the surrounding region since the 1930s. It extended a flourishing program in the University Archives that, by the mid-1980s, had assembled collections of faculty papers and university records in the sciences and engineering. For software history, relevant collections in the University Archives include the papers of Ed Feigenbaum, John McCarthy, George and Alexandra Forsythe, Donald Knuth, and many others, as well as records of the Center for Information Technology (Stanford's computation center), the BALLOTS project papers (an early project in the area of library automation and database technology), the ACME Project collection (a collaboration of Edward Feigenbaum and Joshua Lederberg that led to path-breaking software in the field of expert systems such as MYCIN and DENDRAL), and the Heuristic Programming Project. As the Department of Computer Science, founded in 1965, has become perhaps the leading university program in its field, the University Archives has, by preserving records of its programs and faculty papers, grown in importance for the history of computing.

By 1984, it had become clear that the explosive growth of Silicon Valley not only dominated regional development, but that it was a forerunner of other highly concentrated techno-scientific regions. Because of the close connections between Stanford and specific business ventures located in Silicon Valley, the University Archives already owned significant collections relevant to its historical development. It was a logical step for the Department of Special Collections and University Archives to move forward and actively collect records of Silicon Valley enterprises and individuals not directly tied to Stanford. It appeared that no other institution would invest resources to locate and preserve archival materials documenting research and business growth characteristic of Silicon Valley industries. As a result of the decision to move forward with this program, Stanford has acquired substantial company and laboratory records, such as those of Fairchild Semiconductor Corporation, the American Association for Artificial Intelligence, the System Development Foundation, SRI Laboratories under the direction of Douglas Engelbart and Charles Rosen, and Apple Computer. After the parameters of our project had been established, we proceeded to work with faculty who were known to have contacts in Silicon Valley industry, such as Edward Feigenbaum and, more recently, Doug Brutlag. Another vector from Stanford out to Silicon Valley along the path of software, one of particular interest to this meeting, has been followed in digital typography, with the acquisition of the Euler Project papers of Hermann Zapf and the voluminous papers of Donald Knuth.

A new twist in the Silicon Valley Project has been the acquisition of software in various forms, accompanied by research projects that seek to tell the story of the Silicon Valley in its own medium. In the first instance, the libraries have acquired materials such as data tapes from Engelbart's ARC projects, hard-disk images along with collections of personal papers such as those of Jef Raskin and Mark Weiser, email archives, streamed media and digitally taped audio and video interviews, electronic versions of student papers, and packaged commercial software, such as the Stephen F. Cabrinety Collection in the History of Microcomputing, which includes one of the world's largest collections of early computer and video games. Each of these formats requires special strategies for evaluating, recovering, stabilizing, possibly reformatting, and indexing content. In some cases, the strategies do not yet exist; in others, we have embarked on special projects to test techniques for ensuring that future historians will have access to their contents and creation. For example, in the case of computer game software, Tim Lenoir and I are heading a project called "How They Got Game: The History and Culture of Computer Games and Simulations," funded by the Stanford Humanities Laboratories. As part of this project, the results of which will be entirely Web based, we are evaluating a three-pronged approach to the documentation of game software: streamed video of gameplay, location and preservation of source code, and scanned images of related packaging, marketing materials, and documentation. Note that our efforts steer relatively clear of emulators, metadata packaging, and the preservation of hardware, techniques currently at the center of contention among museum curators, archivists, and librarians about best practices for the long-term preservation of digital documents.

Why Is Access to Software History Collections Difficult?

Although I have certainly left out more topics than I have covered, I would like to conclude with a few remarks about the role of special collections in the preservation of software history. As we have seen, at both Stanford and other institutions, the archival impulse in the history of computing began with paper-based records and documentation. The printed guides cited earlier list the personal papers of computer scientists held in manuscript collections and archives, oral histories, and corporate records. Early computers have been saved by museums such as the Computer History Center and the Science Museum in London, and libraries have saved collections of documentation, technical reports, and the early computing literature. At Stanford, as elsewhere, manuscript, ephemera, and, to a lesser extent, book collections in the history of computing have landed in special collections and archives as an extension of earlier patterns of collecting practice. As the nature of this documentation shifts from paper to electromagnetic storage media, issues of access and technological complexity are calling this habit into question.

Access to software collections is the first problem. The mission of departments of special collections, especially in university libraries, includes not just preservation, but also satisfying the access requirements of users of these materials. Traditional models of access focused on the service desk and reading room as means of mediating complex systems of indexing and identification of materials, as well as supervised reading, fall apart in delivery contexts shaped by computer hardware and virtual libraries of born-digital materials. This is a problem not just for software history, but also for every field of cultural inquiry. Literary drafts, correspondence, graphics media, data, and images created in the 1990s are more likely to reside on disk or in networks than on paper, and the trend, as an optimistic stockbroker might say, is upward.

This issue of access to digital documents and software strikes me as urgently requiring new institutional and curatorial models. Let us consider again the divergent roles of archives, libraries, and museums. W. Boyd Raymond argues in an article on how electronic information is reshaping the roles of these institutions that "the functional differentiation of libraries, museums, and archives as reflected in different institutional practices, physical locations, and the specialist work of professional cadres of personnel is a relatively recent phenomenon. This functional differentiation was a response to the exigencies of managing different kinds of collections as these have grown in size and have had to respond to the needs and interests of an ever enlarging body of actual and prospective users." Raymond's view is that individual scholars continue to favor the ideal of "personal cabinet of curiosities" finely tuned to specific research, an ideal that considers the format of artifacts and media as irrelevant, while stressing content. This was the "undifferentiated past" that these institutions hold in common.[13]

The often-synonymous usage of Special Collections and Rare Books and Manuscripts as designations of library programs will change as a result of collections of digital media and software. This will be a permanent change, and we cannot expect the traditional special collections community to come up with all the answers. One possibility is a kind of functional consolidation of media collections, digital libraries, and software archives. The creation of such cabinets of media curiosities would assemble specialists in curatorial domains that are now separated, while cutting off the uncontrolled extension of established departments of special collections to digital materials and refocusing their attention on the venerated realms of rare books and manuscripts. Still, as Swade has noted in his writings on collecting software, it is tempting to lay aside theoretical problems of proper custody for software and worry, instead, about the work. The conundrum here is that while the relationship of software to hardware, its storage on physical media, or its association with artifacts such as disks, computers, and boxes, might lead one to think of software as fit for the museum, requirements of scholarly access such as identifying and locating sources, standards of indexing and metadata creation, and maintenance of collections for retrieval and interpretation seem more in line with the capabilities and programs of libraries and archival repositories. In short, ad hoc decisions about curatorial responsibility may well have long-term implications for future scholarly work.

Kittler's admonition that "there is no software" provides little relief to archivists and librarians who discover that there is more of it than they can handle. And yet, the separation of physical media from content offers the glimmer of hope that the hard work of software history might be accomplished through a mixture of new organizational models, new technological skills, and established practices, as well as a reconvergence of museum, library, and archival curatorship.

NOTES

1. Friedrich Kittler, "There Is No Software," *C-Theory: Theory, Technical, Culture* 32 (October 18, 1995), http://www.ctheory.com/article/a032.html.

2. "The Computer for the 21st Century," *Scientific American* (1991). I am using the draft Weiser posted at http://www.ubiq.com/hypertext/weiser/SciAmDraft3.html.

3. "Ubiquitous Computing," http://www.ubiq.com/hypertext/weiser/UbiHome.html.

4. Tim Lenoir, "All but War Is Simulation: The Military-Entertainment Complex," *Configurations* 8 (2000): 289.

5. Luanne Johnson, "A View from the Sixties: How the Software Industry Began," *IEEE Annals of the History of Computing* 20, no. 1 (1998): 36–42. Johnson's article provides a summary of this development.

6. Martin Campbell-Kelly, abstract of "Development and Structure of the International Software Industry, 1950–1990," Conference on "History of Software Engineering," Schloß Dagstuhl, August 26–30, 1996, http://www.dagstuhl.de/DATA/Reports/9635/campbell-kelly.html.

7. Paul Edwards, *The Closed World: Computers and the Politics of Discourse in Cold War America* (Cambridge, MA: MIT Press, 1996). Quoted from excerpt of chapter 1 at http://www.si.umich.edu/~pne/cwpref.htm.

8. Bill Gates, *The Road Ahead* (New York: Viking, 1995).

9. Doron Swade, "Collecting Software: Preserving Information in an Object-Centred Culture," in *Electronic Information Resources and Historians: European Perspectives*, ed. Seamus Ross and Edward Higgs (St. Katharinen, Germany: Scripta Mercaturae, 1993), 94.

10. James W. Cortada, preface to *Archives of Data-Processing History: A Guide to Major U.S. Collections* (New York: Greenwood, 1990), ix.

11. From the version provided by the Software History Center at http://www.softwarehistory.org/.

12. This brochure was later reproduced in the *IEEE Annals for the History of Computing* 2 (January 1980). The text of this brochure is available via the Web site of the Software History Center, http://www.softwarehistory.org/.

13. W. Boyd Raymond, "Electronic Information and the Functional Integration of Libraries, Museums, and Archives," in *Electronic Information Resources and Historians: European Perspectives*, ed. Seamus Ross and Edward Higgs (St. Katharinen, Germany: Scripta Mercaturae, 1993), 227–243, esp. 232.

2

Shall We Play a Game

Thoughts on the Computer Game Archive of the Future

The Importance of the Topic

Now that the *Pong* and *Pac-Man* generations have aged and few Stanford undergraduates can remember a world without Mario, computer games have suddenly grown up. It is a good time to think about the place of games among media and how we can preserve their history.

Many would argue that the "medium of the video game" (the title of a recent book)[1] should be a prominent stop on any tour of late 20th-century mediascape, certainly of popular culture and art. The conventional approach cites statistics that measure the commercial success of computer games. Sales of computer and video games in the United States alone, including hardware and accessories, exceeded $10 billion in 2001. By comparison, box office receipts in the US movie industry reached $8.35 billion, itself a record total. Global sales of hardware and software are expected to exceed $30 billion this year, while developed markets for PC or video games in Japan, Korea, Germany, and the UK rivaling the US dollar figures alone do not say enough about the attention our society devotes to computer games. Generally, games take more time to experience than media such as books and films. The publishers of *Half-Life: Counter-Strike*, the most popular networked multiplayer game, report some 3.4 billion player-minutes per month in mid-2002, exceeding estimates based on Nielsen ratings for even the highest-rated US television shows. *Counter-Strike* frags *Friends* in the battle for screen time. About 1.5 billion movie tickets will be sold this year. We can calculate from this

This essay originally was presented at the conference "Bits of Culture: New Projects Linking the Preservation and Study of Interactive Media," Stanford University, 2002.

statistic that on average less than 15 percent of the US population goes to the movies in a given week (down from 46 percent after World War II, by the way). By comparison, the Interactive Digital Software Association reports that roughly 60 percent of the US population played "interactive games on a regular basis" in the year 2000. Yes, video and computer games have got our attention.[2]

There are reasons other than net profits and eyeball time for taking computer games seriously. Games are shedding their reputation for merely providing mindless amusement for young boys. Henry Jenkins suggests that it is time to think about video games as the "art form for the digital age." Some of you may find this thought difficult to reconcile with faded memories of *Pong* and *Pokemon*, or the quietly muttered thanks as your kids stare vacantly into a Game Boy instead of fighting in the back seat. Allow Jenkins to rap your knuckles as he observes that such reactions "tell us more about our contemporary notion of art—as arid and stuffy, as the property of an educated and economic elite, as cut off from everyday experience—than they tell us about games."[3] Indeed, anyone familiar with the narrative aspirations of interactive fiction, the social worlds of MUDs and massively multiplayer games, the technical mastery of id's first-person shooters, the visual storytelling of the *Final Fantasy* series, the strategic depth of Sid Meier's *Civilization*, or Will Wright's open-ended simulations from *SimCity* to *The Sims* knows that game design has matured. Computer games cover an astonishing breadth of entertainment, simulation, artistic, competitive, and narrative experiences.

Computer games are beginning to put their stamp on academic media and cultural studies. It is sufficient for now to cite Jay David Bolter and Richard Grusin's *Remediation*—which is becoming to new media what Elizabeth Eisenstein's *The Printing Press as an Agent of Change* was to print.[4] *Remediation* begins its list of the essential new media with a chapter on computer games. The bibliography of game studies is growing rapidly, fed by scholars in literary, media and cultural studies, and the social sciences, as well as game designers. Let us agree then that computer games are and will be taken seriously. Today I want to talk about the problem of ensuring that this new medium will have a history, one that future scholars can write about critically.

In considering the construction of computer game archives, I will organize my thoughts into three sections: (1) What are some challenges we face in building historical collections of software and interactive media? (2) How will the characteristics of computer games define historical collections and who will be the curators of these collections? (3) What can we do now to advance the computer games archives of the future?

The Challenge of Preserving the History of Software and Interactive Media

The idea of playing games on computers is about as old as the computer itself. Initially, the payoffs expected from this activity were closely related to the study of computation. For example, the mathematician and engineer Claude Shannon proposed in 1950 that computers could be programmed to play chess, leading him to wonder if machines could think. Shannon's proposal stimulated decades of research on chess- and checkers-playing programs by computer scientists working in the field of artificial intelligence. Similarly, games have proven their value as a way of experimenting with computer technology and stretching its capabilities. Pioneering early games such as William Higinbotham's *Tennis for Two* (1958) and MIT's *Spacewar!* (1962) were designed to demonstrate what computers could do. Since the 1970s, connections leading from the lab to the living room have usually been mediated by military or industrial R&D and commercial game developers. Indeed, Tim Lenoir and I would argue that understanding the relationship among these groups is an important and revealing research topic. Games exploit technology; they revel in pushing the technological envelope of personal computers, arcade and television consoles. Since the mid-1990s (one might say, since *DOOM*), it has been games primarily that have driven the need for high-end personal computers and 3D graphics boards, for example. I will mention a few implications of the sturdy technology requirements for games when I talk about emulators later on.

In part, computer and video games are a subcategory of software, and it is instructive to contemplate the work of historians of software in ten or a hundred years. Since the 1970s, the emancipation of software from the closed and bundled world of computer engineering has led to our dependence on computers. While the scope of software's history has thus grown, our personal intimacy with evolving information technologies has intensified. In the last quarter century, personal computer technology, graphical interfaces, networking, productivity software, electronic entertainment, the Internet, and the World Wide Web have made software part of our cultural environment, thus profoundly raising the stakes for being able to tell its history, our history.

It is important in this regard to recall the contributions of Douglas Engelbart, Ted Nelson, Alan Kay, and others in the 1960s and 1970s. They laid a foundation for digital media of communication and entertainment, not just for productivity and information management. They did this by establishing the computer as a communication machine rather than primarily a calculation engine. The monumental expansion of the nature of software that followed has led more recently to convergences of media and technology that push software (and its historians) into nearly every medium of entertainment, art, recreation, and storytelling. Software has become a condition of our lives; culture is embedded in the computer as much as the computer is embedded in culture. About ten years ago, the computer scientist Mark Weiser described the then future omnipresence of software in an article titled "The Computer for the 21st Century" published in *Scientific American*. This essay introduced Weiser's research program, which he dubbed "ubiquitous computing," to the magazine's technologically literate readership, always eager to read about plausible visions of the future. Much of Weiser's argument hinged on a straightforward observation, but one that nonetheless turned his views in an unexpected direction: "The most profound technologies are those that disappear. They weave themselves into the fabric of everyday life until they are indistinguishable from it." He believed it significant not that in his future computers would outnumber people, but that they would have to become "invisible" in order to become useful. As he phrased it a few years after publishing the *Scientific American* article, the "highest ideal is to make a computer so imbedded, so fitting, so natural, that we use it without even thinking about it."[5] We have barely begun to address the long-term preservation of this ubiquitous, yet invisible technology.

Archives of Computer Game History: Collections and Curators

The Nature of the Medium

As we tag along behind Bolter and Grusin down the path of remediation, let us not limit our gaze to cinema, literary texts, and television as the role models for computer games. They are, of course, games. We should keep this in mind when we enter video games in the genre catalog of new media. In *The Study of Games* (1971), the essential treatise of the anthropology of games, Elliott Avedon and Brian Sutton-Smith asked: "What are games? Are they things in the sense of artifacts? Are they behavioral models, or simulations of social situations? Are they vestiges of ancient rituals, or magical rites? It is difficult and even curious when one tries to answer the question 'what are games,' since it is assumed that games are many things and at the same time specific games are different from one another—but are they?"[6]

They are asking two questions here. The first concerns the essential nature of games. Allow me to oversimplify and reduce the options to two: either games are fixed objects—perhaps authored texts or built artifacts—or, alternatively, they are the experiences generated by a framework of rules, codes, or stories and expressed through interaction, competition, or play. Text or performance? Artifact or activity? I will return to this part of their question when I talk about preservation strategies. The second question is whether there are structural similarities among games. Answering this question is out of scope for us today. However, it is worth noting that the authors of *The Study of Games* conclude with "seven elements in games" distilled from studies by psychologists, mathematicians and others. These structural elements include "procedures for action," "roles of participant," "participant interaction patterns," and the like, taking games away from the notion that they are stable artifacts or texts. These elements underscore the importance of documenting interactivity as a historical phenomenon, something that predated computers (an obvious statement to any player of *Diplomacy* or *Dungeons and Dragons*). As fashionable as it has become to discuss games as cinematic or as narratives, let us not forget that actions and responses are fundamental. Games provide a structure within which players do something—whether the games is baseball, *D&D*, or *Myst*—and this structure is not likely to be a linear narrative. Computer games provide a tailor-made opportunity to study interactivity.

The active, performative aspect of games provides a special challenge for documentation strategies. As a thought experiment, think for a moment about the case of a game like basketball. Let's look at some records and try to choose between texts, artifacts, or records of performance. Do any of these sources alone tell us enough about the nature of the game? Don't we need all three?

Computer and video games are both dynamic and interactive. Dynamic, in the sense that no matter how linear, how narrative-driven the game, each instance of it being played results in a different set of experiences. The text is never the same. The interactivity of games is the sine qua non of the definition of this new medium, the aspect without which computer games would lose their identity. Chris Crawford, the dean of American game designers, put it this way: "Interactivity is not about objects, it's about actions. Yet our thought processes push us towards objects, not actions. This explains why everybody is wasting so much time talking about 'content.' Content is a noun! We don't need content; we need process, relationship, action, verb."[7] Randall Packer, a media artist and museum curator, has described a similar quality in interactive art, particularly net-based hypermedia, noting the degree to which computer games embrace the loss of fixed content: "While theater begins with the notion of the suspension of disbelief, interactive art picks up where theater (and film) leave off with branching, user-driven non-linear narrative. The letting go of authorial control has been the big dilemma of interactive works as an art and/or entertainment medium, games being the exception."[8]

The loss of authorial control is not a dilemma for game designers, because they embrace it as the foundation of a medium created by actions as much as content. Games exist somewhere between the text and the experience, confounding preservation strategies that rely on notions of content fixity taken from other media. Hardware and software objects alone cannot document the medium of the computer game. What is saved by preserving consoles, hardware, and software alone, without recording gameplay?

The nature of computer games as software leads to another sort of variability of content, one that does not yet apply so much to console and arcade-based games as it does to computer-based games. I am referring to what Lev Manovich has called the new "cultural economy" of game design, introduced in 1993 with the release of id Software's *DOOM*.[9] When a computer game is released today, it is as much a set of design tools as a finished design. Since the advent of level editors and the increasing popularity of modified games—"mods"—many designers ship game engines accompanied by a game, levels, and maps designed in-house. (Game engines are the software platforms for handling graphics, game physics, artificial intelligence, the structure of game levels and file formats, editors, etc.) Then level, scenario, and mod designers take over, creating their own games. Manovich contrasts modifiable games to the more traditional characteristics of a game like *Myst*: "More similar to a traditional artwork than to a piece of software: something to behold and admire, rather than take apart and modify." The contemporary game scene pulses with the energy of new content developers using game engines to design new games (such as *Dungeon Siege* "siegelets"), modding games and game engines, skinning characters, working up freeware utilities, changing art assets or even conflating media by using game engines to produce movie-like game experiences, such as game movies and machinima. *Counter-Strike*, the most popular Internet-based game of all, is of course a multiplayer modification of single-player *Half-Life*, demonstrating how mainstream the mod economy of game design has become.

Curatorship Issues

Curatorship of interactive digital media collections confronts the growing volume, diversity, and importance of impermanent and invisible software, even for its own preservation. The cat we are trying to put in the bag is ripping our heirloom luggage to shreds. The history of software diverges from the history of print culture, not just in the impermanence of its media—an issue on which the dust has not yet settled—but in the flexibility of its use and capacity for converging previously separated realms: texts, stories, audio-visual experiences, interactive simulations, data processing, records management, and metadata applications such as indexing, among them. Traditional institutions and professional identities provide uncertain guidance in deciding who should be responsible for the custodial care of software and new media collections, given such a diverse range of contents and associated knowledge.

As Doron Swade points out from the perspective of a museum curator, "Some software is already bespoke: archivists and librarians have 'owned' certain categories of electronic 'document': Digitised source material, catalogues, indexes, and dictionaries, for example. But what are the responsibilities of a museum curator? Unless existing custodial protection can be extended to include software, the first step towards systematic acquisition will have faltered, and a justification for special provision will need to be articulated ab initio in much the same way as film and sound archives emerged as distinct organisational entities outside the object-centred museum."[10] Swade considers the problem as one of "preserving information in an object-centred culture," the title of his essay; that is, he ponders the relevance of artifact collections of software and the various methods of "bit-perfect" replication of their content.

Libraries, and within libraries rare books and manuscript librarians, are coming to grips with related issues that might be described as "preserving information in a text-centred culture." In saying this, I realize that these librarians are often the chief protectors of artifact-centered culture in American libraries. Nonetheless, their raison d'être is the preservation of special categories of original source materials—primarily texts. This is one of the rubs in formulating institutional approaches to the preservation of software and related digital media, for software defines a new relationship between media objects and their content, one that calls into question notions of content preservation that privilege the original object. Current debates about the best methods for preserving software are hung up to some degree on different institutional and professional allegiances to the preservation of objects, as well as data migration, archival functions, evidentiary value, and information content. These issues are not likely to be sorted out in time to make serious commitments even to the stabilization, let alone the long-term preservation, of digital content and software. Brewster Kahle's Internet Archive demonstrates what can be done outside these institutional constraints, but it is an exceptional case, rather than providing a new rule.[11]

The historical preservation of games and other interactive multimedia will depend on the development of new models of curatorship and collections. Jürgen Claus, now professor of media art at the Kunsthochschule für Medien in Cologne, introduced this topic in his 1985 essay "Expansion of Media Art: What Will Remain of the Electronic Age?" He said simply, "If media art does constitute a new stage of development, then we have to ask for adequate spaces to display and store this art, that is, we have to ask for media museums."[12] Claus insisted that "the Museum must not be relieved of its duty of being a place of reference for works of remaining value. Certainly, film, photography, video, disc, tape, etc. are media to store events of art. Where should they be collected, examined, and passed on if not in an adequate museum, that is, a media museum?"[13] More recently, Matthew Kirschenbaum, a literary scholar who has written particularly about hypertext, has asked what it means to treat electronic texts such as Michael Joyce's *afternoon* as a textual artifact "subject to material and historical forms of understanding."[14] He calls into question the duality that sees printed texts as durable and fixed, electronic texts as "volatile and unstable." Perhaps Eisenstein's notion of the "fixity of print" has obscured counter approaches best exemplified in such disparate scholarly worlds as editorial and reader studies.

Kirschenbaum points out the equally dangerous position of a postmodernist embracing of the ephemeral qualities of electronic media. He concludes that a kind of bibliographic/textual scholarship can be applied to the "authorial effort to create links, guard fields, and so forth," as long as the network of code, technology, and documentation that underlies the creation of hypermedia is preserved. I would add to this that these artifacts cannot be adequately interpreted without establishing contexts of design, creation, and technology, and for this we need documentation, texts, source code, artwork, and so on. A brochure published by the History of Computing Committee of the American Federation of Information Processing Societies (AFIPS) over two decades ago recommended: "If we are to fully understand the *process* of computer and computing development as well as the end results, it is imperative that the following material be preserved: correspondence; working papers; unpublished reports; obsolete manuals; key program listings used to debug and improve important software; hardware and componentry engineering drawings; financial records; and associated documents and artifacts."[15] The media museum will be equal parts museum, library, and archives.

Access to historical collections of digital files and software strikes me as urgently requiring new institutional and curatorial models. The roles of archives, libraries, and museums will converge rapidly in the realm of new media collections such as the computer game archives of the future. W. Boyd Raymond has written about this convergence, arguing that electronic information is reshaping the roles of these institutions. He points out that "the functional differentiation of libraries, museums and archives as reflected in different institutional practices, physical locations, and the specialist work of professional cadres of personnel is a relatively recent phenomenon. This functional differentiation was a response to the exigencies of managing different kinds of collections as these have grown in size and have had to respond to the needs and interests of an ever enlarging body of actual and prospective users." Raymond's view is that individual scholars continue to favor the ideal of a "personal cabinet of curiosities" finely tuned to specific research, an ideal that considers format of artifacts and media as irrelevant, while stressing content. This was the "undifferentiated past" that these institutions hold in common.[16] Computer game archives, as cabinets of new media curiosities, will need to consolidate access to media collections, digital libraries, and software archives in this way.

A Plan for Action

I have argued that in the realm of digital, interactive software and multimedia, preservation strategies limited to the long-term stabilization of fixed content provide only partial solutions, whether this fixed element is defined as a kind of text, an artifact, or software code itself. Similarly, existing institutional and curatorial models—museums, libraries, special collections departments, and archives—provide only partial models for the care and feeding of the kinds of collections we will need to build.

It seems to me that as we begin building the computer game archives of the future in the present, we should be guided by three notions. The first is careful revision of institutional and curatorial roles for historical new media collections. The second is to begin definition now of the technical foundations of these archives, focusing on how to capture and preserve the "look and feel" of interactive media, interactivity itself, and the social and personal experiences made possible by computer games. At the same time, we should keep our eyes on the ball of preserving existing documentation, hardware, data, and metadata in the full variety of their formats. The third notion is simply that the lynchpin of all that follows will be to solve these problems in collaborative, multi-institutional projects. No single institution owns the resources or expertise under one roof to go it alone in this realm. But a network of museums, libraries, and new digital repositories (such as the Digital Game Archive or the Internet Archive) would make a good start. Moreover, without the cooperation of industry groups such as the International Game Developers Association (IGDA), game designers, and publishers, we cannot make progress on the sticky social, business, and legal issues that hinder the building of safe harbor collections relying on voluntary participation and provide access to collections to as wide a community of users as possible.

Now I will propose five salient tasks and challenges facing such an alliance of interested parties:

1. Build emulation test beds.

We are only at the beginning of the technology debate about the best way to preserve bits of culture—born-digital content and software. Strategies based on migration, hardware and media preservation, emulation, and Rothenberg's "encapsulation" have been debated at great length, though there are few tests of the options. In fact, the application of any one of these methods depends to some degree on one or more of the others. For example, it is difficult to imagine an emulation strategy that does require migration of emulation software to new hardware platforms. It is likely that all of these methods will play some role in the preservation of software history, but at the same time we can agree that in the case of games, a method that preserves "look and feel" is particularly attractive. However, as noted at the beginning of this talk, the technology requirements for running games are evolving rapidly to take advantage of new means for enhancing graphics and gameplay. Emulation design must take account not just of operating systems and software but also the requirements built into games for audio and graphics boards, 3D programming interfaces (application programming interfaces such as OpenGL and Direct3D), networking protocols (and speeds!), displays, and controllers. For this reason, I propose that we fund and implement a very small number of complete test beds to develop emulation as a long-term archival strategy. "Complete" means that these test beds not only deliver technology but also architect repositories that address intellectual property rights, access, cost control, and collection development. These problems are complex, and I propose the following mix of test beds: (1) a system whose software is largely or completely in the public domain (such as the Vectrex); (2) a single console platform and publisher (such as Atari's titles for the 2600 or Nintendo's for the NES); (3) the games published as computer or personal computer software by a single publisher or in a single genre (Infocom's interactive fiction titles, Strategic Simulation Inc.'s military simulations, or id's 3D shooters come to mind); (4) a game produced outside the commercial sphere (such as MUD or games produced by the PLATO project).

2. Build a game performance archive.

Games are about interactivity, and as Crawford has told us, interactivity is about actions, not just content. I do not mean to underestimate the difficulties of documenting performance, but the nature of computer game technology is conducive to creating collections that document gameplay. They are played on screens and computers, opening up the possibility of video capture. Even better, a number of competitive multiplayer games, such as *Command & Conquer*, *Warcraft*, and some shooters, are capable of displaying spectator modes or saving game replays, although playing these replays will require either conversion to digital video files or emulation software.

The new genre of machinima provides a model. Machinima films borrow from the computer graphics technology developed for games, often using game engines such as *Quake* to make films inexpensively on personal computers. They are then distributed over the Internet either as game files or in versions rendered to display on another computer. The Web site machinima.com describes the work of the *Quake* done Quick team, a maker of machinima films, as "to get through *Quake* and its various mission packs, additional levels and various difficulty settings as fast as possible, and then to make an enjoyable Machinima film of the 'demo' recordings of their runs." Note that in its dependence on game engines, a performance archive of game replays, game movies, and machinima depends to some extent on the success of emulation, unless all such formats are to be captured and encoded as digital video files. I propose that a working group formulate a strategy for creating, collecting, and securing rights to collect game performance, then work with a digital repository to build a small demonstration collection within one year.

3. Launch an initiative to build archives of design document, source code, art assets, and ancillary documentation of the history of game design. Even if the application of concepts of authorship is problematic with respect to games, computer games are designed, coded, drawn, built, tested, and published. Design and programming history are important. Games designed by luminaries such as Shigeru Miyamoto, Sid Meier, Will Wright, or Peter Molyneux have their own distinctive style and content. Others, notably the Infocom team or John Carmack, have shaped the technology of game design. At present, it is difficult to write a history of game design and technology because documentation for the work of game designers—particularly for arcade and video game console games—is sorely lacking. I propose that a working group design a documentation survey. This group should draw on experts in the IGDA, the Game Developers Conference, or other industry groups, together with academic repositories developing programs of new media studies (Stanford, MIT, Carnegie Mellon, Georgia Tech). This survey strategy should be tested by applying it to a short list of significant game designs and designers.

4. Stabilize representative artifact collections in museums and archives. Build collections of packaged game software, documentation, and marketing materials to accompany these materials.

This is probably the area in which the most progress has already been made. The collection of game software in Stanford's Stephen M. Cabrinety Collection in the History of Microcomputing (along with materials in the Martin Gardner papers and Apple history collection at Stanford) and the marvelous hardware collections of the Computer History Museum provide two examples of significant manuscript and artifact repositories. Conservators and preservation specialists, some of them in this audience, are already hard at work as they seek to ensure the long-term availability of these collections. In the worst-case scenario, if game software can no longer be run in one hundred years, these collections will still have value as archival resources for assessing the impact and marketing of games and evidence of their structure and play through documentation. They will also prod historians of technology and media to encourage their students to recreate lost technology, much as the Science Museum of London has done for Charles Babbage's difference engine.

5. Collaborate.

Again, no single institution can carry out these tasks. This fact alone demands a collaborative approach. However, there are other reasons that competent institutions must join together to build archives of computer game history. Some of you realized that lay historians of computer games, in Web sites and communities of game players, have already contributed to the first two tasks on this list by making emulators and game movies available. So let's see if we can enlist these pioneers in the effort to create more permanent historical resources. The participation of the game industry, museums, and academic institutions in this project can help to defuse the adversarial relationship between, say, the emulation community and publishers by developing mutually acceptable practices with respect to intellectual property and access.

James Cortada, an IBM executive and historian, made an interesting point in the preface to *Archives of Data-Processing History*, published in 1990: "The first group of individuals to recognize a new subject area consists usually of participants followed closely after by students of the

> field and finally, if belatedly, by librarians and archivists. It is very frustrating to historians of a new subject, because it takes time for libraries to build collections or to amass documentary evidence to support significant historical research. This situation is clearly the case with the history of information processing."[17]

I hope we can begin work on archives of computer games that anticipate the research needs of the future.

NOTES

1. Mark J. P. Wolf, ed., *The Medium of the Video Game* (Austin: University of Texas Press, 2001).

2. The IDSA statistics were gathered from a study. See Interactive Digital Software Association, *Essential Facts about the Computer and Video Game Industry* (Washington, DC: IDSA, 2000), 5. See also Khanh T. L. Tran, "U.S. Videogame Industry Posts Record Sales," *Wall Street Journal*, February 7, 2002; Valve LLC, "Valve Unveils Steam at 2002 Game Developer's Conference," press release, March 21, 2002; Sharon Waxman, "Hollywood's Great Escapism: 2001 Box Office Receipts Set a Record," *Washington Post*, January 4, 2002; Anne Valdespino, "The Big Screen Keeps Pulling Us In," *Los Angeles Times*, July 1, 2002.

3. Henry Jenkins, "Art Form for the Digital Age," *Technology Review* (September–October 2000), http://www.techreview.com/articles/octoo/viewpoint.htm.

4. Jay David Bolter and Richard Grusin, *Remediation: Understanding New Media* (Cambridge, MA: MIT Press, 1999), see esp. "Computer Games," 88–103.

5. Mark Weiser, "The Computer for the Twenty-First Century," *Scientific American* (September 1991): 94–100.

6. Elliott Avedon and Brian Sutton-Smith, *The Study of Games* (New York: Wiley, 1971): 419.

7. Chris Crawford, "Why Is Interactivity So Hard?" *Interactive Entertainment Design* 9 (1995–1996), http://www.erasmatazz.com/library/JCGD_Volume_9/Why_so_Hard.html.

8. Randall Packer, "Net Art as Theater of the Senses: A HyperTour of Jodi and Grammatron," in *Beyond Interface: Net Art and the Art of the Net*, 1998, http://www.archimuse.com/mw98/beyond_interface/bi_frpacker.html. This is the introduction to Mark Amerika's Grammatron, a work of hypertext Web art.

9. The term is found in Lev Manovich, "Navigable Space," 1998, http://jupiter.ucsd.edu/~manovich/docs/navigable_space.doc..

10. Doron Swade, "Collecting Software: Preserving Information in an Object-Centred Culture," in *Electronic Information Resources and Historians: European Perspectives*, ed. Seamus Ross and Edward Higgs (St. Katharinen: Scripta Mercaturae, 1993), 94.

11. My talk was followed by Brewster Kahle's presentation on the Internet Archive, which is located at http://www.archive.org/.

12. Jürgen Claus, "Expansion of Media Art: What Will Remain of the Electronic Age?" [1985?], reprinted in *Ars Electronica: Facing the Future, a Survey of Two Decades*, ed. Timothy Druckrey (Cambridge, MA: MIT Press, 1999).

13. Claus, "Expansion of Media Art."

14. Matthew Kirschenbaum, "Materiality and Matter and Stuff: What Electronic Texts Are Made Of," *ebr* 12 (2002), http://www.altx.com/ebr/riposte/rip12/rip12kir.htm.

15. "Preserving Computer-Related Source Materials," 1979. This brochure was later reproduced in the *IEEE Annals for the History of Computing* 2 (January 1980). The text of this brochure is available via the Web site of the Software History Center at http://www.softwarehistory.org/.

16. W. Boyd Raymond, "Electronic Information and the Functional Integration of Libraries, Museums, and Archives," in *Electronic Information Resources and Historians: European Perspectives*, ed. Seamus Ross and Edward Higgs (St. Katharinen: Scripta Mercaturae, 1993), 227–243, esp. 232.

17. James W. Cortada, preface to *Archives of Data-Processing History: A Guide to Major U.S. Collections* (New York: Greenwood, 1990): ix.

3

Video Capture

Machinima, Documentation, and the History of Virtual Worlds

In March 2010, Sam Pierce, a.k.a. "Sedrin," of the *World of Warcraft*–based machinima group SlashDance, released *Yesterday's News*. This roughly three-minute-long short was created for Blizzard Entertainment's Rise to Power machinima competition, sponsored by Alienware, a high-end personal computer maker acquired by Dell in 2006. The contest was organized to encourage the "best fan-made machinima set in Azeroth, focusing on the rise to power of various characters, heroes and villains from *World of Warcraft*" (Blizzard Entertainment, 2010). The competition attracted some of the best-known *World of Warcraft* machinimators, including Baron Soosdon, Olibith, and SlashDance. Pierce's short, however, turned the theme of glorifying heroes on its head, creating instead a nostalgic, sentimental piece in the style of a silent movie about loss and the past. The video opens with the work of a Dwarf miner and stonecrafter who creates the "meeting stones" used by *World of Warcraft* adventurers to port quickly to the "instances" where groups of players would band together to complete quests and gather loot. He reads a newspaper dated patch 3.3, with the headline, "New LFG Tool: Meeting Stones Obsolete?" When he then visits his lovingly created stones, they are all unused, barren, and survive in the game merely as mute vestiges of an inaccessible past. Another crumpled newspaper reveals a similar story about the obsolescence of flight paths, though the miner is at least able now to share his own sense of loss and bewilderment with the former proprietor of that operation.

Originally published as "Video Capture: Machinima, Documentation, and the History of Virtual Worlds," in *The Machinima Reader*, ed. Henry Lowood and Michael Nitsche (Cambridge, MA: MIT Press, 2011), 3–22.

Yesterday's News points to the possibility of machinima not only as a vehicle for commentary on virtual worlds and the communities of players that inhabit them but also as a source of historical documentation about those worlds and communities as they disappear and fade into memory. It takes its place alongside other projects that explore similar themes having to do with loss in virtual world history, from *Exploration: The Movie* and *Noggaholic: The Movie* (Tobias "Dopefish" Lundmark, 2005) about the exploration of normally inaccessible areas in *World of Warcraft*, through Tristan Pope's *For Honor* (2006), a sly reportage about the impact of Battlegrounds instances in *World of Warcraft* on social interaction and activity in the game, to *EA-Land: The Final Countdown* (How They Got Game, 2008), raw video capture of the last few minutes of *EA-Land* before its final shutdown on August 1, 2008. All of these productions underscore the twofold sense in which machinima is at its heart not only about performance but also about documentation: (1) it is founded on technologies of capture through which in-game assets and performance are redeployed and reworked, and (2) it creates historical documentation that captures aspects of the spaces, events, and activities through the lens of a player's view of the game world. *Yesterday's News* recalls not just the obsolete activities and artifacts of *World of Warcraft* but also links its storytelling to the increasingly vestigial medium of the newspaper.

Figure 3.1. SlashDance, *Yesterday's News*

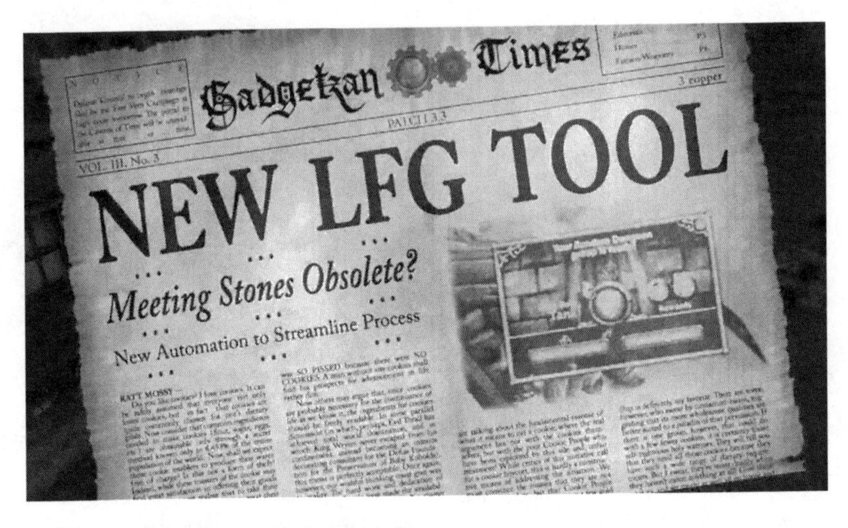

Figure 3.2. SlashDance, *Yesterday's News*

This hint of a linkage between machinima and the recorded chronicle sets the stage for the central thought of this essay, namely, that the three primary methods for making machinima during its brief history—code, capture, and compositing—match up neatly with three ways to document the history of virtual worlds.[1] These linkages are provocative for thinking about what we can do to save and preserve the history of virtual worlds in their early days. As it turns out, they also suggest how we might begin to think about machinima as a documentary medium. It is not really surprising that we should consider this possibility. Historically, the development of machinima has been closely tied to the technologies and practices of game replay and screen capture.[2] Machinima is usually produced with the software used to play games. Seen as a new medium emerging from digital game technology, it can be described as having been co-created by developers and players, yet its development as a vital part of game culture has been dominated by the innovations, creativity, and participation of players. This rootedness in game culture enhances the value of machinima as a method for capturing footage about events and activities that take place in virtual worlds.

Linking the curation of virtual world history to its "capture" along the lines of production methods established by machinima begs the question of what it is that will be captured and documented. The interactive, immersive, and performative aspects of digital games and virtual worlds take the challenges faced by media archivists and historians beyond software preservation to problems that demand new documentation and preservation strategies. Indeed, the social and performative aspects of online worlds lead us away from thinking about the preservation problem solely in terms of the software and proprietary data that support and flow out of these worlds. When we emphasize the *history* of these worlds, our attention turns to events, actions, and activities. Future historians of virtual worlds will want to understand what people—players, residents, users—actually did in early virtual worlds, partly at the level of social and personal experiences, partly in terms of historical events such as political protests or artistic performances, and partly to understand the issues of identity, law, economics, ethnography, and governance that connect the virtual world to real-world activities. Clearly, there is more to the documentation of virtual world history than bit-perfect conservation of software and data.

Machinima and replay both have stimulated cultures of performance and spectatorship within digital game culture. Players enjoy game-based replay movies as game film for following and learning from superior players and thus improving their own player skills. They watch machinima videos for entertainment but also understand them as a mix of commentary, paratext, and insider-joke referencing of specific games, usually those with which a particular piece of machinima was itself created. As spectators of replays or machinima, players deconstruct what they see as a mix of skills, actions, play, performance, and tricks that go into making these game-based videos. The importance of this indexing of machinima with respect to gameplay is at the heart of its potential application for documenting the history of virtual and game worlds. In other words, the documentary value of machinima is not merely a reflection of its potential use as a neutral capture technology; its value as being rooted in game culture is not at all about the objectivity of capture technologies but derives from machinima being in a sense "tainted" by its intimate connection with game environments and gameplay. Efforts to verify and interpret game-based replays and movies have been an integral part of their circulation through player communities since at least the mid-1990s, when games such as id Software's *DOOM* and *Quake* offered robust systems for play recording and the emerging Internet provided a foundation for building networked player communities and file distribution. Replay and machinima, in short, have been at the historical center of player culture, and the production techniques on which they are based inevitably inform the mix of methods available for documenting that culture.

The use of machinima to produce documentation about game and virtual worlds naturally depends on how these movies are produced. More than that, each of the three primary modes of machinima production—code (demo), screen capture, and compositing of game assets—provides a different take on the recording of events and experiences in virtual worlds, giving us three ways of thinking about methods for their documentation that curators and archivists of virtual world history can use to practical effect.

Take 1. Perfect Capture: Demo Recording and Replay

Demo recording was the first form of widely circulated game replays associated with first-person shooter (FPS) games and remains the primary mode for viewing recorded competitive games in genres such as real-time strategy. Strictly speaking, *DOOM* and *Quake* demos or *Warcraft III* replays are not really movies. Rather, they are sequences of commands or scripts that tell the game engine what to do, essentially by repeating the effects of keyboard and mouse input in the same sequence as executed by the player when playing a game. One consequence of the separation of game engine from asset files in John Carmack's canonical solution to the problem of game architecture was that the demo or "intro" movie was stored in a discrete file with its own format, the .lmp (pronounced "lump") file in *DOOM* and the .dem (demo) file in *Quake*. This was a game asset viewed when a player started up the game. Players could also record their own game sessions and play them back inside these games by saving, then loading and running, demo files. In Carmack's design architecture, making these movies required no hacking of the game engine, in effect creating a new performance space separate from the game software but one that also allowed for the distribution of archived performance in the form of demo files that could be viewed by players with the identical software configurations. This convention was carried forward from *DOOM* to many other games, such as *Quake/Hexen*, *Half-Life*, and *Unreal Tournament*. Replays captured as demo movies have been circulated as skills demonstrations, as recordings of significant competitive events (i.e., Thresh's defeat of Billox at Quakedelica in 1998, circulated as a recammed highlight reel), as a form of advertisement or recruiting pitch for player clans and tournaments, and for many other purposes. Demonstrating skills through competitive player performance was the primary motivation for creating the first game movies, such as replays and speed runs; learning about gameplay by viewing these movies depended on the development of practices for spectatorship, witnessing, and certification. The result was the full utilization of this new game-based performance space.

Demo recording and editing also became the basis for the first project that we would today recognize as machinima, initially known as *Quake* movies. The Rangers' *Diary of a Camper* resembled the demo movies of *DOOM* gameplay, with short bursts of frantic action punctuated by flying blood and bits of body parts. Yet *Diary of a Camper* broke with the *DOOM* demo movie in one essential respect: the movie was not seen from the first-person perspective of a player but instead established the independence of the spectator's view from that of any player-actor. It turned out that this *Quake* clan's motto, "Rangers Lead the Way," could be applied to their coding skills as well. They had devised their own programming hacks for editing the *Quake* demo format. The resulting camera view might perhaps be seen as establishing a kind of emancipation from gameplay, perhaps even as a neutral or objective position from which to document gameplay, but this claim hardly squares with the deep dependence of *Quake* movies on game data as well as on programming and player skills, or the fact that *Quake* movies and machinima were often "shot" from a particular player's point of view during an in-game recording session.

Thinking about demos and replays in terms of documenting virtual world history emphasizes another aspect of these recordings as a technique for recording in-game events and actions. A notion that plays into the preservation discussion and is particularly relevant here is the potentially perfect reproduction of digital data. Digital personae, avatars, and player characters are ultimately all bits of data on a machine. If we can acquire and preserve access to these data, should it not be possible to copy the bits forever? Is that how we can solve the problem of preserving the history of virtual worlds? If so, the first step toward this solution is accepting that everything that happens in virtual worlds as software systems is in some sense reducible to data, and the second step asserts that, as a practical matter, it should be possible to capture these data perfectly and preserve them without loss. A paradigm for the perfect capture of activity in the form of data is exactly the recording of in-game activity such as game replay files and demos.

Consider the example of Chris Crosby, a.k.a. "NoSkill." He was among the first wave of *DOOM* players to be recognized by other players as a "Doomgod," a moniker given to exceptionally skilled players. An active player from about 1994 to 1996, he was killed in a car crash in 2001. His memorial site on the Web, like many others, depicts a young man in the prime of life, with his infant son in his arms. The site also offers a number of demo files for downloading, originally recorded from games he played between May 1995 and April 1996 (NoSkill Memorial Site 2004). After a visitor downloads Chris Crosby's demo files from his memorial site and plays these files inside the correct version of this old game, originally published toward the end of 1993, she in effect is able to see a now-obsolete game through the eyes of a dead player. NoSkill comes back to life as the replay file activates the game engine to carry out the exact sequence of actions performed by the now-dead player. Moreover, because we are using an essentially "dead" game to produce this replay, we are also engaging in an act of software preservation and resurrection. The result is that for this FPS, it is possible to see a historical game as played—and seen—through NoSkill's eyes. The player is dead, but his avatar in some sense lives on through this act of perfect reproduction, accessible to any future historians of the game. Yet we cannot help but contrast the potentially infinite repetition and perfect reproduction of his gameplay to the fading memories of his life, and death. His replays alone are mute with respect to his motivation for playing or his experiences as a player. What we are seeing may represent the perfect reproduction of game states and events, but it is not historical documentation. The perfect capture of gameplay represented by the replay is a remarkable act of software and data preservation. Yet as we begin to conduct early work on preservation of games and virtual worlds, approaching this work in terms of software preservation and the perfect capture mode exemplified by game replays will lead to a barren exercise with respect to the documentation of the events and activities—the history—that have occurred in these worlds.

Let us assume that we are able to capture every bit from a virtual world server, everything from 3D models to account information; that we are able to reverse engineer or disable authentication and log-in controls after the original server is no longer live; and that we have received permission from every rights holder, ranging from game developers to third-party developers and players, to copy, store, and use what they created, to show and even inhabit their avatars, and to reveal their identity and activities. The likelihood of all this actually happening is near zero, of course, but let us assume it can be done. If we then could leap the additional hurdle of synchronizing every

Figure 3.3. NoSkill replay from *DOOM*

state or version of the software with the matching states of the server's databases, it might be possible to run a simulation of the virtual world as an archival time machine, flying around on a magic carpet in spectator mode without interacting with events run by the game engine and player data. This would be an act of perfect capture for a virtual world, much like replay for a digital game or the Internet Archive's Wayback Machine for a historical Web site. This is virtual world history as a demo movie.

Take 2. Screen Capture and Documentation

As tempting as the notion of perfect capture of events in game worlds may be, the dependence of replays on game software hardwires documentation to the vexing problem of software preservation. Demos and replay files consist of saved sequences of instructions from a previously played game that, when executed by the game software, show the same game from the perspective of the original players and, for some games, in-game spectators. Unlike video files captured from the screen or video-card output, demos or replays allow different views and settings as permitted by the game software and produce the best visual quality that the software will produce without degradation or the need to compress files to save storage space. However, all this is possible only when a running version of the game engine is available to view these replays. Not only that, the version used to view the demo or replay nearly always must correspond exactly to the version that was played when it was created. Therefore, any decision about which version of the game will be preserved determines which replay or demo files will be viewable in the future. Likewise, any decision about which demos or replays are historically significant in terms of game culture or history presupposes preservation of the appropriate version of the game software. Treatment of the software artifact affects documentation, and selection of documentation affects treatment of the software artifact. At least in the realm of virtual world or digital game history, the separation of these treatment decisions into specialized areas or departments may lead to disastrous consequences for future archivists and historians. Here we might say that Carmack's separation of game engine from data has turned against the historian, who now needs to preserve and synchronize both in order to make sense of either.

Future historians and others interested in the history of virtual worlds will be intensely curious about the inhabitants of early virtual worlds. They will want to know about the things people were doing in virtual worlds, why they were doing them, and what their activities meant to them. The possibility of perfect event capture with respect to digital data, with the game replay or the demo serving as a paradigm for perfectly reproducing the past, suggests that it will even be possible to track the activities of the earliest virtual world inhabitants by seeing through their eyes. Yet from a historian's point of view, perfect capture is not a perfect solution to the problem of how to document the history of virtual worlds. Besides having to preserve the software and data needed to produce such a replay, the notion of the perfection of this mode of capture is misleading, for it must be placed alongside the very real possibility of "perfect loss" in digital spaces. Even if we save every bit of a virtual world, its software, and the data associated with it and stored on its servers, along with a replay of every moment as seen by players, it may still be the case that we have completely lost its history. The essential problem with this approach is that it leaves out the identification and preservation of historical documentation, and these sources are rarely to be found in the data inside game and virtual worlds or on the servers that support them.

The same is true for any form of machinima that merely reproduces or replays events.

So, back to machinima. The hybrid nature of this medium as a found or ready-made technology carried important implications as it broke out beyond *Quake* movies and the demo format into a variety of media spaces less closely tied to the hard-core computer game culture of FPS games. This move was determined in part by changes made by id Software to the game software that consequently, if unintentionally, made the demo format less accessible as a means for making machinima. Significantly, those machinima projects that subsequently became popular with larger audiences relied on a mode of production quite different from *Quake's* demo movies. Instead of capture in the game, editing with special tools, recompilation into the demo format, and playback inside the game, these movies followed a path pioneered by Tritin's *Quad God*. They captured what was displayed on the screen (or perhaps more accurately, the graphics card) as video. Machinima based on video capture could no longer be edited as code; it could only be produced by editing video footage using nonlinear video editing software, compressing the edited tracks, then distributing and viewing the resulting movies in digital video formats. Instead of FPS games, popular console games and massively multiplayer online games became the dominant modes of machinima production based on screen capture rather than replay. This was machinima "for the rest of us," no longer tied to hard-core game genres, expert coding of demo files and decoding of game footage, and access to games as playback machines. The breakout title for this mode of production was Rooster Teeth's *Red vs. Blue*, a comedy series based on the Xbox games *Halo* and *Halo 2* that ran for five seasons (one hundred episodes) between 2003 and 2007. During roughly the same period, hundreds if not thousands of game movies and machinima works were produced in massively multiplayer games (*World of Warcraft*), virtual worlds (*Second Life*), or games that exported directly to video (*The Movies*); in each case, the predominant mode of production was direct capture from the screen image.

The essential difference between code and capture for machinima was that the new "orientation of machinima towards the film format came with a price: The game engine lost its value as replay engine and remained only a production tool."[3] Machinima was thus cut off from its previous dependence on game technology in postproduction and distribution; the game world provided only the production environment. Michael Nitsche underlines a consequence of this separation that has implications for machinima as a documentary medium for virtual worlds: "This is a paradigm shift from the recording of the event (in a demo) to the recording of a viewpoint to the event (in a screen capture)—from a new game-based logging format to the established production of moving images as successive still renderings."[4]

In a post on the *Terra Nova* blog titled "The History of Virtual Worlds," historian Timothy Burke remarked on the difficulties of carrying out qualitative research on this subject, especially when historians lack personal experiences in these environments. One commenter, Greg Lastowka, responded that "actually, it's far easier to get the data on *everything* happening in virtual worlds and to keep it forever"—essentially, the notion of perfect capture. Burke replied to this comment by noting the limitations of data generated and stored on a server. These "proprietary" data of virtual worlds encompass what is owned, or present on the servers that support that world, but even if historians have access to all of it, their utility for the interpretation of specific events is quite limited. As Burke put it, "I think the one thing that *isn't* in the proprietary data is the history of unusual or defining episodes or events in the life of particular virtual worlds. The narrative history, the event history, of any given virtual world, may in fact be obscured by the kinds of god's-eye view data that developers have. After all, they often don't know what is happening at the subjective level of experience within communities, or have to react to it after it's happened. (Say, when players stage a protest)."[5] Thus, focusing on preservation of what Burke calls proprietary data matches up poorly with the likely needs of future scholars of virtual worlds. The problem for historical research is that a complete set of software with a matching trove of all the data associated with a virtual world's server cannot be interpreted without contextual information. Context and personal perspective are needed to supplement historical documentation in demo mode. For a contemporary participant in this history, such as Burke, personal knowledge or interviews can provide guidance in the selection of events and fill gaps in their interpretation. The essential problem, however, is the identification and preservation of historical documentation, and these sources are rarely to be found inside virtual worlds or on the servers that support them.

Machinima based on screen capture provides a kind of documentary recording of events that take place in virtual worlds. It is also dependent on a personal perspective, not just in the sense that the final piece is carefully edited and produced, but also because unlike demo recordings, the camera view is fixed by the original recording, often corresponding to the view of a particular player (usually the director or an assigned "cameraman"). While any kind of machinima production can be edited to produce a narration of historical events in the sense of a documentary film, machinima based on screen capture removes the temptation introduced by the demo analogy of considering "perfect capture" as the documentation mode for virtual world history. The sifting of historical data is not a bad thing. Nor is the introduction of subjectivity in the creation of documentation about events, as opposed to merely recording the events as data. As Hayden White has argued, "histories gain part of their explanatory effect by their success in making stories out of *mere* chronicles." He calls this process "emplotment" and notes that the task of historians is "to make sense of the historical record, which is fragmentary and always incomplete."[6] In a sense compatible with this view of documentation and history, machinima as screen capture supplements the demo format not only as a new basis for game-based movie-making but as a different take on the importance of selectivity and perspective in the use of machinima to produce historical documentation. Collections of virtual world videos, along with other forms of documentation external to virtual worlds, such as blogs, wikis, player-created Web sites, maps, and many other forms of documentation, provide information about player activities that cannot be extracted even from perfectly preserved game software and data.

Douglas Gayeton's *Molotov Alva and His Search for the Creator: A* Second Life *Odyssey* (2007) exemplifies the conflation of documentary film-making and machinima as point-of-view documentation of virtual world history (Au 2007, 2008). Commissioned by the Dutch production company Submarine, it was distributed through a variety of channels, including the Dutch television channel VPRO, mobile phones via Mini-movies, and Cinemax via broadcast rights purchased by the cable channel HBO in the United States. The unprecedented success of *Molotov Alva* as a machinima property belied its mysterious origins as a project. It was originally released as a series of video diaries supposedly created by a *Second Life* character called Molotov Alva. The first of these diaries appeared on YouTube on March 2, 2007. It immediately drew a substantial viewership as an original exploration of virtual worlds set up as this *Second Life* resident's disappearance from "real life" into *Second Life*, followed by his quest to discover the meaning of what he found there. From the beginning, the video destabilized boundaries between documentary, chronicle, and fiction. Gayeton steadfastly insisted that he had produced a documentary about Molotov Alva's disappearance and reemergence in *Second Life*, but eventually revealed that he had created the character himself to produce this work. At the same time, many of the scenes were recorded literally as cinema verité; Gayeton, unfamiliar with machinima production techniques, simply pointed a high-definition camera at his computer monitor, a technique he eventually called "RumpleVision" after the name of the converted farmhouse in which he set up his "studio." Gayeton was aware that part of his work was to document history that would otherwise be lost; he realized that he had "documented something that's never gonna exist again" (Au 2008). Indeed, most of the locations captured through his character's Odyssey have since been removed from *Second Life* without a trace other than documents such as Gayeton's video.

Gayeton's unorthodox production technology sets his work against the perfect capture of demo and replay in several revealing and important ways. His high-definition camera, set three feet away from his monitor, is separated in every way from game software and proprietary data. There can be no confusion about the status of these images as personal, selected, and indexical. It is a point of view derived from literally pointing at the screen, not a direct recording from an in-game camera or imagery generated from gameplay data. This separation underscores the potential of machinima as a means for capturing perspective and context, as ethnography, documentary, and history rather than an exact recording of historical events in virtual spaces. In this sense, such point-of-view documentation is exactly, as Burke proposes, about what software and game data cannot deliver, such as motivations, personal accounts, and situated experiences. In short, it is about the meaning of events and activities.

Gayeton's idiosyncratic documentary points the way to machinima as documentation, but hardly as a rigid method for recording the past. Machinima pieces ranging from Tristan Pope's *Not Just Another Love Story*[7] to raw footage such as the *Final Countdown* video that documents the last minutes of *EA-Land* show us other ways in which personal screen capture provides viewers with thought-provoking documents for interpreting events and activities that take place in virtual worlds. In short, both fictional and nonfictional machinima can contribute to a documentation project by emphasizing point of view rather than perfect data capture.

Figure 3.4. Douglas Gayeton, director, *Molotov Alva and His Search for the Creator*

Take 3. Asset Compositing: Models and Artifacts of Virtual Worlds

Machinima creators do not just use digital games as a production technology. They work inside game spaces and software to scout locations and to build or find sets, artwork, and animation. A third mode of machinima production, neither demo nor screen capture, depends on the direct use of game assets such as models and maps.

This form of machinima production became popular among the many machinima creators who turned to the immensely popular massively multiplayer game *World of Warcraft* as a platform for their work. Server-based games such as *World of Warcraft* deny direct access to code and data outside their use in-game; thus, demo recording is not an option.[8] Initially, *World of Warcraft*–based moviemaking was limited to edited screen captures, as was typical for networked, server-based games. Early *World of Warcraft* movies

Figure 3.5. How They Got Game project, *EA-Land: The Final Countdown*

were created as live performances grabbed via screen capture, then cleverly edited, perhaps with voice-overs or painstakingly lip-synched dialog. These movies rapidly became an essential part of the player culture around the game as it grew in popularity. As the subscriber base grew, so did the demand for new gameplay movies and machinima, which were distributed via sites such as warcraftmovies.com geared to the growing player community. Tristan Pope played on this relationship between machinima and game community in his controversial *Not Just Another Love Story.*[9] At the time this movie was created, Pope could only use video capture to produce his work. He called attention to this dependence of *World of Warcraft* machinima on the game itself when he defended his controversial depiction of sexual imagery in this work by arguing that he had only "executed what the Pixels in *WoW* suggest."[10] But soon another technique for making machinima emerged that provided its makers with more independence from the pixels on the screen.

By early 2006, new tools such as John (Darjk) Steele's WoW Model Viewer opened up the game's model database to *World of Warcraft* players, giving them direct access to every character and equipment model, particle effect, animation, and other game asset. Initially, players used the viewer to do things like dress up characters or show off items they had received as in-game loot. Before long, machinima creators realized they were able now to composite models and maps (viewed with the appropriately titled Map Viewer), mix them with live in-game performance, and composite these elements to make movies. Some went further by editing models or creating new animations and added them to the mix. For example, Deeprun Goldwin Michler's *The Man Who Can* (2006) depicts a character who joyfully escapes the limitations imposed by Blizzard's pixels; he proves that he can by dancing with new moves animated by Michler. Machinima projects such as Tristan Pope's own *Switcher* series (2005–2006), Jason Choi's *Edge of Remorse* (2006), Mike Spiff Booth's *Code Monkey* (2006), Myndflame's (Clint and Derek Hackleman's) *Zinwrath: The Movie* (2005), or Nicholas "Snoman" Fetcko's *Wandering Dreamscape* (Snoman 2007a) used new tools such as the Model Viewer or hand-cut elements of game footage to make machinima by compositing scenes and postproduction editing.

One can trace the evolution of techniques from in-game video capture to model editing and compositing most clearly through the progression of specific machinima series and artists, such as Martin Falch's *Tales of the Past* trilogy (2005–2007), Myndflame's body of work, or Joshua Diltz's *Rise of the*

Figure 3.6. *World of Warcraft* Model Viewer

Living Dead trilogy (2005–2006). Armed with access to game models, effects, and other assets, *World of Warcraft* movies broke open creative (but not necessarily legal) constraints on the use of online games and their artistic assets. Machinima makers now used these games more freely to make movies that were no longer limited to what they could accomplish with in-game puppeteering. Compositing game assets resembled demo in the sense that it depended on game data, but at the same time this technique primarily called on artistic prowess in realms such as 3D model, animation, or video editing rather than coding (demo) or gameplay and puppeteering (screen capture) skills. Terran Gregory of Rufus Cubed probably expressed a typical attitude when he called the Model Viewer "the 'Key' to unlocking WoW Machinima as a whole." He praised its ability to empower the machinima artist to realize his or her vision; the Model Viewer "was the great virtual socioeconomic equalizer that truly allowed *WoW* movie makers to experience all of the freedoms that are inherent to Machinima."[11] By 2009, Naughty Dog's *Uncharted 2* acknowledged the use of such tools by releasing its game with a Cinema Mode that included green screen compositing and built-in video editing.

Works such as Snoman's *Wandering Dreamscape* and Baron Soosdon's *I'm So Sick* (2007) point to a possible role for asset composition not only as a third path for machinima but also as a model for digital archaeology in virtual worlds. Snoman made *Wandering Dreamscape*, the first in a series of *Dreamscape* pieces, as a "tribute to the Nogg-aholic community" (Snoman 2007a) and its primary video creator, Tobias "Dopefish" Lundmark. The Nogg-aholics guild specialized in an activity that can be described as a combination of spelunking, exploration, excavation, and archaeology in terms appropriate for a digital environment: searching out and documenting unfinished or generally inaccessible locations in *World of Warcraft*. (The name of the guild was derived from an elixir that, when imbibed, causes a game character to undergo one of several random changes or effects.) As explorers, the guild members typically ventured into spaces that Blizzard as developer of the game was not ready to release to its players, such as the alpha versions of upcoming expansion areas. Machinima videos created by members such as Dopefish documented what they found, seeking to provide information to the player community about these spaces. As one commentator put it, these and other *World of Warcraft* explorers tried to "document everything that was suddenly new and uncertain" about the game world (Howgego 2008).

Snoman's tributes to the Nogg-aholic community not only construct an "alternate *Warcraft*" but also explicitly show "what is possible when you have the tools and proper knowledge of the *WoW* data" (Snoman 2007a, 2007b). *Wandering Dreamscape*, for example, deploys over a dozen model changes to transform landscapes into alternative virtual realities, the Model Viewer, FRAPS-based video capture, and Soosdon's own landscape model editing to create the Barrens Biohazard area by mixing in elements from other areas in the game world. Like Snoman, in *I'm So Sick* Soosdon also mixes in elements from other games, such as *Unreal Tournament 2004*, *Bioshock*, and *Half-Life 2*, and at one point even shows a *World of Warcraft* model playing in front of a poster for another machinima piece, *Edge of Remorse* (Riot Films, 2006). These projects link the activity of exploring and, in a sense, digging out unrevealed spaces to get access to game data, and a project of documenting through machinima what these data reveal.

If with demo and replay we verify functionality and precisely track events in game worlds, and if with video capture we reveal points of view and the meaning associated with these events, then asset capture and editing allow the recreation of environments for the purpose of discovering what was hidden, lost, or inaccessible below the surface. In short, these techniques hint at an archaeology of virtual worlds. What might a resource built on this premise look like as a historical tool? The newfound power realized by *World of Warcraft* machinima makers amounted to pulling assets out of the game and importing them into the artist's workspace. We return again to the assets and content that go into the creation of a virtual world: models, maps,

Figure 3.7. Baron Soosdon, Barrens Biohazard

geometries, textures, and so on. We do not yet know how future scholars will visualize, analyze, and understand these artifacts in a digital repository consisting of data files and metadata. One way of guessing how future historians might access these materials is suggested by the administrator modes that we have in early examples of digital repositories today, in which the archival collection is presented essentially as a file directory. The use case for access to these materials is that of reconstruction from these data files, such as an installation package or a set of models and textures.

Here is an alternative model for access to the artifacts of a past world: museums. A natural history museum, for example, houses models and suitable spaces for these models. If we go there to see dinosaur skeletons, they are likely to be depicted in front of a diorama that takes the visitor to the prehistoric savannah. The access model in such a museum is based on the visual arrangement of artifacts, and it is reinforced by immersion in a simulation of the historical world of the artifacts. It should be possible to do something similar with 3D artifacts from virtual and game worlds. The importing of models to make machinima provides a clue to how we might go about this project and, in essence, construct digital dioramas about lost virtual worlds.

The first step is to think of objects such as maps and models in virtual worlds as historical artifacts. Historians often distinguish between artifacts and documentation. Both are objects that have survived from the past and can be used by historians as primary sources. According to the *Oxford English Dictionary*, an artifact is an "object made or modified by human workmanship," while a document "furnishes evidence or information."[12] As the practical work of machinima artists described above suggests, tools such as the Model Viewer and Map Viewer make it possible to export 3D models and maps from the software environments in which they support digital games and virtual worlds to another piece of software. The destination for historical software objects of this sort could just as well be another virtual world; this target environment could be a virtual world operated as a library or archive managed by a cultural repository. In this new kind of repository, historical artifacts from virtual worlds would be stored, retrieved, and investigated as 3D objects. This means that just like the machinima maker grabbing models from *World of Warcraft*, a curator would move the original geometry and texture information that defines 3D objects as archival assets from their original environments into such a repository. For example, after the necessary transcoding, an exhibition created on a *Second Life* "island" might be imported into the repository, or perhaps a level from a historical game such as id Software's *Quake* might be created from an original installation. Of course, it would be necessary to validate the authenticity of these objects, if the purpose of this process was to support historical work rather than make machinima.

Maps—also known as levels, zones, and by other names—are among the most important artifacts in game development and player cultures. We have already seen evidence in the work of the Nogg-aholics concerning the importance of documenting the exploration of game worlds. Whereas in the real world, a map is ordinarily a 2D representation of an area, in the design of game and virtual worlds it primarily means the area itself, including the objects and challenges embedded in it. Thus a map is an artifact created by the developers of a game or modified by players. A huge part of player culture encompasses players' efforts to analyze these spaces, recreate them as mods in games other than the ones in which they were originally created, or build viewers and projections to better visualize how to optimize their gameplay in these spaces. As artifacts in a digital repository built with virtual world technology, historical maps would not just be artifacts, they might also provide spaces in which to site other objects and documentation—such as models, screenshots, videos, or documentation—that provide information about what took place in these settings. An interesting quality of virtual and game worlds is that many of them can be navigated by in-world coordinate systems, much like real-world cartography. Two well-known examples are the Second Life URL (SLURL) in *Second Life* and the UI coordinate system in *World of Warcraft*.

Just as we can mash up data by attaching GPS coordinates to real-world maps, photographs, and other media, or a creative player can composite models and maps to create a new machinima piece, these virtual world coordinate systems might make it possible to match documentation we have assembled in our virtual world collections not only with locations in virtual worlds but also with each other. Metadata schemes based on the Dublin Core standard already provide a "coverage" element for individual objects, and as the Dublin Core specification tells us, this element can be applied "for the use of multiple classification schemes to further qualify the incoming information" such as latitude and longitude or other "native coordinate representations."[13] In other words, it is possible to "tag" objects such as an island exported from *Second Life* and a group of machinima videos about events that occurred on that island with the SLURL that locates the island in the digital world. Or demo code, replay movies, and machinima created in a particular

Quake level could be exhibited as documentation in the very space—an artifact created from original game data—in which they were created. Although further development work is necessary to realize this vision, it is certainly possible to export levels (maps) from *Quake* to the open VRML format, from which unaltered geometries and textures can then be moved to other environments, just as Soosdon or Snoman use the Model Viewer or Map Viewer to move a model of an Orc Shaman from *World of Warcraft* into software tools like Milkshape and 3ds Max, so they can edit them. When the pipeline to a virtual repository is completed, it will be possible to drop in and see 3D objects with the same geometries and textures they were given in the original game. In fact, these artifacts will be created from certified copies of original game data used to produce them in the first place.[14] If the Nogg-aholics and their ilk are the spelunkers of virtual worlds, such a resource would be more like an excavation site, where traces of vanished worlds would be revealed and recreated through maps and models.

Conclusion: *Machinima est omnis divisa in partes tres*

Future access environments for digital repositories will need to consider how to provide scholars with access both to data and artifacts from environments such as virtual worlds and to documentation about these artifacts and worlds. Moving artifacts from virtual worlds to historical collections in a manner inspired by asset extraction and compositing in machinima creation is a new way to think about digital repositories as 3D environments filled with 3D objects, rather than simply as bitstreams organized as massive collections of files. Yet it does not replace the "perfect capture" of demo plus replay or documentation that provides context and meaning. Future historians of virtual worlds will not want to be constrained by previous generations' preferences for one form of historical evidence over others; they will want it all. It is our task today to construct a flexible, interdisciplinary approach to virtual world documentation that embraces technical, documentary, and archaeological traditions of work.

I have presented three takes on machinima—demo, screen capture, and asset composition—in parallel with three takes on how to document the history of virtual worlds: replay, POV recording, and asset extraction. As more people spend more time in virtual worlds, the events that take place in those worlds become part of the mixed realities—material and virtual—that the players inhabit and that define who they are. It will not be possible to tell the history of our times without including the history of these places and events. Machinima may well prove to be the documentary medium for recording our experiences, activities, and motivations in virtual worlds. More important for me, however, is that machinima provides compelling evidence of the ability of players to turn digital games into their own creative medium. Historians, curators, and archivists can learn from these examples. Soon it will be our turn to have some fun.

NOTES

1. I use the term "virtual worlds" in this essay to mean multiplayer game worlds.

2. See Henry Lowood, "High-Performance Play: The Making of Machinima," in *Videogames and Art: Intersections and Interactions*, ed. Andy Clarke and Grethe Mitchell, 59–79 (Bristol: Intellect Books, 2007a); and Henry Lowood, "Found Technology: Players as Innovators in the Making of Machiniman," in *Digital Youth, Innovation, and the Unexpected*, ed. Tara McPherson, 165–196 (Cambridge, MA: MIT Press, 2007b).

3. Michael Nitsche, "Claiming Its Space: Machinima," *Dichtung Digital: Journal für digitale Ästhetik* (2007): 37, accessed January 2010, http://www.brown.edu/Research/dichtung-digital/2007/Nitsche/nitsche.htm.

4. Ibid.

5. Timothy Burke, "The History of Virtual Worlds," post and comment on the blog *Terra Nova*, December 1, 2006, http://terranova.blogs.com/terra_nova/2006/12/the_history_of_.html.

6. Hayden White, "The Historical Text as Literary Artifact," *Clio* 3, no. 3 (1974): 280.

7. See Henry Lowood, "Storyline, Dance/Music, or PvP? Game Movies and Community Players in *World of Warcraft*," *Games and Culture* 1 (October 2006): 362–382.

8. Except, of course, to hackers and enthusiasts, who create private servers from *World of Warcraft* code. This technique has been used for machinima production, despite its murky legal status, as it then returns control of game assets to the production team.

9. See Lowood, "Storyline, Dance/Music, or PVP?"

10. Tristan Pope, Crafting Worlds Web site, 2005, accessed September 2005 http://www.craftingworlds.com.

11. Terran Gregory, email to author, December 11, 2006.

12. *Oxford English Dictionary* online, accessed March 2010, http://dictionary.oed.com.

13. Hans Becker, Arthur Chapman, Andrew Daviel, Karen Kaye, Mary Larsgaard, Paul Miller, Doug Nebert, Andrew Prout, and Misha Wolf, Dublin Core Element: Coverage. September 30, 1997, http://www.alexandria.ucsb.edu/historical/www.alexandria.ucsb.edu/docs/metadata/dc.

14. At Stanford, we are currently working with a new, open-source virtual world platform called Sirikata to realize this vision.

4

It Is What It Is, Not What It Was

Let's begin with a question. When did libraries, archives, and museums begin to think about software history collections? The answer: in the late 1970s. The Charles Babbage Institute (CBI) and the History of Computing Committee of the American Federation of Information Processing Societies (AFIPS), soon to be a sponsor of CBI, were both founded in 1978. The AFIPS committee produced a brochure called "Preserving Computer-Related Source Materials." Distributed at the National Computer Conference in 1979, it is the earliest statement I have found about preserving software history. It says,

> If we are to fully understand the process of computer and computing developments as well as the end results, it is imperative that the following material be preserved: correspondence; working papers; unpublished reports; obsolete manuals; key program listings used to debug and improve important software; hardware and componentry engineering drawings; financial records; and associated documents and artifacts. (p. 4)

Mostly paper records. The recommendations say nothing about data files or executable software, only nodding to the museum value of hardware artifacts for "esthetic and sentimental value." The brochure says that artifacts provide "a true picture of the mind of the past, in the same way as the furnishings of a preserved or restored house provides a picture of past society." One year later, CBI received its first significant donation of books and archival documents from George Glaser, a former president of AFIPS. Into the 1980s

Originally published as "It Is What It Is, Not What It Was," *Refractory: A Journal of Entertainment Media* 27 (2016).

history of computing collections meant documentation: archival records, publications, ephemera, and oral histories.

Software preservation trailed documentation and historical projects by a good two decades. The exception was David Bearman, who left the Smithsonian in 1986 to create a company called Archives and Museum Informatics (AHI). He began publishing the *Archival Informatics Newsletter* in 1987 (later called *Archives and Museum Informatics*). As one of its earliest projects, AHI drafted policies and procedures for a "Software Archives" at the Computer History Museum (CHM) then located in Boston. By the end of 1987, Bearman published the first important study of software archives under the title *Collecting Software: A New Challenge for Archives and Museums*.[1]

In his report, Bearman alternated between frustration and inspiration. Based on a telephone survey of companies and institutions, he wrote that "the concept of collecting software for historical research purposes had not occurred to the archivists surveyed; perhaps, in part, because no one ever asks for such documentation!"[2] He learned that nobody he surveyed was planning software archives. Undaunted, he produced a report that carefully considered software collecting as a multi-institutional endeavor, drafting collection policies and selection criteria, use cases, a rough "software thesaurus" to provide terms for organizing a software collection, and a variety of practices and staffing models. Should some institution accept the challenge, here were tools for the job.

Well, here we are, nearly thirty years later. We can say that software archives and digital repositories finally exist. We have made great progress in the last decade with respect to repository technology and collection development. Looking back to the efforts of the 1980s, one persistent issue raised as early as the AFIPS brochure in 1978 is the relationship between collections of historical software and archival documentation about that software. This is an important issue. Indeed, it is today, nearly forty years later, still one of the key decision points for any effort to build research collections aiming to preserve digital heritage or serve historians of software. Another topic that goes back to Bearman's report is a statement of use cases for software history. Who is interested in historical software and what will they do with it? Answers to this fundamental question must continue to drive projects in digital preservation and software history.

As we consider the potential roles to be played by software collections in libraries and museums, we immediately encounter vexing questions about how researchers of the future will use ancient software. Consider that using historical software now in order to experience it in 2014 and running that software in 2014 to learn what it was like when people operated it thirty years ago are two completely different use cases. This will still be true in 2050. This may seem like an obvious point, but it is important to understand its implications. An analogy might help. I am not just talking about the difference between watching *Gone with the Wind* at home on DVD versus watching it in a vintage movie house in a 35 mm print—with or without a live orchestra. Rather I mean the difference between my experience in a vintage movie house today—when I can find one—and the historical experience of, say, my grandfather during the 1930s. My experience is what it is, not what his was. So much of this essay will deal with the complicated problem of enacting a contemporary experience to re-enact a historical experience and what it has to do with software preservation. I will consider three takes on this problem: the historian's, the media archaeologist's, and the re-enactor's.

Take 1. The Historian

Historians enact the past by writing about it. In other words, historians tell stories. This is hardly a revelation. Without meaning to trivialize the point, I cannot resist pointing out that "story" is right there in "hi-story" or that the words for story and history are identical in several languages, including French and German. The connections between storytelling and historical narrative have long been a major theme in writing about the methods of history, that is, historiography. In recent decades, this topic has been mightily influenced by the work of Hayden White, author of the much-discussed *Metahistory: The Historical Imagination in Nineteenth-Century Europe*, published in 1973. White's main point about historians is that History is less about subject matter and source material and more about *how historians write*.

He tells us that historians do not simply arrange events culled from sources in correct chronological order. Such arrangements White calls Annals or Chronicles. The authors of these texts merely compile lists of events. The work of the *historian* begins with the ordering of these events in a different way. Hayden writes in *The Content of the Form* that in historical writing, "the events must be not only registered within the chronological framework of their original occurrence but narrated as well, that is to say, revealed as possessing a structure, an order of meaning, that they do not possess as mere sequence."[3] How do historians do this? They create narrative discourses out of sequential chronicles by making choices. These choices involve the form, effect, and message of their stories. White puts choices about form, for example, into categories such as argument, ideology, and emplotment. There is no need in this essay to review all of the details of every such choice. The important takeaway is that the result of these choices by historians is sense-making through the structure of story elements, use of literary tropes, and emphasis placed on particular ideas. In a word, *plots*. White thus gives us the enactment of history as a form of narrative or emplotment that applies established literary forms such as comedy, satire, and epic.

In his book *Figural Realism: Studies in the Mimesis Effect*, White writes about the "events, persons, structures and processes of the past" that "it is not their pastness that makes them historical. They become historical only in the extent to which they are represented as subjects of a specifically historical kind of writing."[4] It is easy to take away from these ideas that history is a kind of literature. Indeed, this is the most controversial interpretation of White's historiography.

My purpose in bringing Hayden White to your attention is to insist that there is a place in game and software studies for this "historical kind of writing." I mean writing that offers a narrative *interpretation* of something that happened in the past. Game history and software history need more historical writing that has a point beyond adding events to the chronicles of game development or putting down milestones of the history of the game industry. We are only just beginning to see good work that pushes game history forward into historical writing and produces ideas about how these historical narratives will contribute to allied works in fields such as the history of computing or the history of technology more generally.

Allow me one last point about Hayden White as a take on enactment. Clearly, history produces narratives that are human-made and human-readable. They involve assembling story elements and choosing forms. How then do such stories relate to actual historical events, people, and artifacts? Despite White's fondness for literary tropes and plots, he insists that historical narrative is not about imaginary events. If historical methods are applied properly, the resulting narrative according to White is a "simulacrum." He writes in his essay on "The Question of Narrative in Contemporary Historical Theory," that history is a "mimesis of the story *lived* in some region of historical reality, and insofar as it is an accurate imitation, it is to be considered a truthful account thereof."[5] Let's keep this idea of historical mimesis in mind as we move on to takes two and three.

Take 2. The Media Archaeologist

My second take is inspired by the German media archaeologist Wolfgang Ernst. As with Hayden White, my remarks will fall far short of a critical perspective on Ernst's work. I am looking for what he says to me about historical software collections and the enactment of media history.

Hayden White put our attention on narrative; enacting the past is storytelling. Ernst explicitly opposes Media Archaeology to historical narrative. He agrees in *Digital Memory and the Archive*, that "Narrative is the medium of history." By contrast, "the technological reproduction of the past . . . works without any human presence because evidence and authenticity are suddenly provided by the technological apparatus, no longer requiring a human witness and thus eliminating the irony (the insight into the relativity) of the subjective perspective."[6] Irony, it should be noted, is one of White's favorite tropes for historical narrative.

White tells us that historical enactment is given to us as narrative mimesis, with its success given as the *correspondence* of history to some lived reality. Ernst counters by giving us enactment in the form of *playback*.

In an essay called "Telling versus Counting: A Media-Archaeological Point of View," Ernst plays with the notion that, "To *tell* as a transitive verb means 'to count things.'" The contrast with White here relates to the difference in the German words *erzählen* (narrate) and *zählen* (count), but you also find it in English: recount and count. Ernst describes historians as recounters: "Modern historians . . . are obliged not just to order data as in antiquaries but also to propose models of relations between them, to interpret plausible connections between events."[7] In another essay, aptly subtitled "Method and Machine versus the History and Narrative of Media," Ernst adds that mainstream histories of technology and mass media as well as their counterhistories are textual performances that follow "a chronological and narrative ordering of events." He observes succinctly that, "It takes machines to temporarily liberate us from such limitations."[8]

Where do we go with Ernst's declaration in "Telling versus Counting," that "There can be order without stories"? We go, of course, directly to the machines. For Ernst, media machines are transparent in their operation, an advantage denied to historians. We play back historical media on historical machines, and "all of a sudden, the historian's desire to preserve the original sources of the past comes true at the sacrifice of the discursive." We are in that moment directly in contact with the past.

In "Method and Machine," Ernst offers the concept of "media irony" as a response to White's trope of historical irony. He says:

> Media irony (the awareness of the media as coproducers of cultural content, with the medium evidently part of the message) is a technological modification of Hayden White's notion that "every discourse is always as much about discourse itself as it is about the objects that make up its subject matter."[9]

As opposed to recounting, counting in Ernst's view has to do with the encoding and decoding of signals by media machines. Naturally, humans created these machines. This might be considered as another irony, because humans have thereby "created a discontinuity with their own cultural regime." We are in a realm that replaces narrative with playback as a form of direct access to a past defined by machine sequences rather than historical time.[10]

Ernst draws implications from media archaeology for his closely connected notion of the multimedia archive. In "Method and Machine," he says, "With digital archives, there is, in principle, no more delay between memory and the present but rather the technical option of immediate feedback, turning all present data into archival entries and vice versa." In "Telling versus Counting," he portrays "a truly multimedia archive that stores images using an image-based method and sound in its own medium. . . . And finally, for the first time in media history, one can archive a technological dispositive in its own medium."[11] Not only is the enactment of history based on playback inherently non-discursive, but the very structure of historical knowledge is written by machines.

With this as background, we can turn to the concrete manifestation of Ernst's ideas about the Multimedia Archive. This is the lab he has created in Berlin. The Web site for Ernst's lab describes the Media Archaeological Fundus (MAF) as "a collection of various electromechanical and mechanical artefacts as they developed throughout time. Its aim is to provide a perspective that may inspire modern thinking about technology and media within its epistemological implications beyond bare historiography."[12] Ernst explained the intention behind the MAF in an interview with Lori Emerson as deriving from the need to experience media "in performative ways." So he created an assemblage of media and media technologies that could be operated, touched, manipulated, and studied directly. He said in this interview, "Such items need to be displayed in action to reveal their media essentiality (otherwise a medium like a TV set is nothing but a piece of furniture)."[13] Here is media archaeology's indirect response to the 1979 AFIPS brochure's suggestion that historical artifacts serve a purpose similar to furnishings in a preserved house.

The media-archaeological take on enacting history depends on access to artifacts and, in its strongest form, on their operation. Even when its engagement with media history is reduced to texts, these must be "tested against the material evidence." This is the use case for playback as an enactment of software history.

Take 3. The Re-Enactor

Authenticity is an important concept for digital preservation. A key feature of any digital archive over the preservation life cycle of its documents and software objects is auditing and verification of authenticity, as in any archive. Access also involves authenticity, as any discussion of emulation or virtualization will bring up the question of fidelity to an historical experience of using software.

John Walker (of AutoDesk and Virtual Reality fame) created a workshop called Fourmilab to work on personal projects such as an online museum "celebrating" Charles Babbage's Analytical Engine. This computer programming heritage work includes historical documents and a Java-based emulator of the Engine. Walker says, "Since we're fortunate enough to live in a world where Babbage's dream has been belatedly realised, albeit in silicon rather than brass, we can not only read about The Analytical Engine but *experience* it for ourselves." The authenticity of this experience—whatever that means for a machine that never existed—is important to Walker. In a 4,500-word essay titled, "Is the Emulator Authentic," he tells us that, "In order to be useful, an emulator program must be *authentic*—it must faithfully replicate the behaviour of the machine it is emulating." By extension, the authenticity of a preserved version of the computer game *DOOM* in a digital repository could be audited by verifying that it can properly run a *DOOM* demo file. The same is true for Microsoft Word and a historical document in the Word format. This is a machine-centered notion of authenticity; we used it in the second Preserving Virtual Worlds project as a solution to the significant properties problem for software.[14]

All well and good. However, I want to address a different authenticity. Rather than judging authenticity in terms of playback, I would like to ask what authenticity means for the experience of *using* software. Another way of putting this question is to ask what we are looking for in the re-enactment of historical software use. So we need to think about historical re-enactment.

I am not a historical re-enactor, at least not the kind you are thinking of. I have never participated in the live recreation or performance of a historical event. Since I have been playing historical simulations—a category of board games—for most of my life, perhaps you could say that I re-enact being a historical military officer by staring at maps and moving units around on them. It's not the same thing as wearing period uniforms and living the life, however.

Anyway, I need a re-enactor. In his 1998 book *Confederates in the Attic*, Tony Horwitz described historical re-enactment in its relationship to lived heritage. His participant-journalist reportage begins at a chance encounter with a group of "hard-core" Confederate re-enactors. Their conversation leads Horwitz on a year-long voyage through the American South. A featured character in *Confederates in the Attic* is the re-enactor Robert Lee Hodge, a waiter turned Confederate officer. He took Horwitz under his wing and provided basic training in re-enactment. Hodge even became a minor celebrity due to his role in the book.

Hodge teaches Horwitz the difference between hard-core and farby (i.e., more casual) re-enactment. He tells Horwitz about dieting to look sufficiently gaunt and malnourished, the basics of "bloating" to resemble a corpse on the battlefield, what to wear, what not to wear, what to eat, what not to eat, and so on. It's remarkable how little time he spends on martial basics. One moment sticks out for me. During the night after a hard day of campaigning Horwitz finds himself in the authentic situation of being wet, cold, and hungry. He lacks a blanket, so he is given basic instruction in the sleeping technique of the Confederate infantryman: "spooning." According to the reenactor Scott Cross, "Spooning is an old term for bundling up together in bed like spoons placed together in the silver chest."[15] Lacking adequate bedding and exposed to the elements, soldiers bunched up to keep warm. So that's what Horwitz does, not as an act of mimesis or performance per se, but in order to re-experience the reality of Civil War infantrymen.

It interested me that of all the re-enactment activities Horwitz put himself through, spooning reveals a deeper commitment to authenticity than any of the combat performances he describes. It's uncomfortable and awkward, so requires dedication and persistence. Sleep becomes self-conscious, not just in order to stick with the activity, but because the point of it is to recapture a past experience of sleeping on the battlefield. Since greater numbers of participants are needed for re-enacting a battle than sleep, more farbs (the less dedicated re-enactors) show up and thus the general level of engagement declines. During staged battles, spectators, scripting, confusion, and accidents all interfere with the experience. Immersion breaks whenever dead soldiers pop up on the command, "resurrect." In other words, performance takes over primacy from the effort to re-experience. It is likely that many farbs dressed up for battle are content to find a hotel to sleep in.

Specific attention to the details of daily life might be a reflection of recent historical work that emphasizes social and cultural histories of the Civil War period, rather than combat histories. But that's not my takeaway from the spooning re-enactors. Rather, it's the standard of authenticity that goes beyond performance of a specific event (such as a battle) to include life experience *as a whole*. Horvitz recalled that,

> Between gulps of coffee—which the men insisted on drinking from their own tin cups rather than our ceramic mugs—Cool and his comrades explained the distinction. Hardcores didn't just dress up and shoot blanks. They sought absolute fidelity to the 1860s: its homespun clothing, antique speech patterns, sparse diet and simple utensils. Adhered to properly, this fundamentalism produced a time travel high, or what hardcores called a "period rush."[16]

Stephen Gapps, an Australian curator, historian, and re-enactor, has spoken of the "extraordinary lengths" re-enactors go to "acquire and animate the look and feel of history." Hard-core is not just about marching, shooting, and swordplay. I wonder what a "period rush" might be for the experience of playing *Pitfall!* in the mid-21st century. Shag rugs? Ambient New Wave radio? Caffeine-free cola? Will future re-enactors of historical software seek this level of experiential fidelity? Gapps, again: "Although reenactors invoke the standard of authenticity, they also understand that it is elusive—worth striving for, but never really attainable."[17]

Re-enactment offers a take on born-digital heritage that proposes a commitment to lived experience. I see some similarity here with the correspondence to lived *historical* experience in White's striving for a discursive mimesis.

Yet, like media archaeology, re-enactment puts performance above discourse, though it is the performance of humans rather than machines.

Playing Pitfalls

We now have three different ways to think about potential uses of historical software and born-digital documentation. I will shift my historian's hat to one side of my head now and slide up my curator's cap. If we consider these takes as use cases, do they help us decide how to allocate resources to acquire, preserve, describe, and provide access to digital collections?

In May 2013, the National Digital Information Infrastructure and Preservation Program (NDIIPP) of the US Library of Congress (henceforth: LC) held a conference called Preserving.exe. The agenda was to articulate the "problems and opportunities of software preservation." In my contribution to the LC conference report issued a few months later, I described three "lures of software preservation."[18] These are potential pitfalls as we move from software collections to digital repositories and from there to programs of access to software collections. The second half of this paper will be an attempt to introduce the three lures of software preservation to the three takes on historical enactment.

1. The Lure of the Screen

Let's begin with the Lure of the Screen. This is the idea that what counts in digital media is what is delivered to the screen. This lure pops up in software preservation when we evaluate significant properties of software as surface properties (graphics, audio, haptics, etc.).

This lure of the screen is related to what media studies scholars such as Nick Montfort, Mark Sample, and Matt Kirschenbaum have dubbed (in various but related contexts) "screen essentialism." If the significant properties of software are all surface properties, then our perception of interaction with software tells us all we need to know. We check graphics, audio, responses to our use of controllers, etc., and if they look and act as they should, we have succeeded in preserving an executable version of historical software. These properties are arguably the properties that designers consider as the focus of user interaction and they are the easiest to inspect and verify directly.

The second Preserving Virtual Worlds project was concerned primarily with identifying significant properties of interactive game software. On the basis of several case sets and interviews with developers and other stakeholders, we concluded that isolating surface properties, such as image colorspace as one example, while significant for other media such as static images, is *not* a particularly useful approach to take for game software. With interactive software, significance appears to be variable and contextual, as one would expect from a medium in which content is expressed through a mixture of design and play, procedurality, and emergence. It is especially important that software abstraction levels are not "visible" on the surface of play. It is difficult if not impossible to monitor procedural aspects of game design and mechanics, programming, and technology by inspecting properties expressed on the screen.

The preservation life cycle for software is likely to include data migration. Access to migrated software will probably occur through emulation. How do we know when our experience of this software is affected by these practices? One answer is that we audit significant properties, and as we now know, it will be difficult to predict which characteristics are significant. An alternative or companion approach for auditing the operation of historical software is to verify the execution of data files. The integrity of the software can be evaluated by comparison to documented disk images or file signatures such as hashes or checksums. However, when data migration or delivery environments change the software or its execution environment, this method is inadequate. We must evaluate software performance. Instead of asking whether the software "looks right," we can check if it runs verified data sets that meet the specifications of the original software. Examples range from word processing documents to saved game and replay files. Of course, visual inspection of the content plays a role in verifying execution by the software engine; failure will not always be clearly indicated by crashes or error messages. Eliminating screen *essentialism* does not erase surface properties altogether.

The three takes compel us to think about the screen problem in different ways. First, the Historian is not troubled by screen essentialism. His construction of a narrative mimesis invokes a selection of source materials that may or may not involve close reading of personal gameplay, let alone focus on surface properties. On the other hand, the Re-enactor's use of software might lead repositories to fret about what the user sees, hears, and feels. It makes sense with this use case to think about the re-enactment as occurring at the interface. If a repository aims to deliver a re-enacted screen experience, it will need to delve deeply into questions of significant properties and their preservation.

Screen essentialism is also a potential problem for repositories that follow the path of Media Archaeology. It is unclear to me how a research site like the MAF would respond to digital preservation practices based on data migration and emulation. Can repositories meet the requirements of media archaeologists without making a commitment to preservation of working historical hardware to enable playback from original media? It's not just that correspondence to surface characteristics is a significant property for media archaeologists. Nor is the Lure of the Screen a criticism of Media Archaeology. I propose instead that it is a research problem. Ernst's vision of a Multimedia Archive is based on the idea that media archaeology moves beyond playback to reveal mechanisms of counting. This machine operation clearly is not a surface characteristic. Ernst would argue, I think, that this counting is missed by an account of what is seen on the screen. So let's assign the task of accounting for counting to the Media Archaeologist, which means showing us how abstraction layers in software below the surface can be revealed, audited, and studied.

2. The Lure of the Authentic Experience

I have already said quite a bit about authenticity. Let me explain now why I am skeptical about an authentic experience of historical software, and why this is an important problem for software collections.

Everyone in game or software studies knows about emulation. Emulation projects struggle to recreate an authentic experience of operating a piece of software such as playing a game. Authenticity here means that the use experience today *is* like it *was*. The Lure of the Authentic Experience tells digital repositories at minimum not to preserve software in a manner that would interfere with the production of such experiences. At maximum, repositories *deliver* authentic experiences, whether on-site or online. A tall order. In the minimum case, the repository provides software and collects hardware specifications, drivers, or support programs. The documentation provides software and hardware specifications. Researchers use this documentation to reconstruct the historical look-and-feel of software to which they have access. In the maximum case, the repository designs and builds access environments. Using the software authentically would then probably mean a trip to the library or museum with historical or bespoke hardware. The reading room becomes the site of the experience.

I am not happy to debunk the Authentic Experience. Authenticity is a concept fraught not just with intellectual issues, but with registers ranging from nostalgia and fandom to immersion and fun. It is a minefield. The first problem is perhaps an academic point, but nonetheless important: authenticity is always constructed. Whose lived experience counts as "authentic" and how has it been documented? Is the best source a developer's design notes? The memory of someone who used the software when it was released? A marketing video? The researcher's self-reflexive use in a library or museum? If a game was designed for kids in 1985, do you have to find a kid to play it in 2050? In the case of software with a long history, such as *Breakout* or Microsoft Word, how do we account for the fact that the software was used on a variety of platforms—do repositories have to account for all of them? For example, does the playing of *DOOM* "death match" require peer-to-peer networking on a local area network, a mouse-and-keyboard control configuration, and a CRT display? There are documented cases of different configurations of hardware: track-balls, hacks that enabled multiplayer via TCPIP, monitors of various shapes and sizes, and so on. Which differences matter?

A second problem is that the Authentic Experience is not always that useful to the researcher, especially the researcher studying how historical software executes under the hood. The emulated version of a software program often compensates for its lack of authenticity by offering real-time information about system states and code execution. A trade-off for losing authenticity thus occurs when the emulator shows the underlying machine operation, the counting, if you will. What questions will historians of technology, practitioners of code studies, or game scholars ask about historical software? I suspect that many researchers will be as interested in how the software works as in a personal experience deemed authentic. As for more casual appreciation, the Guggenheim's *Seeing Double* exhibition and Margaret Hedstrom's studies of emulation suggest that exhibition visitors actually *prefer* reworked or updated experiences of historical software.[19]

This is not to say that original artifacts—both physical and "virtual"—will not be a necessary part of the research process. Access to original technology provides evidence regarding its constraints and affordances. I put this to you not as a "one size fits all" decision but as an area of institutional choice based on objectives and resources.

The Re-enactor, of course, is deeply committed to the Authentic Experience. If all we offer is emulation, what do we say to him, besides "sorry." Few digital repositories will be preoccupied with delivering authentic experiences as part of their core activity. The majority are likely to consider a better use of limited resources to be ensuring that validated software artifacts and contextual information are available on a case-by-case basis to researchers who do the work of re-enactment. Re-enactors will make use of documentation. Horwitz credits Robert Lee Hodge with an enormous amount of research time spent at the National Archives and Library of Congress. Many hours of research with photographs and documents stand behind his re-enactments. In short, repositories should let re-enactors be the re-enactors.

Consider this scenario for software re-enactment. You are playing an Atari VCS game with the open-source Stella emulator. It bothers you that viewing the game on your LCD display differs from the experience with a 1980s-era television set. You are motivated by this realization to contribute code to the Stella project for emulating a historical display. It is theoretically possible that you could assemble everything needed to create an experience that satisfies you—an old television, adapters, an original VCS, the software, etc. (Let's not worry about the shag rug and the lava lamp.) You can create this personal experience on your own, then write code that matches it. My question: Is the result less "authentic" if you relied on historical documentation such as video, screenshots, technical specifications, and other evidence available in a repository to describe the original experience? My point is that repositories can cooperatively support research by re-enactors who create *their* version of the experience. Digital repositories should consider the Authentic Experience as more of a research problem than a repository problem.

3. The Lure of the Executable

The Lure of the Executable evaluates software preservation in terms of success at building collections of software that can be executed on-demand by researchers.

Why do we collect historical software? Of course, the reason is that computers, software, and digital data have had a profound impact on virtually every aspect of recent history.

What should we collect? David Bearman's answer in 1987 was the "software *archive*." He distinguished this archive from what I will call the software *library*. The archive assembles documentation; the library provides historical software. The archive was a popular choice in the early days. Margaret Hedstrom reported that attendees at the 1990 Arden Conference on the Preservation of Microcomputer Software "debated whether it was necessary to preserve software itself in order to provide a sense of 'touch and feel' or whether the history of software development could be documented with more traditional records."[20] In 2002, the Smithsonian's David Allison wrote about collecting historical software in museums that "supporting materials are often more valuable for historical study than code itself. They provide contextual information that is critical to evaluating the historical significance of the software products." He concluded that operating software is not a high priority for historical museums.[21]

Again, institutional resources are not as limitless as the things we would like to do with software. Curators must prioritize among collections and services. The choice between software archive and library is not strictly binary, but choices still must be made.

I spend quite a bit of my professional life in software preservation projects. The end-product of these projects is at least in part the library of executable historical software. I understand the Lure of the Executable and the reasons that compel digital repositories to build collections of verified historical software that can be executed on-demand by researchers. This is the Holy Grail of digital curation with respect to software history. What could possibly be wrong with this mission, if it can be executed? As I have argued on other occasions there are several problems to consider. Let me give you two. The first is that software does not tell the user very much about how it has previously been used. In the best case, application software in its original use environment might display a record of files created by previous users, such as a list of recently opened files found in many productivity titles like Microsoft Office. The more typical situation is that software is freshly installed from data files in the repository and thus completely lacks information about its biography, for want of a better term.

The second, related problem is fundamental. Documentation that is a prerequisite for historical studies of software is rarely located in software. It is more accurate to say that this documentation *surrounds* software in development archives (including source code) and records of use and reception. It is important to understand that this is not just a problem for historical research. Documentation is also a problem for repositories. If contextual information such as software dependencies or descriptions of relationships among objects is not available to the repository and all the retired software engineers who knew the software inside and out are gone—it may be impossible to get old software to run.

Historians, of course, will usually be satisfied with the Archive. Given limited resources, is it reasonable to expect that the institutions responsible for historical collections of documentation will be able to reconcile such traditional uses with other methods of understanding historical computing systems? The Re-enactor will want to run software, and the Media Archaeologist will not just want access to a software library, but to original media and hardware in working order. These are tall orders for institutional repositories such as libraries and archives, though possibly a better fit to the museum or digital history center.

In *Best Before: Videogames, Supersession and Obsolescence*, James Newman is not optimistic about software preservation and he describes how the marketing of software has in some ways made this a near impossibility. He is not as pessimistic about video game history, however. In a section of his book provocatively called "Let Videogames Die," he argues that a documentary approach to gameplay might be a more pragmatic enterprise than the effort to preserve playable games. He sees this as a "shift away from conceiving of play as the *outcome* of preservation to a position that acknowledges play as an indivisible part of the object of preservation."[22] In other words, what happens when we record contemporary use of software to create historical documentation of that use? Does this activity potentially reduce the need for services that provide for use at any given time in the future? This strikes me as a plausible historical use case, but not one for re-enactment or media archaeology.

Software archives or software libraries? That is the question. Is it nobler to collect documentation or to suffer the slings and arrows of outrageous software installations? The case for documentation is strong. The consensus among library and museum curators (including myself) is almost certainly that documents from source code to screenshots are a clear win for historical studies of software. Historians, however, will not be the only visitors to the archive. But there are other reasons to collect documentation. One of the most important reasons, which I briefly noted above, is that software preservation requires such documentation. In other words, successful software preservation activities are dependent upon technical, contextual, and rights documentation. And of course, documents tell re-enactors how software was used and can help media archaeologists figure out what their machines are showing or telling them. But does documentation replace the software library? Is it sufficient to build archives of software history without libraries of historical software? As we have seen, this question was raised nearly forty years ago and remains relevant today. My wish is that this question of the relationship between documentation and software as key components of digital heritage work stir conversation among librarians, historians, archivists and museum curators. This conversation must consider that there is likely to be a broad palette of use cases such as the historian, media archaeologist and re-enactor, as well as many others not mentioned here. It is unlikely that any one institution can respond to every one of these use cases. Instead, the more likely result is a network of participating repositories, each of which will define priorities and allocate resources according to both their specific institutional contexts and an informed understanding of the capabilities of partner institutions.

NOTES

1. David Bearman, *Collecting Software: A New Challenge for Archives and Museums*, Archival Informatics Technical Report #2 (Spring 1987); see also David Bearman, "What Are/Is Informatics? And Especially, What/Who Is Archives and Museum Informatics?" *Archival Informatics Newsletter* 1, no.1 (Spring 1987): 8.

2. Bearman, *Collecting Software*, 25–26.

3. Hayden White, *The Content of the Form: Narrative Discourse and Historical Representation* (Baltimore: Johns Hopkins University Press, 1987), 5.

4. Hayden White, *Figural Realism: Studies in the Mimesis Effect* (Baltimore: Johns Hopkins University Press, 2000), 2.

5. Hayden White, "The Question of Narrative in Contemporary Historical Theory," *History and Theory* 23, no. 1 (February 1984): 3

6. Wolfgang Ernst, *Digital Memory and the Archive* (Minneapolis: University of Minnesota Press, 2012), 1053–1055.

7. Ernst, *Digital Memory and the Archive*, 2652–2653.

8. Ernst, *Digital Memory and the Archive*, 1080–1084.

9. Ernst, *Digital Memory and the Archive*, 1029–1032.

10. Ernst, *Digital Memory and the Archive*, 1342–1343.

11. Ernst, *Digital Memory and the Archive*, 1745–1746, 2527–2529.

12. Media Archaeological Fundus, January 21, 2016, http://www.medienwissenschaft.hu-berlin.de/medientheorien/fundus/media-archaeological-fundus.

13. Trevor Owens, "Archives, Materiality and the 'Agency of the Machine': An Interview with Wolfgang Ernst," *The Signal: Digital Preservation*, February 8, 2013, http://blogs.loc.gov/digitalpreservation/2013/02/archives-materiality-and-agency-of-the-machine-an-interview-with-wolfgang-ernst/.

14. John Walker, introduction to *The Analytical Engine: The First Computer*, Fourmilab, March 21, 2016, http://www.fourmilab.ch/babbage/; John Walker, "The Analytical Engine: Is the Emulator Authentic?" Fourmilab, March 21, 2016, http://www.fourmilab.ch/babbage/authentic.html.

15. Tony Horwitz, *Confederates in the Attic: Dispatches from the Unfinished Civil War* (New York: Pantheon Books), 1998.

16. Horwitz, *Confederates in the Attic*, 153–157.

17. Stephen Gapps, "Mobile Monuments: A View of Historical Reenactment and Authenticity from inside the Costume Cupboard of History," *Rethinking History: The Journal of Theory and Practice* 13, no. 3 (2009): 397.

18. Henry Lowood, "The Lures of Software Preservation," *Preserving.exe: Toward a National Strategy for Software Preservation* (October 2013): 4–11, http://www.digitalpreservation.gov/multimedia/documents/PreservingEXE_report_final101 813.pdf.

19. Margaret L. Hedstrom, Christopher A. Lee, Judith S. Olson, and Clifford A. Lampe, "'The Old Version Flickers More': Digital Preservation from the User's Perspective," *American Archivist* 69, no. 1 (Spring–Summer 2006): 159–187; Caitlin Jones, "Seeing Double: Emulation in Theory and Practice. The Erl King Study," Paper presented to the Electronic Media Group, June 14, 2004, Electronic Media Group.

20. Margaret L. Hedstrom and David Bearman, "Preservation of Microcomputer Software: A Symposium," *Archives and Museum Informatics* 4, no. 1 (Spring 1990): 10.

21. David K. Allison, "Preserving Software in History Museums: A Material Culture Approach," in *History of Computing: Software Issues*, ed. Ulf Hashagen, Reinhard Keil-Slawik, and Arthur L. Norberg (Berlin: Springer, 2002), 263–265; Cf. Len Shustek, "What Should We Collect to Preserve the History of Software?" *IEEE Annals of the History of Computing* 28 (October–December 2006): 110–112.

22. James Newman, *Best Before: Videogames, Supersession and Obsolescence* (London: Routledge, 2012), 160.

5

Screen Capture and Replay

Documenting Gameplay as Performance

Digital game performance involves interaction between two systems. It is liminal, in the sense that it occurs at a place of contact between humans and machines. Digital games encourage us to explore what happens in this space, which Alexander Galloway in *The Interface Effect* describes as an interaction "perched there, on the mediating thresholds of self and world."[1] Such claims lead us to consider human-machine co-performance as a particular problem for performance studies, as well as game, software, or computer design studies. In other words, we would like to know more about digital play as performance. In order to do that, we will need methods for archiving and reactivating play, whether as a research object or for other purposes, such as training or spectatorship. Where do we begin? We propose to divide this problem, like Gaul, into three parts. First, we consider how capturing game performance has, for want of a better phrase, played out through technologies of *replay*. Second, we review historical terms and technologies of replay in sports and sports media, thereby focusing on an ongoing tension between re-viewing and re-doing past performance. Third, we will track the implications of defining gameplay performance as a sequence of actions associated with *interactions* of a player or players with a game system. Finally, we conclude with a proposal for a performance capture tool that could provide a method for saving, playing back, and analyzing past gameplay that takes

Originally published as "Screen Capture and Replay: Documenting Gameplay as Performance" (with Eric Kaltman and Joseph Osborn), in *Histories of Performance Documentation: Museum, Artistic, and Scholarly Practices*, ed. Gabriella Giannacchi and Jonah Westerman (New York: Routledge, 2017), 149–164.

neither the player nor the game system as primary, but instead considers game performance as a production of their interaction.

Perfect Capture

Chris Crosby, known as NoSkill, was among the first highly skilled players of the multiplayer game *DOOM* (id Software, 1993) to receive the accolade "Doomgod." Before e-sports, he personified a new kind of competitive performance. Its arenas were local area networks or game servers hosted somewhere on the emerging Internet, along with forums and channels for supporting game communities. Crosby's reign as a Doomgod was brief, lasting roughly from 1994 to 1996; in 2001, he died in a car accident. In thinking about how to capture or replay such superlative performance, we must consider the methods available at the time and their capacity to deliver a sense of that performance to us as viewers more than two decades later.

The documentation formats that we use to revisit theatrical or athletic performances include a wide range of media: texts (scripts and playbooks), video and audio recordings, narrative accounts in newspaper reviews, analytical summaries in box scores, spectators' recollections and camera photos, and more. Researchers, coaches, fans, and others can recall historical performance through such documentation. These ways of re-experiencing performance take place in a new venue under new circumstances, such as on a display screen or printed artifact or in a researcher's imagination. Videotaping NoSkill's performances or writing about one of his games would have provided such documentation. However, like many activities that take place in digital environments, his gameplay can also be re-experienced in another way. It could be reactivated in a sense quite different from point-of-view recording or a narrative account. In addition to photographs and other documents associated with his lived life, visitors to Crosby's online Memorial Site also find a collection of *DOOM* replay files. These digital artifacts make it possible to recreate matches in which he played between May 1995 and April 1996 (NoSkill Memorial Site 2004). The replays are stored as "demos" or "lump" files (from the .lmp file extension for files holding "lumps" of *DOOM* data); loading these data into the correct version of the game instructs the game "engine" to execute a sequence of commands. These commands mimic the input control commands generated by Crosby's actions during every "tic" or time frame of a historical game. Additional metadata in the file exactly describe the conditions under which the game engine will re-execute these commands, such as the game version, map, and difficulty level. A demo is not a video recording, although such a recording (or "video capture") might seem to show the same imagery on a computer screen. The demo replay is a script that reactivates a game system and produces a perfectly rendered re-performance by the machine alone, provided that the instructions are executed by a copy or emulation of the same version of the game. This is reactivation, not documentation, of historical performance.

DOOM introduced the computer game genre later known as the "first-person shooter," a fast-action, shoot-'em-up format played from a first-person perspective. Demo replay therefore provides an eerie paradigm for performance documentation. When we load NoSkill's files into *DOOM*, we reactivate a dead game—in the sense that the version and computing system he used to play the game have long been obsolete—*and* reanimate the gameplay of a deceased player. Not only do the game and the gameplay come back to life; we watch through NoSkill's eyes. Is this performance capture so perfect that it not only enacts a historical performance, but also puts us in touch with the experiences of the original player?

Gameplay preservation has long been characterized by the metaphor of *capture*, whether it takes the form of screenshots, replays, video, or other means. The process involves capturing data, whether the data consist of scripts that represent discrete actions, as in the demo file, or streams of time-stamped audio and visual information that record what we see on the screen and hear in our headphones. As a metaphor, capture is a freighted word. Capturing a performance might mean that we are trapping a perfect reproduction of game events, or it might emphasize putting away a player's subjective view of an event for later liberation by a game engine. As historians, we are of course intrigued by the possibility of perfect reproduction of historical activities suggested by replaying NoSkill's matches and seeing more or less what he saw, at least in terms of images flashing on the screen. We expect that future historians will be thankful for access to replays as experiences of past performances in digital spaces. We are nevertheless convinced that data preservation and replay without contextual documentation fall short of satisfying their research needs because data alone reveal little about the events and motivations that created, surrounded and informed the use of these data.[2] Speaking of digital gameplay as performances that might be captured obligates us to dig into the implications of such capture.

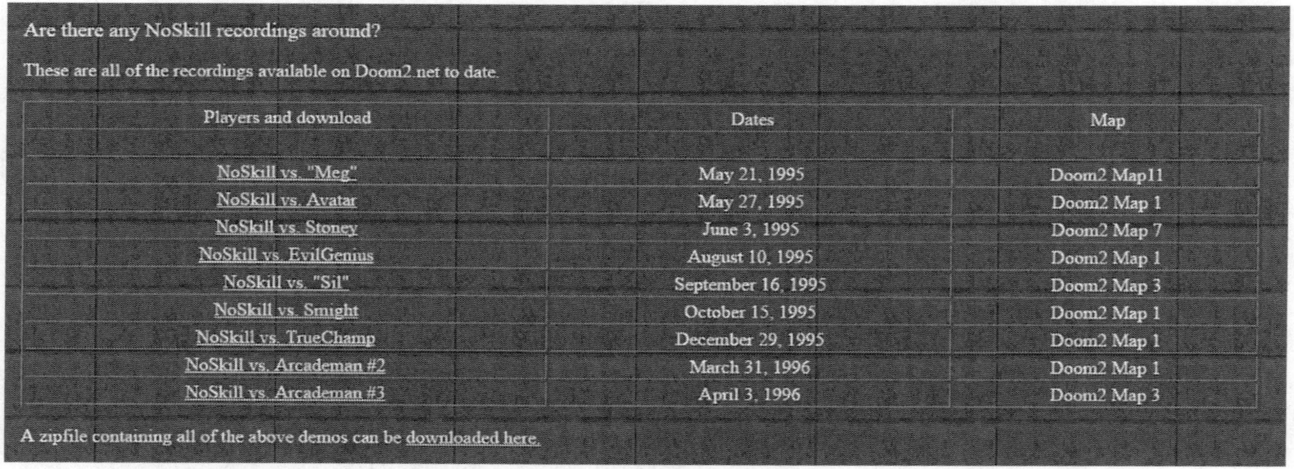

Are there any NoSkill recordings around?

These are all of the recordings available on Doom2.net to date.

Players and download	Dates	Map
NoSkill vs. "Meg"	May 21, 1995	Doom2 Map11
NoSkill vs. Avatar	May 27, 1995	Doom2 Map 1
NoSkill vs. Stoney	June 3, 1995	Doom2 Map 7
NoSkill vs. EvilGenius	August 10, 1995	Doom2 Map 1
NoSkill vs. "Sil"	September 16, 1995	Doom2 Map 3
NoSkill vs. Smight	October 15, 1995	Doom2 Map 1
NoSkill vs. TrueChamp	December 29, 1995	Doom2 Map 1
NoSkill vs. Arcademan #2	March 31, 1996	Doom2 Map 1
NoSkill vs. Arcademan #3	April 3, 1996	Doom2 Map 3

A zipfile containing all of the above demos can be downloaded here.

Figure 5.1. Replay files on the NoSkill Memorial Site. Internet Archive, captured January 14, 2002

Replay History

Capturing digital gameplay has generally involved either saving game data or recording what is visible on-screen to screenshots or video files. These methods offer several options. In the case of saved data, for example, we will distinguish among replays, "game saves," and "memory saves." All of these methods produce data files that can be stored, then described by adding metadata for later discovery. Specific kinds of computer applications, such as a game engine for a demo replay or a media player for recorded video, execute these files. The "captured" performance is thus re-enacted.

Replay and recording predate digital games, of course. The connection between game replay and television replay is particularly resonant. Creative programming (e.g., ABC Television's *Wide World of Sports*, f. 1961), innovative production technologies and a growing appetite for televised sporting events created a media climate that encouraged new ways to watch live matches and athletic competitions on TV. Replay in particular became a familiar concept to television viewers during the 1960s in the form of "instant replay," which was used almost exclusively during televised sports events. US television viewers first encountered it during the live broadcast of the annual Army-Navy football game by CBS Sports in 1965. Two years later, Ampex Corporation in California released the HS-100 "Color Slow-Motion Sports Recorder/Reproducer." Its name highlights replay as both capture and re-production. Ampex engineers worked closely with television producers such as Roone Arledge at ABC Sports on video recording technologies such as the HS-100 that might enhance live broadcasts. Producers used Ampex's magnetic disk-based video recording technology not just to revisit a performance highlight but also to slow down or "freeze" the action, allowing in-booth commentators to explain the buildup of a play by applying "slow motion" or "stop motion" effects during a replay, techniques also developed by Ampex.

Viewers accepted this new way of watching live performance, and instant replay became a staple of live sports coverage. Along with other technical innovations, it profoundly altered sports spectatorship. Viewers learned how to integrate recorded (replayed) segments and live broadcasts. As a consequence, the privilege associated with physical presence at a sporting event was diminished. As broadcast teams learned to use instant replay, slow motion, telestrators (real-time video markers), and more, "live spectatorship, with a distanced view and lacking replay, could seem to be missing an ele-

Figure 5.2. Ampex HS-100 Sports "Recorder/Reproducer." Product brochure, circa 1967. © Stanford University, image provided courtesy of Department of Special Collections, Stanford University Libraries

ment. No longer was the spectacle staged simply for those who were there; now it was, increasingly, stage-managed for the television viewer."[3] Instant replay shifted attitudes about live and mediated performance. The concept became familiar enough for the American football player Jerry Kramer to use *Instant Replay* as the title of his diary of the 1967 football season.[4] Less than a decade later, Marshall McLuhan praised instant replay "as a very powerful and important new development in our time." He concluded that there was no longer much reason to prefer live over recorded performance, whether of sports or a symphony.[5]

As instant replay changed the experience of watching sports, it also redefined "replay" for the conduct of games and matches. Before the 1960s, replay usually meant playing again; after *instant*, televised replay, its use invoked viewing again. The earlier sense of replay was that a new performance would occur; it did not refer to recording or archiving a previously played game or event within a game. For example, in May 1935 the Harvard baseball team, as a sign of good sportsmanship, agreed to "replay" a game played during the previous month against the rival Princeton team in response to a "debated interpretation of rules." This was replay as do-over, with a new match erasing the results of a previous competition ("Harvard Sportsmanship" 1935). A similar meaning carried over from sports to games in the form of the pinball replay introduced during the late 1930s. In this case, replay referred to the reward from a winning game that entitled the winner to a free play on the same machine. This kind of replay substituted for earlier forms of payoff prizes (including cash) that ran afoul of gambling laws in various US jurisdictions, a complicated episode of game and legal history that does not concern us here.[6] What is worth noting is that, as in sports, pinball replay referred to playing a new game or continuing gameplay, rather than revisiting a previously played session. After television's successful introduction of replay technology the older prospective meaning of replay as playing again gave way to the retrospective idea of watching again.

The examples we have recounted illustrate a general tension in replay between capturing and re-performing, or between *review* and *redo*, that predates replay as associated with digital games. On the one hand, we have a meaning associated with recording, with playing back rather than playing again. On the other, we have a meaning that describes a new event. Teams meet again, players set up the pieces for another go, or pinball machines yield a free play to a victorious player.

As a form of performance capture for digital games, replay has assimilated the tension between redo and review in a way that matches the particular affordances of game systems with respect to data and image capture. The history of performance capture for digital games raises the question, what do we mean by game-based performance? Richard Schechner, in his influential introduction to performance studies, reveals many different ways to understand the statement that "performances mark identities, bend time, reshape and adorn the body, and tell stories. Performances—of art, ritual, or ordinary life—are made of 'twice-behaved behaviors,' performed actions that people train to do, that they practice and rehearse."[7] He arranges numerous kinds of performance along a spectrum that allows for multiple modes of interaction among various performance activities: play, games, sports, popular entertainments, performing arts, daily life, and ritual.[8] The range and diversity of performance as understood by Schechner and others[9] is of course a compelling idea; we are interested especially in Schechner's emphasis on play (along with ritual) as anchoring components of any kind of performance. This approach leads Schechner to consider play, including games and sports, as a component of a more general framework of performance studies. As he puts it, "one definition of performance might be: ritualized behavior conditioned/permeated by play."[10] A game, in particular, is a structured form of play consisting of a sequence of "basic physical units" that Schechner calls "play acts."

In their magisterial study of game design, Salen and Zimmerman break play activity into three categories: gameplay, ludic activities, and "being playful." Gameplay, in particular, encompasses the "formalized interaction" of a player with a game system. As the term suggests, it takes place only through and in games, while ludic activities and playfulness involve less structured modes of play. As Salen and Zimmerman put it, gameplay is the "experience of a game set into motion through the participation of players" and includes a range of forms that include competitive, athletic, and performative play activities.[11] We will consider the "play acts" of game-based performance as the expression of gameplay, that is, as a sequence of actions associated with the interaction of a player or players with a game system.

Our perspective on game-based performance as it relates to digital games thus involves both the player as enactor and the active system (platform + game software). Capturing this performance involves more than recording human activity. Archiving player activities outside this interaction—facial expressions, bodily movements, conversation with others in the living room, etc.—produces contextual documentation that helps us to understand performance, but we do not consider these actions here as game performance. This does not lessen the importance of such documentation, particularly with regard to questions such as the "feelings and moods of the players and the observers."[12] However, these documents tell us about para-performance rather than play acts per se. Focusing on player interaction with a game system emphasizes the liminal nature of game-based performances. The interaction between enactor and system occurs at a boundary between the player and the game system, and it produces data that are captured there, such as replay scripts based on controller inputs, records of system memory, or video captured at the display screen. We are looking for systems that capture these data in a manner that makes it possible to reproduce, view, and cite game performance as a sequence of play acts.

We would be delighted to write a clean definition of performance capture based on the specific criterion of time-based interaction between player and system. However, our definition thus far skips over consideration of the difference between replays and saved games. Is a saved game a form of performance capture? The term "save" refers to a mix of methods and data formats for documenting the state of a game system at some point in time during gameplay, such that the saved data can be reloaded into a compatible system to produce the same state at a later time.[13] Saved games have been produced by save systems for many home console games, often tied to specific memory cards and game data formats, as PC game files (for saved games or replays) valid only for specific versions of a game title, or as maps of memory state information generated by emulators. We distinguish two broad classes of saves: "game saves" and "memory saves." We will have more to say below about these methods. With respect to performance capture, we consider saves as preserving the result of play performance rather than performance per se; saves give us a particular moment, often the concluding moment, of gameplay, but tell us little about the interactive play acts that occurred. Of course, this distinction is not absolute, as replay might be understood as the sequential reproduction of a series of game states, and saved games often provide opportunities for continuing a previously played game. On the whole, we associate replays with re-viewing previously played (captured) games, and saves with re-doing them (re-performance rather than performance capture).

The history of capturing game-based performance weaves through a remarkable variety of practices. This variety responded to technical constraints associated with specific platforms and games, modes of performance (e.g., single-player vs. competitive multiplayer games), and uses for archived game performance such as re-doing and re-viewing. This history includes methods used for pre-digital games. Samuel Tobin has written that, "Games were being saved long before they went digital."[14] He points to examples such as play-by-mail games, chess books, newspaper bridge problems, scribbled notes about the state of an interrupted game on writing pads, and clips for holding together sets of cards in a particular order. Some of these methods involved recording game states (such as the point at which one intended to continue playing), but others came closer to later game replay by summarizing sequences of already played actions. The latter group includes chess books with move-by-move summaries of a played game, and the "after action reports" of wargamers. Kirschenbaum notes that these "AARs" were "narrative retellings of events that unfolded during gameplay" and sometimes offered "procedural records of moves."[15] Players of manual (non-digital) games developed methods of performance capture that tracked specific elements of gameplay that they considered central to their experience.

Saves and replays of digital games from the era of *Spacewar!* and *Adventure* during the 1960s and 1970s, or of arcade consoles, and early home video game systems during the 1980s did not circulate between users at the time for the simple reason that provisions were not made in early game systems to create them. What we do have in the way of documentation from this early period does not match up well with our criterion that game-based performance capture must reproduce interactions between players and game systems. Television coverage and reportage about the computer game *Spacewar!* informs us about players and environments like computer laboratories and arcades around these interactions.[16] Saved paper tapes and printouts deliver historical software, including source code, but we learn little from these records about gameplay during the 1960s and 1970s. Some transcripts of text-based games such as *Adventure* and *MUD1* survive as printouts of sequences of typed player commands and computer responses. Jerz, for example, cites a transcript of a short segment of gameplay from the Crowther-Woods version of *Adventure* dating from 1977, and Richard Bartle has preserved short passages recording player interactions in *MUD1*, probably from the late 1970s.[17] These transcripts give us text output from conversations between players and game systems as well as among players and thus might be the closest thing we have to replays for this period. In their historical study of the early dissemination of *Spacewar!*, Monnens and Goldberg consider the paucity of relevant documents from the period between 1962 and 1972 by speculating that, "neither *Spacewar!*'s creators nor its players had any idea of its historical significance, perhaps even that they were witnessing the origins of something extraordinary."[18] This observation might also explain why little attention seems to have been given to saving data related to gameplay.

As digital games became more popular during the 1980s, greater attention to replay by developers gave players more options for saving game states and performances. Some games published during the 1980s stored information about played games in order to play again from a specified point in a previously played game. For example, a computer game called *Chess Partner* (c. 1983)[19] stored information that made it possible for a player to recreate her chess game from the beginning move; she could then delete an unsuccessful last move and redo it. *Chess Partner* exemplifies performance capture both for reviewing and redoing; as with many of the more robust replay systems of the 1990s forward, the goal of such replay was educational. Reviewing played games became a standard method for improving one's own performance. Yet this was not the only purpose. Several sports games during the 1980s were inspired by television's instant replay. Electronic Arts' *One on One*, released in 1983, rewarded a particularly good play by showing it to the player again immediately afterwards, but without capturing information for playback after the game was completed. Saving games for continuation at a later time also became a common practice by the end of the 1980s. Game systems began to offer ready methods for storing save files, such as using floppy drives on home and personal computers (and some game consoles of this era) or the battery-powered save functionality provided by some game cartridges for the Nintendo Entertainment System console. The capability of preserving game states for continuation of gameplay aided the success of long-form games, some lasting dozens of hours or more, such as Nintendo's *Legend of Zelda* (1986). As we noted earlier, however, such game saves usually served replay in the sense of continuing or redoing play rather than reviewing it.

By the end of the 1980s, a mix of increasing attention to competitive and real-time games and improvements in technology encouraged closer attention to the ways in which gameplay could be recorded and studied. Dani Bunten Berry's head-to-head multiplayer game *Modem Wars* (Electronic Arts, 1988) emphasized player performance from the perspective of how "each person had their own specialized style of play." Recognizing this as a mode of performance conducive to spectatorship, she designed the game to make it possible to create "game film" from stored data associated with a played game, expecting that players would use the "opportunity the game films offered to [. . .] create stories out of the intense and ephemeral experience" of the game (Berry, 1992). Berry's games, including *Command HQ* (Microprose, 1990) and *Global Conquest* (1992), brought competitive player performance and spectatorship to real-time strategic gameplay. Unfortunately, technical limitations may have limited the use and distribution of this replay technology, as replay files associated with these games have proven difficult to locate. Over the course of the next decade, replay (including demo replay) and screen capture began to be used extensively for recording game performance, particularly in online, competitive games played on personal computers and the networks that connected them.

Most replay formats since the mid-1990s have defined ways of producing linear, time-stamped logs of play actions recorded in a proprietary or game-specific data format for later execution by a copy of the same version of the original game. These play actions usually involve a player's interaction with the game system, such as moving at a specific rate, pressing a controller button, or picking up an object in the game world. NoSkill's replays were recorded as *DOOM* demos stored as lump files. Nitsche describes demo replays as "traces of play, recorded in-game action" that also "can be manipulated and moved independently from the main game engine to be played back on different computers."[20] Introduced as a feature of *DOOM*, during the 1990s such replays became a vital part of the culture of competitive, multiplayer games, especially first-person shooters and real-time strategy games. As Berry's vision of turning strategy gaming into a space for social performance had predicted, networked players made movies that documented their prowess for other players to watch.[21]

Space does not permit a full accounting of all the practices and games associated with active communities of replay creators and spectators. Instead, we would like to emphasize three historical points that position replays in a larger context of play, spectatorship and community-building associated with digital games. First, replay could also be pre-play. *DOOM* had followed a tradition set by "attract mode" screens in arcades by running a demo of co-designer John Romero's own gameplay as the game started up. Thus demos were sometimes called "intro" movies. Loading a demo file to prepare the player recalls usages such as game film (the term appropriated by Berry) for sports. Second, the use of replay as rehearsal echoes our previous hints at the educational use of game replays. These educational uses recall other ways of thinking about the demo (or demonstration) as a method for showing or showing off. Demos in this sense ranged from the computer program as prototype or demonstration (e.g., *Spacewar!* 1962, or Douglas Engelbart "Mother of all Demos," 1969) to the "single-standing pieces of real-time animation" called cracktros and from there to the "demoscene" that required "remarkably optimized code to render out largely noninteractive animation sequences."[22] As one veteran of *DOOM*'s rather different demo scene pointed out, "Use of demos for their educational value has been going on since almost the beginning." *DOOM* demos as demonstrations of skill by admired players such as NoSkill, XoLeRaS, and Smight circulated widely, supporting study by "a new player who wants to get better."[23]

Third, using replays to improve skills took the viewer beyond spectatorship. Game replays were interactive in a way that televised replays could not offer. When a game engine was activated to reproduce a previously played game based on information stored in the replay file, it did more than produce a recording. Since the game engine interpreted and played back scripts stored in the proprietary replay format, features of the game were available while viewing. A replay spectator might choose which player's view to follow or even move the camera, sometimes by interacting with in-game menus or those made available in replay mode. This active spectatorship bled into another practice made possible by the demo format: editing the replay scripts. Once the demo format was understood, players learned tricks such as "recamming" (changing the camera perspective) first associated with *Quake* speedrun replays. Editing demo files converged with related practices such as modifying game assets to lead from replays to narrative movie-making, thus setting the stage for the game-based video productions first known as *Quake* movies, but later renamed Machinima.

Demo replays produce small script files. These files could be easily distributed even during the early days of the low-bandwidth Web. This advantage was mitigated by the reliance of demo recording on specific game formats. Efforts to broaden the appeal of machinima by releasing these entertainment videos from their binding to specific player communities and game engines led to the adoption of video as a more accessible format. The availability of higher download speeds and better viewer applications by the early 2000s encouraged the use of video, as did the desire of many players to include footage captured on game consoles or in virtual worlds, platforms that did not offer access to demo replay. The gradual shift to video as the format of choice for machinima contributed to the increased use of screen capture for performance documentation, which meant routing audiovisual game output directly to raw video files (or later, to compressed video formats). Capturing video directly from a display screen (or more likely, a PC graphics card) differs from the demo or replay as performance documentation in several respects, beginning with the interface between player and system that it monitors. Unlike replay, which tracks play acts as primitive commands generated by control devices and interpreted by game engines, video capture records gameplay as a view on performance. It is also produced at a different interface: the screen.

These differences—actions vs. view, controller vs. screen—produced contrasting takes on performance capture recorded, say, as *DOOM* demos versus video recordings. Techniques for capturing gameplay to video date back to the 1980s, and even earlier if we include the use of film and video cameras to document arcade gameplay. For example, Sharp and Nintendo collaborated on the production of a device called the Famicom Titler, released in Japan in 1989; the Titler was essentially a Nintendo (Famicom) game console that produced color RGB video and output S-Video signals directly to an external device such as a video camera; it also included a few basic editing and audio recording capabilities. While the Titler had little impact outside Japan, these features emphasized a use of replay that has become a prominent feature of more recent screen capture technologies: the encapsulation of experiences. Aptly named, the Titler enhanced a spectatorial point-of-view produced at the screen with subtitles and commentary. By the late 1990s, the intimate connection of video capture and game-based movie-making in the form of machinima, better software applications for producing and playing back video on computers, and increasing access to broadband for file transfer of large video files favored increasing preference for video as a replay format. Video capture differs from demo replay not only because it produces documentation that is not interactive, but also because it leaves out machine states. Both of these differences condition video replay as preservation of a specific and unchanging point-of-view rather than detailed understanding of game actions.

A Performance Capture Tool

The various approaches to documenting game performance, whether the work of game designers or players, have led us to consider how game studies might benefit from access to replays and saves. How might a research tool providing access to recorded gameplay work? We begin with our insistence on the replay as documenting interactions between the player and game system. As a digital game is played, a player and a game system engage in a feedback loop. The player provides inputs to the system via a controller, keyboard, or other device, and the system interprets this input as it relates to the state of the game in its memory. The system then responds via visual, aural, or, in some cases, haptic output to which the player again responds.

Due to the nature of this feedback loop, the player and system are simultaneously spectators and performers. The player's input performance is a reaction to the output performance of the system, and vice versa. Performance capture of digital games based on video capture of either the visual output of the system or the physicality of the player's input is limited in ways that video recordings of a theatrical or sports performance might not be. Videos of digital games are not interactive and do not recreate the original conditions of the performance, they merely deliver the images and sounds of a past instance of play. Because playing a digital game is conditioned by the actions and reactions of a digital system, the performance of that system must also be captured as a digital object.

Digital game systems are always managing a specific part of computer memory during gameplay, in accordance with the rules of the game's source code. The player's inputs trigger specific transitions from one state of memory to the next, and these transitions usually correspond to changes of audiovisual output. In "replays" based on screen capture, these audiovisual outputs are captured through an external source (video camera, microphone, etc.) Demo and other game replay files provide the means for a game system, or a specific engine, to itself replay previous inputs recorded from a player. However, because a system has to be designed to interpret, record, and play back demo files, most games do not support this functionality. On the other hand, many older games can be played today through emulation, where a host computer imitates the processes and architecture of an earlier machine. This circumscription of one system (the target) within another (the host) allows for the host to record all the inputs, internal processes, and outputs of the target. Demo files can reproduce input (if available) and video can record the outputs, but emulation collapses the input/output (redo/review) distinction because it provides a means to capture the performance of the entire system. Since the system is being reproduced inside the host, all the memory transitions, the succession of computational states that manifest into a game performance, are available for recording and reproduction at a later time. This blurs the distinction between redoing a performance and reviewing it, because from the perspective of a system under emulation, the same succession of inputs (and of memory states) will reproduce the same outputs. It is as if the aforementioned Harvard-Princeton game could be replayed on the same field, with the same players (and their internal mind-sets), and the same play conditions resulting in the same plays being made with exactly the same final result.

We propose a citation tool that leverages the capabilities of emulated systems for sharing and reproducing system performances. Specifically, emulators can record inputs for any game, as well as computational state information. Emulation allows for a demo file–like replay of any game, since it is possible to provide an emulator with the same time-indexed inputs as those contained in demo formats. An emulator also allows for the saving of the system's memory at any point in time, which amounts to creating a snapshot of all system processes that can then be referenced and reactivated at a later time. This "memory save" enabled by the emulator is akin to a traditional saved game file, except that the latter must be saved and restored by the original game system. A saved game file is also constrained temporally by the location or availability of save functions within a played game (such as save points or a menu allowing one to "Save"). Emulated memory saves avoid this constraint by, in effect, copying the entire system.

This performance citation tool allows for the management, capture, and replay of system performances and provides a means of linking them to structured descriptions of their provenance (the game and emulator that produced them). For example, using the tool it is possible to load *Super Mario Bros.* (*SMB*) into an emulator for the Nintendo Entertainment System (in this case a program called FCEUX). The tool will record the inputs of a player to the emulator along with a video capture of the resulting gameplay. If at any point some aspect of the game piques the player's interest, say a particular level or an especially difficult jump, he or she can instruct the tool to create a memory save and mark that point for later investigation. If for some reason the performance does not go as planned (maybe after falling prey to a difficult jump) the tool can instruct the emulator to load an earlier memory save and continue play from that point. The tool will also automatically edit the failed jump out of the accompanying video. When the player is satisfied with the performance, they can instruct the tool to save it. This instruction engages its management component.

When recording a system performance the tool accounts for certain crucial bits of information. Any citation of a system performance requires a reference to the specific emulator and specific game that created it. In the *SMB* example, there are many different potential NES emulators and many different versions of the game. A change to either component (emulator or version) might disrupt the ability to reactivate the system performance, so the tool records all this information through a detailed metadata scheme. It

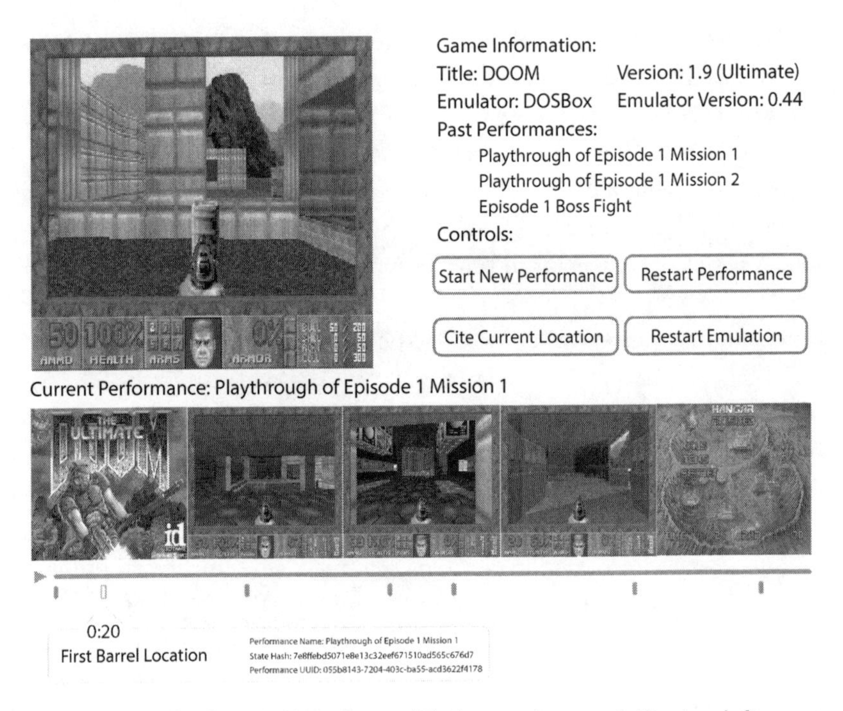

Figure 5.3. A citation tool interface, with the running emulation top left

checks that a playback system configuration matches the original before reactivating a previous performance. Additionally, the tool links the *SMB* performance both to its video recording and to the various memory saves the player made along the way. If a researcher wants to simply view an example of play, the video is available immediately, but if that researcher prefers to experience something seen in the video (maybe a final battle with the evil Bowser needs a closer look), the tool will reactivate the system performance at the closest memory save to the video's current position. Unlike video capture, the tool provides the options of jumping right into the game to continue a particular performance or simply reviewing gameplay segments that they themselves may not have the skill to reach.

Our game performance capture tool records all the relevant data for reactivating a system performance; it is a simple step to encode and share a citation of a performance that is also an entry-point *into* that performance. Because many emulators can be executed inside modern Internet browsers, linking to

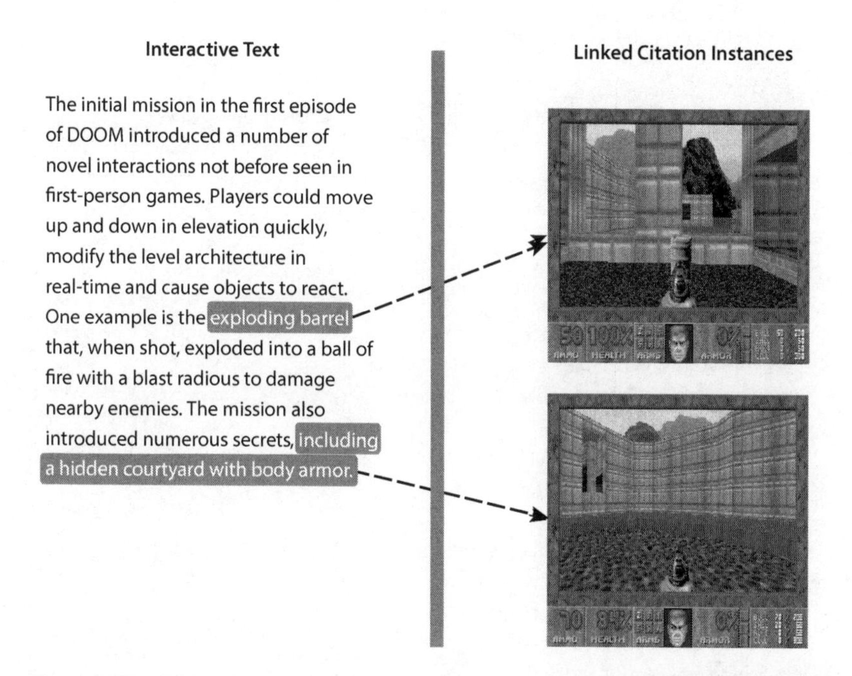

Figure 5.4. Linking from a researcher's text to a running emulation illustrating the text with historical gameplay, which in turn could be played anew from the point cited

an active system performance is now a possibility. We can record our performance, share a link to it with a collaborator, and have that person see what we did or take over where we left off. The tool will also allow for the embedding of multiple system performances on the same page, making comparative analysis much easier. If a specific glitch in *SMB* differs in two versions of the game, a researcher could describe them and provide playable links to supplement the comparison. If a historical player recorded a seminal feat of *SMB* mastery, he or she could allow anyone in the future to see it, to hop in and play along, or maybe even do them one better. System performance citations might thus produce a meaningful dialogue between the gameplay (and players) of the past and future, an affordance that incidentally encourages consistent citation and description of games, game systems, and the performances produced with them. In these ways, we hope to exploit the potential for gameplay performance capture as the ultimate synthesis of historical takes on redoing and reviewing, as well as the interaction of human and system performance.

NOTES

1. Alexander Galloway, *The Internet Effect* (Cambridge, UK: Polity, 2012), 78–79.

2. See Henry Lowood, "Memento Mundi: Are Virtual Worlds History?" in *Digital Media: Technological and Social Challenges of the Interactive World*, ed. Megan Winget and William Aspray (Lanham, MD: Scarecrow Press, 2011), 3–25; Henry Lowood "The Lures of Software Preservation," *Preserving.exe: Toward a National Strategy for Software Preservation* (October 2013): 4–11, http://www.digitalpreservation.gov/multimedia/documents/PreservingEXE_report_final101813.pdf.

3. Garry Whannel, "Television and the Transformation of Sport," *Annals of the American Academy of Political and Social Science* (September 2009): 215.

4. Kramer, Jerry. *Instant Replay: The Green Bay Diary of Jerry Kramer.* New York: World, 1968.

5. Marshall McLuhan, "'It Will Probably End the Motor Car': An Interview with Marshall McLuhan," Pay-TV (August 1976), 27.

6. See "Slot Machines and Pinball Games," *Annals of the American Academy of Political and Social Science* (May 1950), 68–69 (anonymously written); R. J. Urban, "Gambling Today via the 'Free Replay' Pinball Machine," *Marquette Law Review* 98 (Summer 1958), http://scholarship.law.marquette.edu/mulr/vol42/iss1/12.

7. Richard Schechner, *Performance Studies: An Introduction* (London: Routledge, 2002), 22.

8. Schechner, *Performance Studies*, 42.

9. Examples include Marvin Carlson, *Performance: A Critical Introduction* (London: Routledge, 1996); Erving Goffman, *The Presentation of Self in Everyday Life* (Garden City, NY: Doubleday, 1959).

10. Schechner, *Performance Studies*, 79.

11. Katie Salen and Eric Zimmerman, *Rules of Play: Game Design Fundamental* (Cambridge, MA: MIT Press, 2004), 305, 309–310.

12. Schechner, *Performance Studies*, 84.

13. See Samuel Tobin, "Save," in *Debugging Game History: A Critical Lexicon*, ed. Henry Lowood and Raiford Guins (Cambridge, MA: MIT Press, 2016).

14. Tobin, "Save," 385.

15. Matthew Kirschenbaum, "War Stories: Boardgames and (Vast) Procedural Narratives," in *Third Person: Authoring and Exploring Vast Narratives*, ed. Pat Harrigan and Noah Wardrip-Fruon (Cambridge, MA: MIT Press, 2009), 357, 368.

16. See Stewart Brand, "SPACEWAR: Fanatic Life and Symbolic Death Among the Computer Bums," *Rolling Stone*, December 7, 1972, http://www.wheels.org/spacewar/stone/rolling_stone.html; and Devin Monnens and Martin Goldberg, "Space Odyssey: The Long Journey of *Spacewar!*, from MIT to Computer Labs around the World," *Kinephanos* (June special issue 2015), 140.

17. See D. Jerz, "Somewhere Nearby Is Colossal Cave: Examining Will Crowther's Original 'Adventure' in Code and in Kentucky," *Digital Humanities Quarterly* 1, no. 2 (2007), http://www.digitalhumanities.org/dhq/vol/001/2/000009/000009.html; Richard Bartle, *Richard Bartle Papers*, 1979–1997, Stanford University Libraries, 1992.

18. Monnens and Goldberg, "Space Odyssey," 125.

19. See "Biding Your Time with Computerized Chess," *PC Magazine*, September 1983, 449–458.

20. Michael Nitsche, "Demo," in *Debugging Game History: A Critical Lexicon*, ed. Henry Lowood and Raiford Guins (Cambridge, MA: MIT Press, 2016), 103.

21. See Henry Lowood, "'It's Not Easy Being Green': Real-Time Game Performance in *Warcraft*," in *Videogame/Player/Text*, ed. Barry Atkins and Tanya Krzywinska (Manchester, UK: Manchester University Press, 2007): 83–100; Henry Lowood, "*Warcraft* Adventures: Texts, Replay and Machinima in a Game-Based Story World," in *Third Person: Authoring and Exploring Vast Narratives*, ed. Pat Harrigan and Noah Wardrip-Fruin (Cambridge, MA: MIT Press, 2009): 407–427.

22. Nitsche, "Demo," 103.

23. L. Herrmann, email from Laura "BahdKo" Herrmann to Henry Lowood, January 28, 2004.

6

Software Archives and Software Libraries

Defining the Software Collection

Conversations about historical software archives and collections preceded the rise of the Web and the advent of digital libraries, events of the early 1990s that stimulated projects for the preservation of digital objects. The problem of how to document computing hardware and activities emerged even earlier, before much attention was paid to software as a historical object. In this essay, I will focus on collecting software with three points of emphasis. The first is historical. When did cultural institutions in the United States begin to collect software, and what did they collect? This historical background sets up the rest of the essay. Second, my discussion of "software" will be limited generally to formats that have thus far, for better or worse, dominated collections in cultural institutions. This means microcomputer software and digital games of the period from the late 1970s through the early 21st century. I understand the limitations of collections that do not nearly reflect the full range of historical software development, but they do provide a foundation for addressing issues of preservation, access, and exhibition that museums and libraries will face as we move forward. Those issues will be the third point of emphasis. The endgame will be a discussion of the challenges faced by institutional curators in developing priorities for their care of historical software collections.

Originally published as "Software Archives and Software Libraries," in *Challenging Collections: Approaches to the Heritage of Recent Science and Technology*, ed. Alison Boyle and Johannes-Geert Hagmann, Artefacts: Studies in the History of Science and Technology (Washington, DC: Smithsonian Institution Scholarly Press, 2017), 68–86.

Institutions and individual collectors began in the late 1970s to gather archival documentation and build libraries for research on the history of computers. The Charles Babbage Institute (CBI), for example, was founded in 1978. Its efforts dovetailed with those of a History of Computing Committee formed by the American Federation of Information Processing Societies (AFIPS), and AFIPS became a major sponsor of CBI. Around that time the AFIPS committee produced a brochure called "Preserving Computer-Related Source Materials," which was distributed at the National Computer Conference. The brochure provided the following information:

> If we are to fully understand the process of computer and computing developments as well as the end results, it is imperative that the following material be preserved: correspondence; working papers; unpublished reports; obsolete manuals; key program listings used to debug and improve important software; hardware and componentry engineering drawings; financial records; and associated documents and artifacts.

In other words, paper records. The recommendations did not mention the preservation of data files or executable software, although there was a brief nod to the museum value of hardware artifacts for "esthetic and sentimental value" and to provide "a true picture of the mind of the past, in the same way as the furnishings of a preserved or restored house provides a picture of past society." One year later, in 1979, CBI received its first significant donation of books and archival documents from George Glaser, a former president of AFIPS. Into the 1980s institutional projects primarily gathered documentation: archival records, publications, ephemera, and oral histories.[1]

Attention to historical software archiving and preservation followed documentation projects or historical work and to some degree grew out of them. Two noteworthy movers in shifting attention to software as archival objects were David Bearman and Margaret Hedstrom. Bearman left his position as deputy director for information resource management in the Office of Information Resource Management in the Smithsonian Institution in 1986 to create a company called Archives and Museum Informatics. He envisioned this new company as a research organization and a consulting firm. Bearman had written extensively on the application of new information technologies to archival and museum management. He began publishing the *Archival Informatics Newsletter* in 1987 (the title changed to *Archives and Museum Informatics* in 1989). Both the company and its newsletter offered a forum for exchanging information about informatics, particularly about how to automate archives and museums. Informatics did not simply mean institutional change, however. Bearman was also considering software both as a collection object and as an interactive medium that could extend museum programs.[2]

As one of its earliest consultations, Archives and Museum Informatics drafted policies and procedures for a "Software Archives" at the Computer History Museum (CHM) then located in Boston. In 1987, Bearman published the first important study of software as a collection object: *Collecting Software: A New Challenge for Archives and Museums*. Bearman's report provided reasons both for frustration and persistence. He wrote frankly that a telephone survey of "a few computer companies, high-technology firms, universities, professional associations and governments" conducted in 1987 had revealed that "no one seems to have formed a software archive nor is anyone collecting software documentation, except as an inadvertent biproduct of institutional archives." He named only one exception: CBI. Indeed, "the concept of collecting software for historical research purposes had not occurred to the archivists surveyed; perhaps, in part, because no one ever asks for such documentation!"[3] He found neither actions nor intentions to build software archives among those surveyed. Meetings with the CHM, Smithsonian, and CBI had instead convinced him that "no single institution, or even group of institutions as prestigious and well situated as the sponsors [of his CHM report] were, could expect to collect the entire corpus of software related materials."[4] Undaunted, Bearman produced a report that carefully considered software archives as a multi-institutional endeavor, including collection policies and selection criteria, use cases, policies, a rough "software thesaurus" to provide terms for organizing a software collection, and a variety of practices and staffing models. Should an institution accept the challenge, here were some of the tools needed for the job.

Bearman found no signs of life for software archives. However, he took note of important work in related fields. His research revealed progress in archiving of machine-readable data files or archival documents such as email, as well as contributions to archival automation, an area familiar to him through his work at the Smithsonian. There was hardly an overabundance of these studies, yet at least there were reliable and informative publications. One of them was Margaret Hedstrom's *Archives and Manuscripts: Machine-Readable Records*, a practical guide that had been published by the Society of American Archivists in 1984.[5] Trained as a historian of technology and an archivist, Hedstrom was among the first US archivists concerned with electronic records and data, first at the State Historical Society of Wisconsin (1979–1983) and then as director of the Center for Electronic Records at the New York State Archives (1985–1995). The interplay between digital archives and software collecting—or between documentation and software—continues to the present, and Bearman addressed the relationship of these siblings in his report. He was at first skeptical about the relevance of progress in building data archives for the different problem of collecting software, primarily because methods for preserving data in the 1980s involved a model of data migration that broke the connection between the original context of data creation and its eventual storage in a repository.

Bearman's remarks identified a tension between the goals and approaches of data archives and software archives. He was not, however, opposing work that emphasized documentation over software preservation. He was simply pointing out that these are different problems with different solutions. He praised Hedstrom's work and others in later issues of the *Archival Informatics Newsletter* and eventually collaborated with her on several publications and projects. The point was (and remains) that archives, libraries, and museums that chose to build collections for the history of computing and its impact faced a multifaceted challenge. They would have to develop new practices for assessing, collecting, and preserving in numerous and diverse formats and nuance these practices to account for different use scenarios for documents, data, and software. Hedstrom and Bearman provided one argument for linking software archives and digital records in a report coauthored in 1993: "Indeed in the electronic age, custody of archives may require the on-going maintenance of a range of hardware and software and continuing migration of both data and applications, both of which activities are never ending and very expensive."[6] In other words, use of historical documents in digital form may require access to historical software. Writing in the late 1980s, Bearman's case for software archives rested on a simple proposition for the digital archives and libraries of the future: The more participants, the merrier.

Attention to the problem of software preservation took off during the 1990s. There probably are several explanations for this, but the deluge of commercial microcomputer software published during the 1980s surely played a leading role in awakening institutional repositories such as libraries and museums to the problem of software preservation. A specific moment for this awakening might have been the symposium organized by Columbia

University called "Preservation of Microcomputer Software" that took place in March 1990 at Columbia's Arden House Conference Center in the Catskills. The invitation to participants noted that the proposed topic of "Planning a Center for the Preservation of Microcomputer Software" had become "necessary and urgent." The conference addressed topics such as a consortium of collection centers, the establishment of standards and techniques for migration and preservation of electronic data, collecting strategies, and technical problems with executing historical software.[7]

In his paper on the proposed centers, Bernard Galler addressed the kinds of materials that would "ideally [be] available for future study" in a collection devoted to the history of microcomputer software:

- executable code, including the complete environment (hardware, operating system, peripheral devices, and so on) to run it
- source language statements, preferably in machine-readable form but at least on paper
- enough of an environment so that executable code could be produced from the source code, if needed, for whatever system is available for execution
- representative code (source and executable) for each distinct version that appeared and that received a distinctive name or label
- user documentation, as supplied by the creator or vendor
- program logic documentation, as much as is available
- advertising/marketing materials
- environment specifications for both creating and running object code
- user statistics, both numbers of users and their profiles

This is a comprehensive list, running the gamut from source code to executable software and operating environment and on to documentation, marketing, and evidence of use. Galler must have known that this would be a tall order for any repository that one could imagine in 1990; he wrote that "as always, what can be saved will depend on the cost of acquisition, and on the priorities assigned to the kinds of information, just as with the product as a whole."[8] The Arden conference did not result in the creation of such a center, and Bearman described the papers as "largely disappointing" in a review.[9] However, the conference did spur a subsequent project that drew upon its energy: a proposed National Software Archives (NSA). The NSA proposal arose out of a series of meetings convened by David Allison of the Smithsonian Institution's National Museum of American History. The participants included collection curators, historians, and representatives from several companies (Apple, Microsoft, Hewlett-Packard, WordPerfect). The NSA meetings reviewed many of the issues that had been vetted in Bearman's report and the Arden conference, adding participation by software developers and the notion of surveying, identifying, and marking materials held by developers and other institutions for eventual deposit in the NSA. Yet it failed to produce an active collecting program. The same could be said of a similar effort by a Software Task Force of CBI, completed in 1998.[10] These initiatives raised provocative issues about software archives, but they did not provide a model for building collections of historical software.

One of the papers at the Arden conference noted that future historians of computing would require access not just to "the numerous species of academic and business software"; it proposed that a "Center for the Preservation of Microcomputer Software" also include "software for such functions as games, entertainment and personal financial management."[11] This comment was prescient because the road into institutional repositories was paved with packaged consumer (microcomputer) software and digital games. Of course, there are other important classes of software and many domains other than personal use, ranging from computer science and academic software to business and military applications. Also, it is easy to name media used historically for data storage and software execution, such as punch cards, paper tape, and magnetic tape, that one will not run across in a collection of floppy diskettes, game cartridges, and optical discs associated with consumer software. The emphasis on consumer software in the first historical software collections to land in repositories thus provokes a question: Is there a rationale for "preferring" these kinds of software objects in libraries and museums, or was it just an accident of collecting history?

Two software collections will help us to consider this question. The Machine-Readable Collections Reading Room (MRCRR) and the Stephen M. Cabrinety Collection in the History of Microcomputing were made available to researchers by the Library of Congress and the Stanford University Libraries, respectively, under rather different circumstances and at different ends of the 1990s. Both collections were substantial and consisted primarily of software in the categories of microcomputer software intended for consumer use, including games and various forms of entertainment and educational software. They were, as far as I know, the first major collections of historical software that researchers used in controlled reading rooms, that is, within the walls of traditional research repositories. To the extent that these collections represent the first wave of software collecting in the United States, their acquisition emphasized specific kinds of software for specific classes of computers. This specificity, in turn, shaped a conception of not just the software object itself but also how software was organized, packaged, and documented and, in turn, how institutions would describe, preserve, and exhibit it. And yet collections consisting primarily of "software for such functions as games, entertainment and personal financial management" could document only a slice of software history. They omitted important categories of academic, scientific, and business software; computing environments other than microcomputer platforms and game consoles; and means of distribution other than retail stock.

In his annual report for the fiscal year ending September 30, 1988, the US librarian of Congress, James Billington, described a new project of the library:

> In July the Library opened a new reading room, the first of its kind in the nation, as a one-year pilot project. The Machine-Readable Collections Reading Room is intended to be a facility for the study of the design, history, and documentation of software and information data files. It focuses attention on the Library's continuing program to acquire, catalog, and make available to researchers materials in this format. It brings these items together physically and serves to underscore the significance of traditional library materials formatted in new technologies to contemporary society.[12]

The Library of Congress (LC) is the copyright deposit library of the United States, but its software collection in 1988 had been acquired mostly by other means, such as donations, exchanges for copyright registration (not the same process as deposit), and purchases. Indeed, the MRCRR collection featured items acquired by several LC divisions, from Science and Technology to Rare Books and Special Collections. "Machine-readable materials" had been collected since at least the early 1980s by these units, and three different LC committees had considered various aspects of the collection of these materials by 1987.[13] Mandatory deposit had recently been proposed and was still under discussion, with noteworthy resistance from the software industry. In light of the contentious matter of copyright deposit, it is worth noting that representatives from Apple, the Association of American Publishers, IBM,

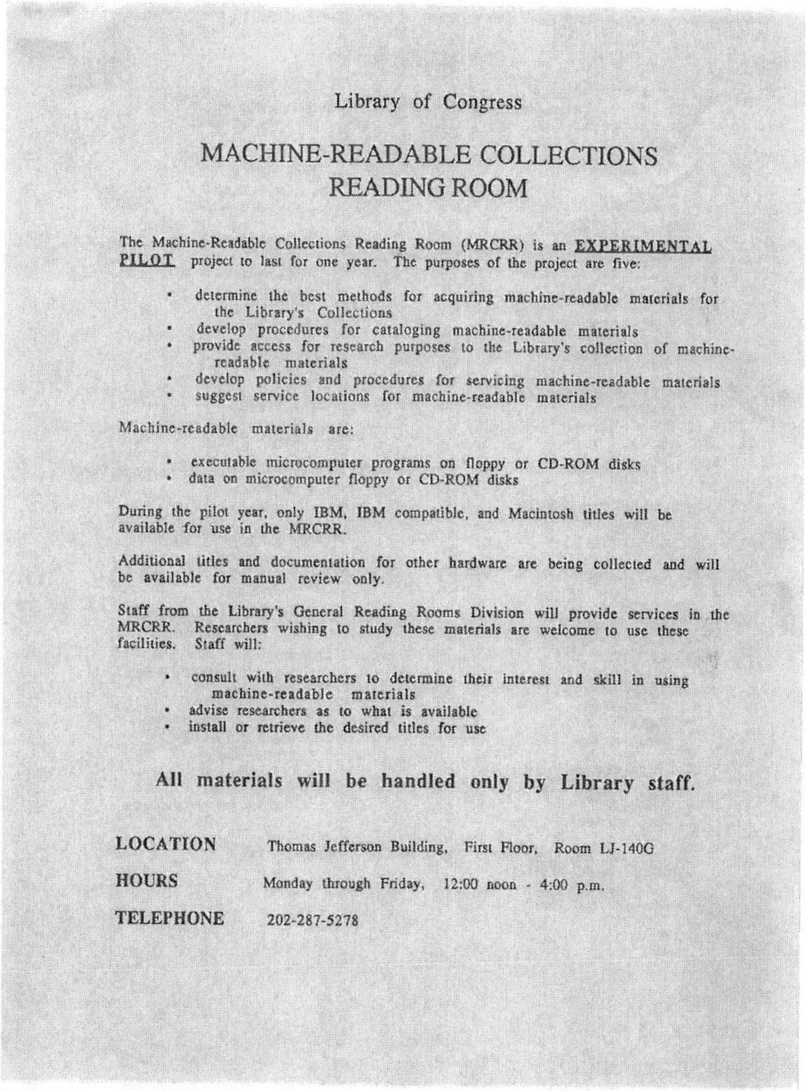

Library of Congress

MACHINE-READABLE COLLECTIONS READING ROOM

The Machine-Readable Collections Reading Room (MRCRR) is an **EXPERIMENTAL PILOT** project to last for one year. The purposes of the project are five:

- determine the best methods for acquiring machine-readable materials for the Library's Collections
- develop procedures for cataloging machine-readable materials
- provide access for research purposes to the Library's collection of machine-readable materials
- develop policies and procedures for servicing machine-readable materials
- suggest service locations for machine-readable materials

Machine-readable materials are:

- executable microcomputer programs on floppy or CD-ROM disks
- data on microcomputer floppy or CD-ROM disks

During the pilot year, only IBM, IBM compatible, and Macintosh titles will be available for use in the MRCRR.

Additional titles and documentation for other hardware are being collected and will be available for manual review only.

Staff from the Library's General Reading Rooms Division will provide services in the MRCRR. Researchers wishing to study these materials are welcome to use these facilities. Staff will:

- consult with researchers to determine their interest and skill in using machine-readable materials
- advise researchers as to what is available
- install or retrieve the desired titles for use

All materials will be handled only by Library staff.

LOCATION Thomas Jefferson Building, First Floor, Room LJ-140G

HOURS Monday through Friday, 12:00 noon - 4:00 p.m.

TELEPHONE 202-287-5278

Figure 6.1. Library of Congress software reading room notice, 1998. Lowood Papers, Stanford. Courtesy of Department of Special Collections and University Archives, Stanford University Libraries, Stanford, California

the Software Publishers Association, and other industry organizations attended a meeting held in May 1988 to discuss the operations of the MRCRR and attended a ribbon-cutting ceremony to launch the project after the meeting.[14]

The collection consisted of "executable programs" and "data" on "microcomputer or CD-ROM discs" and "video discs." Representative titles ranged from productivity titles such as bibliographic managers (ProCite, Sci-Mate) and word processing software (MS Word, WordPerfect, WordStar) to reference databases (*Oxford English Dictionary*, ABI/Inform) and even a category for "windowing" (MS Windows, TopView). Only one entertainment title (arguably) was included in the MRCRR information flyer out of roughly 70 titles: Flight Simulator.[15] The entire collection comprised 1,419 IBM-compatible and Macintosh titles, with the largest number of titles, 861, having been acquired as of July 1989 by copyright registrations between 1978 and 1989.[16] The collection and the service registered a serious tone and purpose. Machine-readable materials could be retrieved, handled, and installed only by library staff. A columnist for the *Computer System News* reported in January 1989 that policies enforced in the room were set to mollify industry concerns about copying through access to software acquired by the proposed mandatory deposit. An MRCRR librarian cited "limitations on the use of software programs," noting that "we don't allow people to do their taxes here or write papers." The author commented that "the librarians spend a lot of time looking over patrons' shoulders, making sure the rules aren't broken."[17] It is difficult to rectify the lofty goals of studying the "design, history, and documentation of software" with the wrangling over copying of software and copyright deposit that constrained the reading room's policies and, clearly, put a damper on research.

A report issued in January 1990 summarized what the Library of Congress learned from its yearlong experiment with providing access to software. The report called for the library to "expand the scope of the collections to include more sophisticated and expensive titles that are leaders in their areas."[18] It identified areas such as "computer-aided design/manufacturing, image processing, mathematics, expert systems and full-text optical disk publications." Today, about twenty-five years after the conclusion of the MRCRR pilot project, the current overview of computer files at the Library of Congress notes an increase from the approximately 1,500 titles held in 1988 to more than 88,000 items, most of them in the Humanities and Social

Sciences Reading Room. The collection scope and formats collected have not changed much, although the materials in the current collection "are usually PC compatible." The original goal of supporting work on the "design, history and documentation" of software has not resonated strongly with the services eventually implemented for the Machine-Readable Collections, but it has not been entirely absent either. The Humanities and Social Sciences Collection has collected "sample software products" selected "for their representation of the industry and as archival artifacts to represent computer and software development." These samples include software intended for use on operating systems and devices that are not included in the collecting policy of the Machine-Readable Collection.[19] Responsibility for preservation of roughly 3,000 entertainment and related educational titles in the collection has been assumed by the Moving Image Section, following its strong commitment to the history of media.[20]

The Cabrinety Collection aligns neatly with patterns of collection building that have for centuries brought private collections from the idiosyncratic to the monumental into museums and libraries. Stephen Cabrinety, the son of a DEC (Digital Equipment Corporation) executive, began collecting software as a teenager and steadily added to his collection right up to his untimely death in 1995, at the age of 29. He wrote programs and founded a company called Superior Software. In 1989, two years after Bearman's conclusion that no institution had yet begun to collect historical software and just as the Arden conference was being planned, Cabrinety set up the Computer History

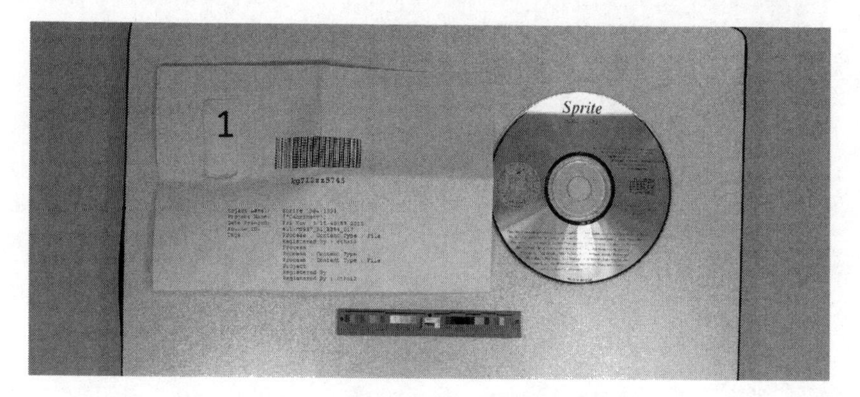

Figure 6.2. An artifact from the CHIPS collection. Copyright Board of Trustees of Stanford University. Image provided courtesy of Department of Special Collections, Stanford Libraries

Institute for the Preservation of Software (CHIPS)—at the ripe age of 23. His mother later described the goal of CHIPS as being "to preserve all software, and as much hardware as possible, so that future generations could visually see how the industry evolved."[21] In the CHIPS business plan completed in May 1989, Cabrinety reported that his collection included more than 18,000 pieces of software and 55 microcomputer systems. The plan also detailed staffing (including historians of computing and a curator) and fundraising plans and anticipated an annual collecting budget of about $150,000. Finally, Cabrinety indicated a preference for locating the new institution in the Boston area, near Washington DC, or in "the world's single largest technological hotspot," Silicon Valley.[22]

Less than a decade later, the CHIPS collection arrived in Silicon Valley. The Stanford University Libraries acquired it from the Cabrinety family in 1997, two years after Stephen's death. Shortly before the gift was finalized, a reporter for the *Los Angeles Times* marveled at the work of the youthful collector, noting that "one of the world's largest vintage software collections is not housed in the Smithsonian, or the Computer Museum in Boston, or even the in-house museum at Microsoft Corp."[23] The remark is not inaccurate, but at the same time it misses the longstanding importance of personal collecting and private initiatives as cornerstones of institutional collections. The Cabrinety Collection at Stanford complements hardware and software with documentation such as publications, ephemera, messages and files downloaded from bulletin board systems (BBS), Stephen Cabrinety's papers, and records of Superior Software and CHIPS. Like other collectors, Cabrinety also pursued specific interests within the larger framework of software history; for example, the collection is rich in various flavors of interactive books and "edutainment" titles, the areas in which Superior Software was active.[24]

Software as a Collection Object

Shortly after Stanford acquired the Cabrinety Collection, Doron Swade (coincidentally) concluded that "despite the formidable obstacles that face a fully-fledged software preservation programme, there is at least one modest but significant programme of software acquisition that is technically achievable and that has affordable resource implications—namely, software for PCs—'shrink-wrapped' consumer software, as well as custom-written special applications software."[25] As we have seen, libraries first dipped their toes in the waters of software collecting in exactly this way. This first generation of collections thus implicitly defined the historical software object as Swade predicted, that is, as executable software carried on a media format packaged inside a physical container, along with a few other inserts such as a manual, publisher catalog, or even extras such as the "feelies" included with Infocom's games.

As we consider the ways in which cultural institutions collect, curate, and preserve software artifacts, both material and virtual, it is worth asking when software became an object or artifact in the sense of something that might find its way into a repository such as a museum or library. Today, we speak easily of software in terms that fit the parlance of artifact curation. Similar terms have been employed in other contexts such as programming for decades and allowed for slippage between the virtual and the material. To give only two examples, "object-oriented" programming emerged during the 1960s and was followed by the graphical interfaces of the 1980s that translated programs and file directories into a system of visual icons that represented trashcans, file folders, and other office objects on a computer display. Notions of software as an object or artifact have since proliferated. The distribution of consumer software (including games) accompanying the rise of home computers, personal computers, and video game consoles opened up the possibility of speaking of software as an artifact with reference to actual physical items such as floppy diskettes and ROM cartridges, as well as retail packaging. In recent years, distribution units, including games, apps, and entertainment media, have joined the mix of "objects." Of course, all of these notions are relevant for libraries, museums, and other cultural repositories already possessing or hoping to acquire collections of historical software.

Despite having these reasons to talk about software at least potentially as a collection object, there is little evidence that anyone did that before the mid-1990s. As we have already seen, there was not much institutional collecting to encourage this manner of speaking. Moreover, it seems likely that museum curators and librarians initially resisted the idea that software could be considered as a collection object. Swade in his 1998 essay has carefully studied the hurdles an "object-centred culture" in museums put before software preservation, considering them at one point as "philosophical misgivings about the materiality of software" in institutions created to preserve physical objects.[26] The words "object" and "artifact" do not appear in Bearman's 1987 report about collecting software, although he does suggest that the "transformation of software into a consumer product" is a historical topic of concern to the software archive. Still, Bearman stresses that a software archive museum is about software history, which requires documentation but does not necessarily require the operation or even ownership of historical software. Conversely, a software library "has as its sole purpose the provision of software for use," not software history per se. In other words, software archives do not need to collect software objects. As for the early collections, Cabrinety refers to "hardware and software artifacts" in the 1989 CHIPS business plan, and he frequently uses the word "artifact" throughout the document.

Before turning to the place of software in the historical archive, it will be helpful to say a few more words about terminology. The early Library of Congress reports use terms like "titles," "collection of software," and "files."[27] When did software become an artifact or collection object? If we turn to Google's NGram Viewer, we can generate trend lines for the use of terms such as "digital object" and "software artifact" in Google's huge corpus of works published between 1980 and 2008. This resulting visualization suggests two tentative conclusions. The first is that use of "digital object" emerged during the mid-1990s and then increased rapidly. Looking at occurrences of this term shows that it was associated with writing about digital libraries, archives, and preservation or related technologies such as the "digital object identifier." In other words, we are dealing here with the vocabulary of data preservation, digital archives, and documentation activities, rather than software preservation. An example in this context is Ross Harvey's "From Digital Artefact to Digital Object," published in the proceedings of a conference on digital preservation sponsored by the National Library of Australia in 1995. Harvey specifically deploys the term "digital object" to describe a method of preserving digital data. His topic is "efforts toward migrating the data (or digital 'object') to new systems as they are introduced," which he contrasts to methods that emphasize preserving original artifacts, meaning media formats such as optical disk.[28]

The second conclusion is that the term of art adopted by digital libraries and archives is "digital object." As already noted, this term's popularity has grown significantly over the past two decades. Roughly concurrently, the term "software object" has declined in use, despite its longer history of being connected to the "object-oriented" programming vocabulary. At least this is what the Ngram Viewer tells us. Related terms such as "software artifact/artefact" and "digital artifact/artefact" have also become more popular, especially since the turn of the century. It seems we are becoming more comfortable with speaking of software as an artifact, and conversely, in contemporary speech our artifacts do not have to be material objects. The proliferation of digital libraries, archives, repositories, commercial stores, and other collections of software and data has encouraged our general acceptance of virtual artifacts. Computing technology long ago inflected the meaning of "virtual" to mean something that does not physically exist but is made by software to appear as if it did exist. Consider "virtual memory," a term that goes back to the 1950s. The popularity of terms like "digital object" and "software artifact" in more recent years reflects the growth of stores of data and software as much as it has benefited from their increasing importance for cultural repositories. These institutions considered how to design and build digital libraries and archives as their personnel (curators, technical staff, etc.) began to realize the necessity of considering the kinds of collections and objects that would be collected and, so to speak, fill virtual shelves and boxes. The pace of this work was accelerated by the increasing accessibility of the Web after the release of the Mosaic browser in 1994 and the National Science Foundation's Digital Libraries Initiative, launched in the same year. It is hardly surprising that cultural repositories both old (Library of Congress, National Library of Australia) and new (Internet Archive) focused their collective attention at about the same time on what came to be seen as digital objects and artifacts. These intellectual shifts during the mid-1990s helped shape institutional commitments that included building and preserving software collections.

Software Archives or Software Libraries?

The early software collections and the rise of digital libraries during the 1990s both begged an essential question: What do we collect, and what do we do with these collections? Jim Bennett's observation about museums that collect scientific instruments applies equally to software collections: "One obvious thing to remember about museum collections is that they show you not what there was but what was collected."[29] Scholarly research, repository practice, and collection curation all are addressing this question of what to collect in one way or another. Studies have been devoted to a range of issues from low-level forensics and data capture to institutional frameworks for acquisition, data transfer, description, and use. They are also redefining curatorial work as it pertains to digital collections.

The writing and projects that paced the emergence of software collections did not resolve several tensions. The first was highlighted in Bearman's 1987 report and contrasted data preservation as a documentation or archival activity and software libraries. A second tension reflected different priorities in forms of software collecting. One line of projects led from electronic records to Web archiving and concentrated on the collection, organization, and migration of vast stores of data and documentation, often created and stored in systems built around larger computing systems: mainframes, minicomputers, or the Advanced Research Projects Agency Network (ARPANET)/Internet. A second, often independent line produced collections that consisted mostly of discretely packaged software produced for use on microcomputers, personal computers, home computers, and game consoles. These two categories of software artifacts produced different ideas about repository design and curatorial activity. Collections featuring one or the other required different approaches toward curation, preservation, access, and use. The Internet Archive and the Department of Special Collections at Stanford, where the Cabrinety Collection is housed, are quite different operations—complementary rather than oppositional, but still different.

Figure 6.3. An example of documentation accompanying historical software. Copyright Board of Trustees of Stanford University. Image provided courtesy of Department of Special Collections and University Archives, Stanford Libraries

Why collect historical software at all? Since Bearman began to write about this question, responses have appealed to the profound impact of computers, software, and digital data on virtually every aspect of recent and contemporary history. Bearman, as we have seen, defined the "software archive" as serving the history of software as a research field, thus distinguishing it from the software library. Hedstrom reported, however, that those present at the Arden conference "debated whether it was necessary to preserve software itself in order to provide a sense of 'touch and feel' or whether the history of software development could be documented with more traditional records."[30] More recently, David Allison of the Smithsonian's National Museum of American History has suggested that "supporting materials are often more valuable for historical study than code itself. They provide contextual information that is critical to evaluating the historical significance of the software products."[31] He too argues that operating historical software is not an aspiration in institutions that serve a historical mission. The needs of historians, of course, do not account for all potential uses of software collections. Computers have been a fact of life for more than a half century. As a result, archives collect historical records in digital formats, and the separation between data archiving and use cases for historical software is not as absolute as a casual reading of Bearman's report would suggest. (Indeed, a close reading shows that he respected their connections.) Yet institutional priorities must be set and resources allocated. As a result, curators prioritize among collections and services that provide access to documentation of software history and those that deliver the experience of using historical software. It is not (and cannot be) a necessarily binary choice between software archive and software library, but choices still have to be made.

More than a decade ago, I wrote an essay called "The Hard Work of Software History." I tried to come to grips with some of the then-emerging difficulties that cultural repositories were beginning to face with collecting software.[32] The positions I have described above that separate the software library and archives as methods for preserving software history struck me then as being "partly stuck on different institutional and professional allegiances to the preservation of objects, data migration, archival functions, evidentiary value, and information content. . . . These issues are not likely to be sorted out before it is necessary to make serious commitments at least to the stabilization, if not the long-term preservation, of digital content and software."[33] Rapid changes in media formats and computing technologies add to the uncertainty associated with the future of software collections. For example, it remains to be seen whether practices associated with collecting packaged software will provide useful methods for collecting software distributed via streaming and software-as-a-service delivery models.[34] In addition to technological evolution, licensing and legal restrictions on both collecting activities and access to collections further constrain curatorial work in this area. The bottom line is that it is not yet clear how curation of software objects and the particular experiences associated with the virtual spaces created by software—whether a game or the white space facing me on the screen as I write these words—will be captured by established library or archival collecting, cataloging, exhibition, and preservation activities.

In May 2013, the National Digital Information Infrastructure and Preservation Program at LC held a conference called "Preserving.exe" with the goal of articulating the "problems and opportunities of software preservation." In my response to the conference included with the LC report issued a few months later, I described three "lures of software preservation," by which I meant potential pitfalls as we begin to connect the dots from software collections to digital repositories and from there to programs and services such as access to collections and exhibitions.[35] In a nutshell, the lure of the screen is the idea that what counts in digital is what is delivered to the screen. With respect to software preservation, the problem comes up when judging success in delivering significant properties of software entirely by auditing interaction with surface properties (graphics, audio) and not accounting either for the variable and contextual nature of these properties or the invisibility of software abstraction levels and procedural aspects of software design such as game mechanics in video games. The second lure is that of the authentic experience, which I consider the mandate that digital repositories at a minimum must be careful to not preserve digital objects in a manner that would interfere with re-enactment of such experiences as are deemed historically authentic. To put it bluntly, I believe that there is no such thing as unconstructed and pure authenticity in this sense, and it is a waste of resources to seek to provide it.

Finally, the third lure is the lure of the executable, and it will conclude my reflections here on historical software collections. As we have seen, the kind of historical resource that Bearman called a software archive had a head start on software libraries. The primacy of documentation was implied in the AFIPS brochure, and it was already being collected by CBI and, indeed, in corporate and university archives by the late 1970s. The first repository collections of software followed more than a decade later and more than twenty years on remain few and far between. As we have seen, these collections have generally been restricted to a limited set of software categories and platforms, and indeed, curators cited above have been skeptical about the potential for much expansion beyond them. Indeed, some have doubted the value of maintaining software collections for historical research at all.

So what is the value of historical software for the history of software? An attractive version of the end product of software preservation is the well-maintained library of executable historical software. The work to reach that product involves a series of tasks from selection and collection through ingest and migration from original media and creation of technical and rights metadata, as well as provenance, descriptive, and contextual information. This workflow is at the heart of what has been called the life cycle model for digital preservation, the most influential statement of which has been that of the Digital Curation Centre.[36] Although the model says little about related physical components such as packaging, manuals, and box inserts, archiving these components can be considered a separate problem, and as projects at the Strong National Museum of Play, the University of Texas, Stanford University, and elsewhere indicate, this work is underway. The point for software preservation is that the lure of the executable compels digital repositories to focus on building collections of verified historical software that can be executed on demand by researchers. This lure could well function for some institutions as a holy grail of digital curation with respect to software history.

What could possibly be wrong with this mission, assuming that it can be executed? I have argued on several occasions that there are at least three problems with the software library. Perhaps the good news is that they are problems of omission, rather than commission. The first problem is that software does not tell the user very much about how it has previously been used. In the best case, such as previously installed software available in its original use environment (not a common occurrence), application software might display a record of files created by users, such as a list of recently opened files found in many productivity titles like Microsoft Office. The more typical case for a software library, however, will be that software is freshly installed from the library and thus completely lacking information about historical use.

The second point might best be illustrated through the example of virtual world software, such as *Second Life* or a game world such as *World of Warcraft*.

Let us assume that we can capture every bit of historical data from a virtual world server and then successfully synchronize all of these data with carefully installed (and patched) software; we can reverse engineer authentication controls, and we can solve the complex problems of ownership and rights. We will have accomplished an act of perfect software and data capture, and of course, we will then preserve all of the associated digital objects. Now we can create a historical time machine through which we can fly through the virtual world at any moment in time of its existence. Of course, historians would applaud this effort, until it becomes evident that all we can do is tour an empty world. In the absence of historical documentation that helps the researcher to understand motivations and reactions of the participants, the payoff will be limited. Such documentation generally will not be found in the recreation of such a virtual environment, content without context.

The third point and perhaps the most fundamental is that the documentation that is a prerequisite for future historical studies of software and digital media such as games and virtual worlds is simply not located in software. It is, in a sense, on both sides of software: the design materials (including source code) that document software development and the archives, both digital and nondigital, that document context, use, and reception. It is important to understand that this is not just a problem for historical research. It is also a problem for repositories, whether they are doing the work of digital preservation or addressing the needs of a researcher requesting access to this software. If contextual information such as software dependencies or descriptions of relationships among objects is not available to the repository and all the retired software engineers who knew the software inside and out are gone, it may be impossible to get old software to run.

In *Best Before: Videogames, Supersession and Obsolescence*, James Newman argues that a host of publishing, retail, marketing, journalistic, and other practices have cultivated a situation "in which the new is decisively privileged and promoted and the old is constructed either as a benchmark by which to measure the progress of the current and forthcoming 'generation' or as a comprehensively worn out, obsolete anachronism to be supplanted by its update, superior remake or replacement."[37] Newman is not optimistic about software preservation. This does not mean that he is pessimistic about every possibility for historical preservation, however. In a section of his book provocatively called "Let Videogames Die," he argues that a documentary approach to gameplay might be a more pragmatic enterprise than the effort to preserve playable games. We might generalize his conclusion about digital games to other classes of software. Newman calls for a "shift away from conceiving of play as the outcome of preservation to a position that acknowledges play as an indivisible part of the object of preservation."[38] Stated more generally, software preservation is not about preserving historical software for enactment as historical research and even less about reenactment of past experiences. Rather, it provides a method for producing documentation about contemporary use for (later) historical purposes. The lure of the executable is not just about following a false idol; recognizing the lure leads to acknowledging that the library of executable software can at best be only a partial solution to the problem of software preservation.

Software archives or software libraries? That is the question. Is it nobler to collect documentation or to suffer the slings and arrows of outrageous software installations? The case for documentation is strong. The consensus among library and museum curators (including myself) who have taken a side is that historical documents are a stone-cold lock to be a winning prerequisite for future historical studies of software. Useful records of software could be virtually anything from source code to screenshots. Moreover, historians will not be the only users of these documents. There is even a software library case to be made for software archives: the success of software preservation will depend on the availability of historical documentation. Documents provide contextual information that can be included in transfer protocols or descriptive metadata and inform long-term preservation activities by documenting software dependencies and relationships among objects or clarifying ownership of intellectual property. Documents tell us how software was used historically, and they can also help tell us what we are supposed to do or what we are looking at when operating ancient software on long-defunct operation systems or platform.

And yet every argument for software archives leaves room for preserving software, whether as artistic or cultural content, for technology studies or for forms of scholarship that treat aspects of digital games and virtual worlds as authored texts or artistic objects. Indeed, this tension between the collected artifact and archival documentation will be familiar to curators working with other object and media formats. The difference posed by software collections is that preservation of the software "artifact" necessarily involves commitment to a long-term cycle of data curation activities, rather than object conservation. It is an expensive proposition, involving skills, approaches, and technologies quite different from those required for the conservation of physical artifacts. The problem of institutional commitments to software collections thus ultimately boils down to the identification of relevant use cases and the allocation of resources. It is virtually impossible that any one institution will be able to do it all, but every collecting institution committed to supporting software history should at least try to do something.

NOTES

1. American Federation of Information Processing Societies, "Preserving Computer-Related Source Materials," 1979, reproduced in *IEEE Annals of the History of Computing* 2 (January–March 1980): 4–6.

2. David Bearman, "What Are/Is Informatics? And Especially, What/Who Is Archives and Museum Informatics?" *Archival Informatics Newsletter* 1, no. 1 (Spring 1987): 8.

3. David Bearman, *Collecting Software: A New Challenge for Archives and Museums*, Archival Informatics Technical Report 2 (Toronto: Archives and Museum Informatics, 1987), 25–26.

4. Bearman, *Collecting Software*, 24.

5. Margaret L. Hedstrom, *Archives and Manuscripts: Machine-Readable Records*, SAA Basic Manual Series (Chicago: Society of American Archivists, 1984).

6. David Bearman and Margaret Hedstrom, "Reinventing Archives for Electronic Records: Alternative Service Delivery Options," in *Electronic Records Management Program Strategies*, ed. Margaret Hedstrom, Archives and Museum Informatics Technical Report 18 (Pittsburgh: Archives and Museum Informatics, 1993), 82–98.

7. Hans Rütimann to Henry Lowood, November 19, 1989, "Planning a Center for the Preservation of Microcomputer Software," Henry Lowood Papers, Department of Special Collections and University Archives, Stanford University Libraries, Stanford, CA.

8. Bernard A. Galler, "A Center for the Preservation of Microcomputer Software: Software Issues," 1990, Henry Lowood Papers.

9. Margaret Hedstrom and David Bearman, "Preservation of Microcomputer Software: A Symposium," *Archives and Museum Informatics* 4, no. 1 (Spring 1990): 10.

10. David K. Allison, "National Software Archives: Exploring the Concept," printed email attachment, from David K. Allison email to Henry Lowood, "Subject: Software Archives Project," November 23, 1993; George Glaser, William T. Coleman III, Paul N. Edwards, Henry Lowood, and Keith Uncapher, "Charles Babbage Foundation Software Task Force, Final Report," November 1998, Henry Lowood Papers.

11. "A Center for the Preservation of Microcomputer Software: Access and Outreach," 1990, 3, Henry Lowood Papers.

12. Library of Congress, Annual Report of the Librarian of Congress for the Fiscal Year Ending September 30, 1988 (Washington, DC: Library of Congress, 1989), 26.

13. Suzanne Thorin, John Kimball, Linda Arret, and Sandra Lawson, "Machine-Readable Collections Reading Room Pilot: Report" (Washington, DC: Library of Congress, 1990), 4; William J. Sittig to William J. Welsh, memorandum, "Report of the Ad Hoc Committee on Selection Policy for Machine-Readable Publications," February 23, 1983; Machine-Readable

Collections Task Force, "Machine Readable Collections: Library of Congress," August 26, 1985; Machine-Readable Collection Committee, "Establishing a Machine-Readable Collection and Pilot Reading Room in the Library of Congress," May 1987, all in Henry Lowood Papers. Most of the materials cited regarding the MRCRR were sent to me in February 1990 by John Kimball, head of the Automation and Reference Collections Section of LC.

14. Thorin et al., "Machine-Readable Collections," 6.

15. Library of Congress, "Machine-Readable Collections Reading Room," Henry Lowood Papers.

16. Thorin et al., "Machine-Readable Collections," 15 and Appendix H: Collections Statistics.

17. Stacey Peterson, "Open to the Public," Computer System News, January 2, 1989, Henry Lowood Papers.

18. Thorin et al., "Machine-Readable Collections," 1.

19. Library of Congress, "Collections Overview: Computer Files," http://www.loc.gov/acq/devpol/colloverviews/computer.pdf.

20. Trevor Owens, "Yes, the Library of Congress Has Video Games: An Interview with David Gibson," *The Signal* (blog), September 26, 2012, http://blogs.loc.gov/digitalpreservation/2012/09/yes-the-library-of-congress-has-video-games-an-interview-with-david-gibson/; Library of Congress, "Moving Image Materials," Library of Congress Collections Policy Statements, November 2008, http://www.loc.gov/acq/devpol/motion.pdf. The section collects video games at the "research level" and selects "via copyright and purchases video games, their associated hardware, and magazines about them that reflect the breadth and depth of gaming culture."

21. Patricia Cabrinety to Henry Lowood, ALS, February 15, 1998, Henry Lowood Papers.

22. Computer History Institute, Inc., business plan, May 1, 1989, 8, Stephen M. Cabrinety Collection in the History of Microcomputing, circa 1975–1995, Department of Special Collections and University Archives, Stanford University Libraries, Stanford, CA.

23. Greg Miller, "Software Trove Is Testament to Its Collector," *Los Angeles Times*, August 12, 1996.

24. For more details on the collection, see "The Collection," http://web.archive.org/web/19961221075312/http://www.clark.net/pub/kinesixd/chipsdesc.html and "We Need Your Help!" http://web.archive.org/web/19961221075305/http://www.clark.net/pub/kinesixd/stevesstory.htm. Both of these websites were prepared by the Cabrinety family after his death and were retrieved by the Internet Archive on December 21, 1996. On the role of collectors in shaping institutional collections, see Bernard Finn, "Collectors and Museums," in *Exposing Electronics*, ed. Bernard S. Finn, Robert Bud, and Helmuth Trischler (Amsterdam: Harwood, 2000), 175–191.

25. Doron Swade, "Preserving Software in an Object-Centred Culture," in *History and Electronic Artefacts*, ed. Edward Higgs (Oxford: Oxford University Press, 1998), 205.

26. Swade, "Preserving Software," 196.

27. Bearman, Collecting Software, 10, 36–38.

28. Ross Harvey, "From Digital Artefact to Digital Object," in *Multimedia Preservation: Capturing the Rainbow; Proceedings of the Second National Conference of the National Preservation Office*, November 28–30, 1995 (Canberra: National Library of Australia, 1996), 202.

29. Jim Bennett, "Scientific Instruments," Department of History and Philosophy of Science, University of Cambridge, 1998, archived August 16, 2002, https://web.archive.org/web/20040406120525/http://www.hps.cam.ac.uk/research/si.html.

30. Hedstrom and Bearman, "Preservation of Microcomputer Software," 10.

31. David K. Allison, "Preserving Software in History Museums: A Material Culture Approach," in *History of Computing: Software Issues*, ed. Ulf Hashagen, Reinhard Keil-Slawik, and Arthur L. Norberg (Berlin: Springer, 2002), 263–265. Another contribution to this topic is offered by Len Shustek, "What Should We Collect to Preserve the History of Software?" *IEEE Annals of the History of Computing* 28 (October–December 2006): 110–112.

32. Henry Lowood, "The Hard Work of Software History," *RBM: A Journal of Rare Books, Manuscripts, and Cultural Heritage* 2 (Fall 2001): 141–161.

33. Lowood, "The Hard Work," 149.

34. Eric Kaltman, "Current Game Preservation Is Not Enough," *How They Got Game* (blog), June 6, 2016, http://web.stanford.edu/group/htgg/cgi-bin/drupal/?q=node/1211.

35. Trevor Owens, "Preserving.exe: Toward a National Strategy for Software Preservation," in *Preserving.exe: Toward a National Strategy for Software Preservation* (Washington, DC: Library of Congress, 2013), 2–3, http://www.digitalpreservation.gov/multimedia/documents/PreservingEXE report final101813.pdf; Henry Lowood, "The Lures of Software Preservation," in *Preserving.exe*, 4–11.

36. See Ross Harvey, *Digital Curation: A How-To-Do-It Manual* (New York: Neil Schuman, 2004).

37. James Newman, *Best Before: Videogames, Supersession and Obsolescence* (London: Routledge, 2012), 121.

38. Newman, *Best Before*, 160.

II GAME HISTORIES AND HISTORIOGRAPHY

Author's Introduction

My graduate training in history of science and technology prepared me to approach the history of games as a subject that writing in technology studies could help me understand. Melvin Kranzberg, the cofounder of the Society for the History of Technology (SHOT) occasionally credited with founding the history of technology as a scholarly discipline in the United States, is perhaps best known for "Kranzberg's Laws," six "truisms deriving from a longtime immersion in the study of the development of technology and its interactions with sociocultural change" that he delivered in the annual SHOT presidential address for 1985.[1] One of these laws stuck with me, the fifth: "All history is relevant, but the history of technology is the most relevant." When I stepped down as SHOT bibliographer, I was given the framed poster that has delivered Kranzberg's Fifth Law to me every day for the last fifteen or so years.

Technological determinism is not what I take from Kranzberg to the history of games. Rather, the lessons that I find useful to port over from histories of technology usually have to do with how human beings interact with other people to use, work on, and transform technologies while these same technologies also work on them.[2] So games are a form of technology, and the history of games is therefore a subfield of the history of technology? No, that is not the point I am trying to make. A closer starting point is the quotation from Brian Sutton-Smith at the head of "Game Studies Now, History of Science Then" (chapter 7 in this volume): "The adaptive problem to which the video game is a response is the computer. The computer is, to this century [the 20th], what printing was to the sixteenth century."[3] Admittedly, this is a broad McLuhanesque take on the impact of media technology on human culture. That is not what attracts me to this statement, however. Rather, it is the idea of games and computers as "adaptive problems." In other words, people individually and collectively respond to games as part of the larger, longer-term impact of computing technology on human cultures. As evidence that this idea has stuck with me, allow me to cite the opening questions of my essay in the book *EA Sports FIFA: Feeling the Game* (2022): "When players crave information about game software that developers do not provide, where do they get that information? How do they share it? And how do these efforts affect the game, the developer, and its players."[4]

While this overview of my previous writings on the history of games has been focused on technology, I cannot help noticing that the "Game Studies Now" essay cited above, written in 2005, refers to the history of science. Indeed, that essay refers to historians like Herbert Butterfield, George Sarton, and Thomas Kuhn, none of whom made much of a dent in technology studies. The missing link is the historian Michael Mahoney. He was the first historian of computing I met who had at some point in life mixed in some serious software programming work. It was fun talking to him about that work, of course, but the deeper connection was probably that he had started out as a historian of early modern mathematics, working on Fermat, Descartes, Barrow, Newton, and the like. His book on Fermat remains to this day an essential point of departure on the French mathematician's life and work.[5] When I first met Mike, probably in the early 1990s, I had not only transitioned from historian to curator, but also from the 18th to the 20th century (history of computing, Silicon Valley Archives, software preservation) as the focus of my historical and curatorial work.[6] I follow Thomas Haigh, who writes in the introduction to his collection of Mahoney's essays that his "historiographic papers published from 1988 to 2008 . . . constitute the most sustained and self-conscious examination so far attempted of the fundamental question hanging over a growing body of work: What is the history of computing a history of?"[7] Rather than dwell on Mahoney's impact, allow me to point to one of the essays that Haigh undoubtedly had in mind. Word for word, it has probably had more impact on focusing my attention on the history of games than anything else. The modest, but far-reaching title of this essay is "The History of Computing in the History of Technology."[8] It would not be far off to say that the general topic in which to place my writing might be "the history of games in the history of computing in the history of technology," a modest branch or subset of the issues raised by Mahoney. The summary of his work in "Game Engines and Game History" (chapter 9 in this volume) perhaps illustrates this statement. The point I will emphasize now paraphrases the concluding sentence of Mahoney's essay: pursued within the larger enterprise of the history of technology (or computing, if you prefer), the history of games will acquire the context of place and time that gives history meaning.

Besides "technology," computer technology specifically, another important word that cuts through most of my historical work is "performance." Jane McGonigal[9] pointed me some years ago to the text that sparked my attention to performance as a broad concept applicable to game studies: Richard Schechner's *Performance Studies: An Introduction*.[10] He devotes a substantial chapter to play, with references to several theorists (e.g., Roger Caillois) who have been widely referenced in game studies, unlike Schechner. We read in *Performance Studies* that one definition of performance could be "ritualized behavior conditioned and/or permeated by play."[11] Reading that the first time was of course interesting, but the important move in Schechner's work was his unrestrained application of "performance" to what seemed like an almost limitless set of activities related to technology, social interactions, culture, politics, sports, and more, not just entertainment spheres such as theater and dance. In other writings, Schechner called this effort to move performance studies beyond its traditional center "the broad spectrum approach."[12] When he began writing about the "broad spectrum" of performance studies in 1988, Schechner called for "treating performative behavior, not just the performing arts, as a subject for serious scholarly study."[13] Reading this call for studies of performance across a range of human activities from a performance studies scholar for whom play was central to the understanding of performance seemed like an invitation to begin to look at performative aspects of playing digital games. My work on the history of machinima and competitive play, outside the scope of the present volume, took off directly from this challenge.[14]

The common element among the chapters in this section is the identification and documentation of ways in which players engage with games, from creating artworks about games to transgressive play and modifying games. Not much in the way of theory, I admit, but I hope the articles hang together in my commitment to working from things players do to ideas about game-based creativity and performance. It is worth noting here that several of these pieces are not about or only partly about digital games. While I have not played digital games my entire life, I have played board games for as long as

I can remember. It was long my wish to write about the history of tabletop games, and indeed there is a glimpse of that interest in the early work of the How They Got Game project at Stanford on wargames and simulations.[15] Still, player creativity cuts across the essays on board games. Examples include methods for playing board games by postal mail during the 1960s and the authoring of scenarios in modular games such as *PanzerBlitz* from the 1970s forward. These player-driven tinkerings with game systems—I would call them technologies—are closely related to the game modifications made possible by id's game engine design. Player-generated mods figured prominently in the game-based responses to the 9/11 attacks. The thread connecting these disparate moments of player agency cuts across game design for tabletops *and* computers. Computerization of wargames, for example, did not fundamentally change the inclination of players to alter and tinker with games. The shift described by James Dunnigan as a move from "mushware" (our brains, where the system is located when players must read, understand, and apply board game rules) to software systems (in which rules are hard-wired, so to speak, in code) significantly altered how players changed games, but the inclination of players to take on that challenge remained. Historical game studies, at least those devoted to the second half of the 20th century forward, will benefit from tracing common elements across different game formats, by which I mean both physical and digital games.

Without meaning to seriously compare myself to Mel Kranzberg, I might propose just one "Lowood's law" of game history that I hope these chapters back up: good histories of games are rarely, if ever, just about games. Banal perhaps, but true.

NOTES

1. Melvin Kranzberg, "Kranzberg's Laws," *Technology and Culture* 27, no. 3 (July 1986): 544–560.

2. Some examples of thoughtful work that explores this topic: Susan J. Douglas, "Some Thoughts on the Question 'How Do New Things Happen?'" *Technology and Culture* 51, no. 2 (2010): 293–304; Susan J. Douglas, "The Turn Within: The Irony of Technology in a Globalized World," *American Quarterly* 58, no. 3 (2006): 619–638; Eric von Hippel, *Democratizing Innovation* (Cambridge, MA: MIT Press, 2005); Kelly Oudshoorn and Trevor Pinch, *How Users Matter: The Co-Construction of Users and Technology* (Cambridge, MA: MIT Press, 2005); Ronald Kline and Trevor Pinch, "Users as Agents of Technological Change: The Social Construction of the Automobile in Rural America," *Technology and Culture* 37, no. 4 (1996): 763–795; Merritt R. Smith and Leo Marx, eds. *Does Technology Drive History? The Dilemma of Technological Determinism* (Cambridge, MA: MIT Press, 1994).

3. Brian Sutton-Smith, *Toys as Culture* (New York: Gardner, 1986), 64.

4. "'Where There Is Smoke, There Is Fire . . .': The FIFA Engine and Its Discontents," in *EA Sports FIFA: Feeling the Game*, ed. Raiford Guins, Henry Lowood, and Carlin Wing (New York: Bloomsbury Academic Press, 2022).

5. Michael S. Mahoney, *The Mathematical Career of Pierre De Fermat, 1601–1665* (Princeton, NJ: Princeton University Press, 1994).

6. From that past life in the 18th and 19th centuries: "The Calculating Forester: Quantification, Cameral Science, and the Emergence of Scientific Forestry Management in Germany," in *The Quantifying Spirit in the Eighteenth Century*, ed. Tore Frängsmyr, J. L. Heilbron, and Robin E. Rider (Berkeley: University of California Press, 1990), 315–342; *Patriotism, Profit, and the Promotion of Science in the German Enlightenment: The Economic and Scientific Societies, 1760–1815* (New York: Garland, 1991); "The New World and the European Catalog of Nature," in *America in European Consciousness, 1493–1750*, ed. Karen O. Kupperman (Raleigh: University of North Carolina Press, 1995), 295–323.

7. Thomas Haigh, "Unexpected Connections, Powerful Precedents, and Big Questions: The Work of Michael S. Mahoney on the History of Computing," in *Histories of Computing*, Michael S. Mahoney and Thomas Haigh (Cambridge, MA: Harvard University Press, 2011), 2.

8. Michael Mahoney, "The History of Computing in the History of Technology," *IEEE Annals of the History of Computing* [then called *Annals of the History of Computing*] 10 (1988): 113–125.

9. Author of *Reality Is Broken: Why Games Make Us Better and How They Can Change the World* (New York: Penguin, 2011); and *Superbetter: The Power of Living Gamefully* (New York: Penguin, 2016).

10. Richard Schechner, *Performance Studies; An Introduction* (London: Routledge, 2002).

11. Schechner, *Performance Studies*, 45.

12. Richard Schechner, "Broadening the Broad Spectrum," *TDR: The Drama Review* 54 (Fall 2010): 7–8, which is a follow-up to his earlier "Performance Studies: The Broad Spectrum Approach," *TDR: The Drama Review* 32 (Autumn 1988): 4–6.

13. Schechner, "Performance Studies," 4.

14. E.g., "Joga Bonito: Beautiful Play, Sports and Digital Games," in *Sports Videogames*, ed. Mia Consalvo, Konstantin Mitgutsch, and Abe Stein (London: Routledge, 2013), 67–86; "*Warcraft* Adventures: Texts, Replay and Machinima in a Game-Based Story World," in *Third Person: Authoring and Exploring Vast Narratives*, ed. Pat Harrigan and Noah Wardrip-Fruin (Cambridge, MA: MIT Press, 2009), 407–427; and (coedited with Michael Nitsche) *The Machinima Reader* (Cambridge, MA: MIT Press, 2011).

15. "Theaters of War: The Military-Entertainment Complex," (with Tim Lenoir) in *Collection, Laboratory, Theater: Scenes of Knowledge in the 17th Century*, ed. Helmar Schramm, Ludger Schwarte, and Jan Lazardzig (Berlin: Walter de Gruyter, 2005), 427–456.

7

Game Studies Now, History of Science Then

Maybe it's a pathetic symptom of some modern malaise in a world lacking things really worth striving for. Perhaps the game is a pure place to get yourself a good spate of solitary willpower in a social world with decreasing options for courageous expression.

—David Sudnow[1]

The adaptive problem to which the video game is a response is the computer. The computer is, to this century, what printing was to the sixteenth century.

—Brian Sutton-Smith[2]

In a recent editorial for the Digital Games Research Association's Web site, Frans Mäyrä (n.d.) depicted the "quiet revolution" in contemporary culture that has created a profound need for the academic discipline of game studies.

> There is an ongoing, mostly silent revolution taking place in our culture and society. . . .
>
> Extension and investment of modern life and energy into digital puzzles and parallel universes presents modern universities with a major challenge. We must take these popular realms seriously, or face loss of both intellectual and social relevance. To meet the demands presented by these changes, there is need for a new discipline.[3]

Originally published as "Game Studies Now, History of Science Then," *Games and Culture* 1 (January 2006): 78–82.

Mäyrä's reasoning struck me as eerily familiar. Coming to grips with a modern cultural revolution by establishing a new academic discipline . . . where have I heard this song before? The notes resonate with themes familiar from my own graduate training in the new academic discipline of the late 1970s: the history of science. You say you want to study a revolution? In the formative years of the history of science, the Scientific Revolution and Industrial Revolution were focal points of research. Many of the seminal writings of the field presented, reconfigured, or debated notions of cultural and intellectual revolutions that could be applied to or derived from the rise of modern science. The most influential book of the history of science's growth years was surely Thomas Kuhn's (1970) *The Structure of Scientific Revolutions*. A twofold conviction motivated the discipline's rapid growth from the mid-1950s through the 1970s: that modern science had revolutionized human affairs and that scientific change itself could be understood as intellectual disjunctions ripe for contextualization rather than a linear progression of discoveries. This growth could be quantified in terms of publications in the *Isis Current Bibliography in the History of Science*, academic positions and programs, membership in professional societies, graduate students, or opportunities for participation in wider forums of discourse.

Mäyrä's (n.d.) passionate appeal for academic game studies echoes convictions that could be found among postwar historians of science. If anything, they were even more vigorous in proclaiming the significance of the revolutions they studied. Consider this often cited passage in Herbert Butterfield's (1958) *The Origins of Modern Science 1300–1800*:

> Since that [scientific] revolution overturned the authority in science not only of the middle ages but of the ancient world—since it ended not only in the eclipse of scholastic philosophy but in the destruction of Aristotelian physics—it outshines everything since the rise of Christianity and reduces the Renaissance and Reformation to the rank of mere episodes, mere internal displacements, within the system of medieval Christendom. Since it changed the character of men's habitual mental operations even in the conduct of the non-material sciences, while transforming the whole diagram of the physical universe and the very texture of human life itself, it looms so large as the real origin both of the modern world and of the modern mentality that our customary periodisation of European history has become an anachronism and an encumbrance.[4]

Maybe we are not willing to go quite this far in making a case for game studies—but we must nonetheless commit ourselves to the notion that games reflect significant changes in contemporary culture and society. Mäyrä (n.d.) and others, such as Sudnow (1983) and Sutton-Smith (1986) in the opening quotes, have suggested a few ways in which this case can be made. Reaching the conviction that games are this important—that they are representative, symptomatic, impact causing—and thus concluding that they deserve scholarly attention is the yeast that will give rise to an academic discipline of game studies.

As compelling as Butterfield's (1958) dramatic conclusion may be for the need to devote attention to the history of science, tracking this momentous revolution in human affairs was not everyone's incentive. Humanists committed to bridging the "two cultures," scientists tracing lineages of invention and the triumphs of the scientific method, and critics focusing on questionable and even threatening impacts of scientific progress all pitched tents in this new academic territory. Translated into terms relevant for game studies, we can see similar concerns in the two cultures of design and critical studies[5] or the continuous proliferation of publications on the positive (serious games, games for health, game-based learning) and negative (violent content, cultural stereotyping, addiction) ramifications of games and gameplay. These parallels may justify the reflections in this article, drawn from my own parallel professional allegiance to the history of science. What can we learn from the elevation of the history of science to an established discipline and profession that might help us understand the situation of game studies? And why are we talking about game studies today in ways similar to the rhetoric that accompanied the history of science in the 1960s and 1970s? I would like to suggest that the growth of history of science then and game studies now has been fueled by similar motivations and strategies.

The history of science is a relatively recent academic discipline. The number of departments, research centers, disciplinary history centers, library collections, and scholarly journals founded in the past fifty years testifies to the proliferation of instruction, research, and publication in history of science. The History of Science Society, the primary professional organization, counts thousands of institutions and individuals among its members. Of course, scholarship set the stage for these formations of an intellectual discipline or professional identity, whether as introductory historical chapters in textbooks or as seminal essays and treatises going back as far as Sir Francis

Bacon (1561–1626). The modern historiography of science began with George Sarton (1884–1956), who in 1912 founded the first (and still leading) scholarly journal in the history of science, *Isis*. His activities included four decades of writing, lecturing, editing, and bibliography at Harvard University and a key role as cofounder of the History of Science Society in 1924. Despite these various contributions, the history of science grew slowly as a discipline and profession during these years. Given the rapid installation of the history of science on American university campuses in the 1960s and 1970s, it is easy to forget that as late as the 1950s, there were fewer than a dozen full-time professional appointments in this field. No institution had granted more than a handful of doctorates, the first not until the late 1930s. On the eve of World War II, membership in the History of Science Society stood at less than one tenth its membership today.

It is instructive to contrast Sarton's vision of the history of science to the positions that later propelled its dramatic expansion. For Sarton and the cadre of historians, scientists, and philosophers attracted to his projects, the history of science proposed a cultural synthesis. Bridging the humanities and sciences by applying the methods of the humanist to the activities of the scientist, Sarton called this fusion the "New Humanism." Through the 1920s and beyond, his goals for the history of science included this cultural ideal alongside the scholarly tools and professional contributions for which he is remembered today. By contrast, the key text of the growth period during the 1960s and 1970s was Thomas Kuhn's (1970) *The Structure of Scientific Revolutions*, which gave us "normal science" and the "paradigm shift." In his introduction, "A Role for History," Kuhn put down the foundation for a "quite different concept of science," neither the succession of scientific achievements found in pedagogy and textbooks nor "the discipline that chronicles both . . . successive increments and the obstacles that have inhibited their accumulations."[6] He shifted the paradigm by proposing a history of science that displays "the historical integrity of science in its own time" and by considering how views of nature depicted in a larger set of contexts and often incommensurable with one another could lead to scientific activity and knowledge.

The conceptual transformation that paced the disciplinary growth of the history of science was not the unifying New Humanism but a dividing methodological controversy. Following Kuhn's (1970) work—and here I pass over predecessors such as Alexandre Koyré and more than a decade of sharp debate that succeeded it—the growth of the history of science as a professional discipline was accompanied by disagreement among historians over the incorporation of historical context, discontinuity, and competition as counterweights to the successive triumphs of an eternal scientific method. As Kuhn noted, contextualization put historians in the position of discounting ideas drawn "partly from scientific training itself." Critical academic work might thus be seen as opposing, if not irrelevant to, scientific practice. Inside history of science, the distinction of "internalist" and "externalist" methods, those focusing on the relationship of ideas and discoveries to each other versus those that situated science in a sociohistorical context, drew a sharp line between camps of historians during the growth years.

The issue that translates from the history of science to game studies in both of their formative stages is a tension between inside and outside. As the sharp lines between internalists and externalists gradually blurred during the 1980s, many historians spoke of penetrating the "black box," of closing in on the details of scientific practice through contextualization of various sorts. Many of these methods challenged the authority, stability, and results of scientific work and by implication, its practitioners, thereby raising the specters of social construction and science studies opposing the scientific enterprise as understood by scientists. In game studies, we face and will continue to face a similar set of problems. If the history of science offers a lesson, it is that a new discipline can grow and mature despite tensions between critical study and practice or even between ludology and narratology, our own riff on the theme of internalist versus externalist and how to dissolve such distinctions.

The point of game studies today is not recognition of the maturity or the midlife crisis of the games industry, or even hoisting the impressive aspirations of modern games as art, business, or technology. Its potential lies in critical engagement with games as symptoms of and impacts on society and culture—contextualization—but in ways that illuminate the structure of revolutions in design and gameplay. Such game studies will speak, eventually, to both academic scholars and enlightened developers; it may be possible to realize the goal of moving inside the black box along mutually traveled paths more successfully than historians of science have done in the eyes of many scientists. If game studies is successful, tones of approval and disapproval, justification and critique, will be recognized and debated in ways that reflect wider issues of the impact of games on society, culture, religion, warfare, and other aspects of life, just as they did for the history of science in recent decades. But as these reflections on the history of science suggest, there is nothing about such divisions that dooms or even threatens the growth and eventual success of our new discipline.

NOTES

1. David Sudnow, *Pilgrim in the Microworld* (New York: Warner Books, 1983), 210–211.

2. Brian Sutton-Smith, *Toys as Culture* (New York: Gardner Press, 1986), 64.

3. Frans Mäyrä, *The Quiet Revolution: Three Theses for the Future of Game Studies*, n.d., http://www.digra.org/hardcore/hc4.

4. Herbert Butterfield, *The Origins of Modern Science* (London: G. Bell and Sons, 1958), 7.

5. See Chris Crawford, *The Two Cultures, Maybe Three* (2004), http://www.igda.org/columns/ivorytower/ivory_May04.php.

6. Thomas S. Kuhn, *The Structure of Scientific Revolutions* 2nd ed. (Chicago: University of Chicago Press, 1970), 2.

8

Video Games in Computer Space

The Complex History of *Pong*

Computer games such as *Spacewar!* and *Adventure* were created in institutions, such as the Massachusetts Institute of Technology, BBN, and Stanford University, that defined the main streams of computing research during the 1960s and 1970s.[1] Telling the stories of these games reveals the emergence of "university games" out of laboratories and research centers.[2] The institutional contexts of *Spacewar* and *Adventure* suggest an important, and at times underappreciated, relationship between exploratory work in computer science and the early history of computer games. Both games grew out of the very institutions that played an essential role in defining time-shared and then networked computing in its early days. Games such as these exemplified the technical mastery of programmers and hardware hackers. These links between games and computing recall Brian Sutton-Smith's argument that games are fundamentally "problems in adaptation" and that computer games specifically address the problem that "is the computer."[3]

A success story that marked the early evolution of the video game, Atari's original *Pong* arcade console betrays few obvious connections to computer technology of the mid-1970s. The prototype's cabinet and circuitry, designed by Al Alcorn, reveal only a modest investment in electronic components, a modified television set, and some ad hoc wiring and parts. Years later, Alcorn made a block diagram of the game's logic for a later generation of computer

Originally published as "Video Games in Computer Space: The Complex History of *Pong*," *IEEE Annals in the History of Computing* 31 (July–September 2009): 5–19.

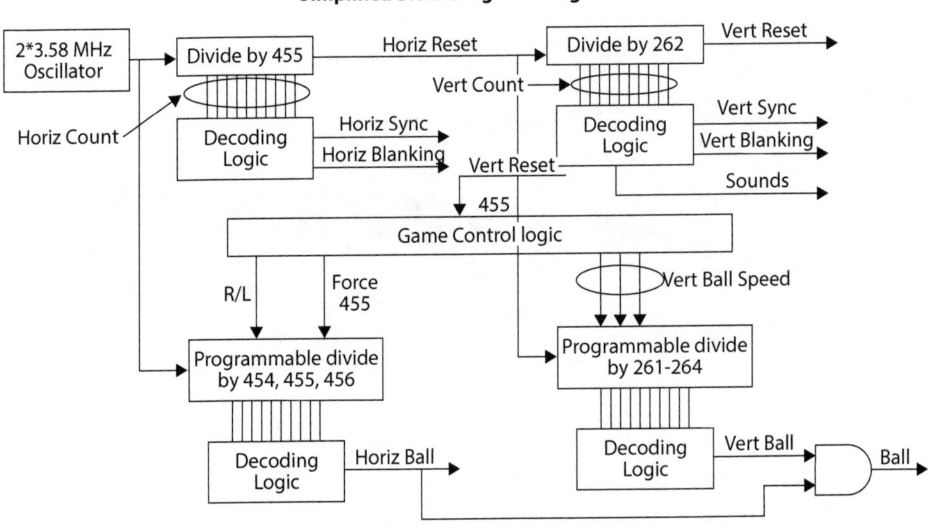

Figure 8.1. Al Alcorn's simplified block diagram for *Pong*. Courtesy Allan Alcorn.

science students less familiar with tricks of transistor-transistor logic (TTL); it shows the circuitry that generated game control functions as well as video signals to produce the game's images and sounds. The original game ran not one line of program code. It did not use a microprocessor or a custom integrated circuit; rather, it was a digital logic design made from components familiar to a television engineer who thoroughly understood the various ways pulse waveforms could be generated and manipulated.

Despite being constructed entirely from television technology, *Pong* is occasionally depicted as a product of the computer age or even as a computer artifact. One of the earliest critical studies of video games, Geoffrey R. Loftus and Elizabeth F. Loftus's *Mind at Play* described *Pong* as being "entirely under the control of a computer," and their version of the video game's "family tree" showed arcade games and digital computing as its parents. Michael Malone wrote in his excellent history of Silicon Valley that *Pong* was put together by a computer programmer.[4] These mischaracterizations of *Pong* reflect a natural, if perhaps careless, assumption about the dawn of the video game. If much of its past—and, as we now know, its future—was bound to the computer, we are tempted to read these connections into every video game artifact. Like the theory of pre-formation in the 18th century, this idea leads us to see a fully formed adult in the germ of origin, a little computer inside every game machine.

The relationship of *Pong*'s creators at Atari to computer technology has not yet been investigated in a manner that illuminates the connections, if any, between early computer games and video games. On the face of it, there would seem to be no particular connection, at least nothing that can be traced in the particular technologies involved; the convergence certainly cannot be found inside the original *Pong* arcade console. The relationship might have played out in other ways, however, that might better be described in terms of influence rather than convergence. The reading of *Pong* as a product of the computer age sidesteps the emergence of the video game out of TV engineering, but it also calls attention to other factors in the development of the new video game technology—what the key figures had in mind, their entrepreneurial aspirations, their specific engineering training, and the impact of subsequent litigation on the story.

Pong was an easy game to play, a definite competitive advantage for an arcade game often placed in bars and restaurants where a patron could play with a drink in one hand. Has its simplicity discouraged serious attention to its history? This first success story of the commercial video game has deflected critical reflection through reduction to a "keep it simple, less is more" narrative. According to this argument, *Pong* succeeded in a manner that requires little explanation; it was easy to learn and fun to play, providing uncomplicated amusement suited to taverns and arcades.[5] End of story.

But what about the antecedents that set up its success? Nolan Bushnell's guiding vision of an electronic arcade game inspired by computer technology and previous work in TV game technology do not figure much in the explanation that *Pong* succeeded only because of its design simplicity. Hindsight and the appeal of a good Silicon Valley success story have perhaps postponed deeper investigation of how the earliest commercial video games were envisioned and built.[6] In fact, *Pong* emerged from a complex of research projects, product design, and business relationships that later figured in heated legal and corporate disputes about intellectual property, priority of invention, and so on through the 1980s.[7]

This history of *Pong* begins with its problematic connection to computer-based games such as *Spacewar*. These games inspired Bushnell's industrial design project, *Computer Space*. Tracking how that inspiration led to *Pong* corrects other accounts of the relationship between the TV game console and computer technology. Claims put forward in the courtroom, for example, attempted to portray computer games as a prior art of TV game technology. These disputes usually had little to do with *Pong* but were about rights and licenses associated with the invention of the television game and the reduction to practice of the relevant technology. In any case, the argument for computer games as prior art failed to make this case persuasively, as evidenced by settlements that benefited the holders for the original Baer/Harrison/Rusch TV game patents, assigned to Sanders Associates [now BAE Systems] and licensed to first Magnavox, and later Philips.[8] As for Atari, it is fair to say that no other company of the early video-game era negotiated as many contradictions and convergences of computer and TV technology, which besides developing *Pong* also manufactured arcade games, dedicated home consoles, programmable game machines, and home computers. The path of invention, innovation, and design that led to *Pong* reveals points both of intersection and disconnection between computer and TV technology during the video game's early development.

Spacewar

The winding path from the computer to *Pong* began with *Spacewar*. Few computer games are linked so tightly with the technical and institutional contexts of digital computing. Steve Russell, Alan Kotok, J. Martin Graetz, and others at MIT created *Spacewar* in February 1962 to demonstrate the new PDP-1 minicomputer and Precision CRT Display Type 30, both donated by the Digital Equipment Corporation (DEC) to the Electrical Engineering Department only months earlier.[9] The *Spacewar* authors were part of the Tech Model Railroad Club (TMRC) on campus, and *Spacewar* became an integral part of that culture. They were unimpressed by the "little pattern-generating programs" that others had made to show off the PDP-1, assuming more could be done with "this display that could do all sorts of good things!"[10]

The group decided that the most interesting demonstration of the computer's capabilities would be "a two-dimensional maneuvering sort of thing, and . . . that naturally the obvious thing to do was spaceships."[11] They were guided by several principles for a good demo program, especially that "it should involve the onlooker in a pleasurable and active way—in short, it should be a game."[12] Russell led the project, influenced by earlier computer programs such as *Tic-Tac-Toe* and Marvin Minsky's Tri-Pos: Three Position Display, better known as the Minskytron. Russell's collaborators contributed code and built control boxes so that players could maneuver virtual spaceships around on the CRT and shoot at their opponents. The code was available to all hackers at MIT, who improved, extended, and modified the game. Peter Samson, offended by the sparse background of empty space, coded "Expensive Planetarium" to portray accurately the stars in the night sky; Dan Edwards worked on gravity calculations; Graetz programmed explosions; and so on. The game superbly showcased the lab's new computer, while stimulating better understanding of new graphics, I/O, and display technology. In April 1962, soon after *Spacewar* was unveiled, J. M. Graetz, the editor of *Decuscope*, wrote that the "use of switches to control apparent motion of displayed objects amply demonstrates the real-time capabilities of the PDP-1." He had visited the computer room and could "verify an excellent performance" for the emerging PDP community.[13]

Spacewar thus demonstrated not only computing technology, but the technical mastery of programmers and hardware hackers as well. It expressed their shared culture and institutional setting. In working out the game, Russell fixed on the popular science fiction novels of Edward Elmer Smith's Lensman series. This was the early Space Age, so it is not surprising that a fan and hacker would place his game in the world of these novels. Smith's writing excelled at portraying action and movement, with spaceships blasting away at each other, so what better homage than a fast-paced shoot-'em-up action game?

Setting it in outer space only required a visual backdrop of flickering stars that was relatively easy to render graphically because the Type 30's display could directly plot stars as points. Having an essentially dark background and a fantasy setting meant that the game's visual treats, spaceships and missiles, could be set in an appropriate visual space without overburdening the hardware. Russell noted also that,

> by picking a world which people weren't familiar with, we could alter a number of parameters of the world in the interests of making a good game and of making it possible to get it onto a computer.[14]

For example, Edwards's gravity calculations were realistic, but the programmers decreed that "photon torpedoes" ignored gravitational attraction to ease the computational task. Collaboration and design flexibility became the project's defining characteristics, much like the nascent culture of the computer lab.

Figure 8.2. ***Spacewar*** **at the Stanford AI Lab. Courtesy of Stanford University Libraries, University Archives, http://infolab.stanford.edu/pub/voy/museum/pictures/AIlab/list.html**

Spacewar was distinctly a product of MIT computing. Like Whirlwind, the TX-0, and the PDP-1, it exemplified the tradition of what Gordon Bell has called "MIT personal computers."[15] The gift of the PDP-1 established DEC as a provider of equipment for academic research, and *Spacewar* returned the favor. Freely distributed via paper tape in the lab, the game was shipped by DEC with PDP computers as a test program to verify their operation after new installations. *Spacewar* became a fixture in university and industrial laboratories of the 1960s and 1970s. A community of programmers and players formed around it as a popular and competitive entertainment, described in Brand's reports of the 1972 *Spacewar* Olympics at Stanford (see figure 8.2). Programmers everywhere added elements to the game or tweaked settings and controls in a local version. *Rolling Stone* reported, "Within weeks of its invention *Spacewar* was spreading across the country to other computer research centers [that] began adding their own wrinkles."[16] This convergence of competitive skill, programming wizardry, and collaborative community characterized hacker culture.

Computer Space

How did *Spacewar* extend its influence from the TMRC hackers to the design of products such as arcade consoles? With the growth of a network of research laboratories funded by DARPA, especially its Information Processing Techniques Office (IPTO), a generation of computer science students was introduced to computers such as DEC's PDP series that ran *Spacewar*. One of these laboratories was at the University of Utah, home of a strong program in computer graphics that DARPA generously funded. One historian of computer graphics has remarked that, "almost every influential person in the modern computer-graphics community either passed through the University of Utah or came into contact with it in some way," while DARPA historians have called its program "especially influential in the birth and development of interactive graphics."[17]

Nolan Bushnell graduated in 1968 from the University of Utah with a degree in electrical engineering. While there, he had access to the program's computers, and like many other students, he often played *Spacewar*. He also held a summer job as an amusement park employee, staffing a pinball arcade and other attractions. This unusual exposure to both carnival and computer culture stimulated his notion of creating a new kind of entertainment arcade filled with *Spacewar*-like games.

After graduating, he moved to California to work for Ampex, a leader in the development of magnetic recording, video, and computer storage technologies. He was now in the hotbed of high-technology entrepreneurship, at the southern boundary of the region Don Hoefler began calling "Silicon Valley U.S.A." in 1971.[18] The first big wave of Silicon Valley start-ups crested between 1967 and 1969, with the founding of National Semiconductor, Intel, Advanced Micro Devices, and many more companies. Bushnell's entrepreneurial imagination responded to this environment. Surrounded by first-rate research engineers and product development teams at Ampex, he thought more about his vision of a *Spacewar* arcade. He was in the right place to ponder the impact of component miniaturization and integration. Moreover, located in Redwood City, Ampex was a short drive from a hotbed of *Spacewar* activity in Stanford University's Artificial Intelligence Laboratory (SAIL), an important center for computer science research.[19] Bushnell revived his enthusiasm for the game and pondered how money could be made in the arcade video game business.

Bushnell's original plan was to create an arcade video game based on *Spacewar* using a "Data General 1600—to have a minicomputer running multiple games."[20] Having played *Spacewar* and used time-sharing systems at Utah and Stanford, Bushnell started out with the ambitious goal of creating an arcade system that utilized a time-sharing environment to display interactive graphics concurrently on several consoles. Although the concept was understandable, it was not feasible. In terms of cost, DEC's 36-bit, multitasking PDP-10/DEC System-10 had begun shipping to computer science departments in 1968, but a complete system cost well in excess of $100,000—far more with displays, disk storage, printers, and hardware peripherals. Still, Bushnell was determined to create a modestly priced computer arcade game inspired by *Spacewar*, a commitment underscored by calling this project *Computer Space*.

Key events related to the story of *Computer Space*—Bushnell's departure from Ampex in March 1971, his partnership with Ted Dabney, the decision to join Nutting Associates, which acquired and manufactured *Computer Space*, and the founding of Atari (originally called Syzygy) in June 1972—have been well documented.[21] Less has been written about the technical and cultural contexts of Bushnell and Dabney's work on *Computer Space*. While still at Ampex, Dabney and Bushnell began to consider how to build an arcade version of the six-figure computing platforms used to play *Spacewar*. In the early 1970s, the steady progress of hardware miniaturization and software innovation sounded the call to deliver computing technology in smaller packages for many applications, so *Computer Space* can be instructively compared to other projects. Microcomputer kits such as the Altair 8800 would not be available for a few years, but digital logic components such as TTL and other ICs had become standard electronics parts.

In November 1971, Intel introduced its first single-chip microprocessor, the 4004. Douglas Engelbart, Ted Nelson, and others had already begun to ponder the impact of computing on human potential. Nelson, for example, called for "computer liberation" in *Computer Lib* and proclaimed that everyone "can and must understand computers NOW."[22] He predicted new applications for a variety of purposes and asked, "Can the public learn, in time, what good and beautiful things are possible" from computer systems?[23] Nelson's manifesto included computer games, for he had observed that "wherever there are graphic displays, there is usually a version of the game *Spacewar*." Citing the 1972 *Rolling Stone* article,[24] Nelson might have been looking through Bushnell's eyes when he remarked that "games with computer programs are universally enjoyed in the computer community."[25] He discussed computer games at about the same time as Bushnell began working on *Computer Space*, specifically mentioning versions of Conway's *The Game of Life*[26] and BASIC programs published by the People's Computer Company, an organization near Stanford University (and Ampex) that sought to bring programming power to the people through recreational and educational software. Nelson provided a voice for those who proposed to move advanced text, graphics, networking, and other computer technologies out of academic laboratories to make them available to everyone. Bushnell took the engineer's route as he thought about building a machine on which anyone could play a version of *Spacewar*.[27]

As Bushnell worked on *Computer Space* during 1971, he might have been aware of other projects like his. In 1969, Rick Blomme had written a two-player version of *Spacewar* for the Programmed Logic for Automatic Teaching Operations (PLATO) time-sharing system at the University of Illinois. It was the first multiplayer game hosted by the PLATO project, which during the 1970s became a hotbed for innovative, networked games.[28]

As he was completing *Computer Space*, Bushnell probably heard about a summer project closer to home; a recently graduated SAIL student, Bill Pitts, and his friend Hugh Tuck built a coin-operated (coin-op) computer game, *The Galaxy Game*, for the newly released PDP-11/20, DEC's first 16-bit computer. DEC had fit the PDP-11 into a relatively small box and listed it for a mere $20,000, hoping to open new markets and applications.[29] Pitts and Tuck formed a company called Computer Recreations, bought the low-end version of the PDP-11 for only $13,000, and converted the PDP-10 version of *Spacewar* for this machine. Including a Hewlett-Packard vector display, wooden cabinet, and other parts, their expenses came to roughly $20,000. In September 1971, they installed it in Stanford's student union, where a later version that supported up to four monitors (eight players) could be found until 1979. *The Galaxy Game* was faithful not only to *Spacewar*, but also to the player community (university students and computer engineers) and to the technical configuration (software code, vector displays, time-sharing, and so on) that produced it.[30]

Bushnell started out on the same course of programming a version of *Spacewar*. Like Pitts and Tuck, his first thought was to purchase an inexpensive minicomputer, maybe the new Data General Nova or the SuperNova.[31] Instead of coupling the computer to expensive monitors, he would link up several game stations equipped with cheap, off-the-shelf TV sets using raster, not vector, graphics.[32] At first, Bushnell knew almost nothing about how TV sets might function as monitors, but Dabney brought him up-to-speed quickly on TV signal generation and related topics. Indeed, Dabney showed Bushnell how to modify a TV using TTL components to move an object around on the screen.[33]

This promising union of technologies—the computer from the university lab and the TV from the home—proved impractical, however. There was little chance of getting the design to work with the equipment available, and no chance to do so at an acceptable cost. According to Bushnell, the burden of providing images to multiple TV monitors would bring a computer to its knees; it would be "blindingly slow" even if Dabney and Bushnell were able to tweak monitors and build circuitry to offload processing from the CPU.[34]

Dabney certainly grasped quickly that computers designed to drive vector displays could not be used to produce raster-scan output for analog TV monitors. Bushnell's telling of the story suggests that they were able to lash up parts of a working system, but it is clear that the original concept was abandoned quickly. According to Al Alcorn, Bushnell's wife was responsible for ordering the computer but considered the price tag for the computer "crazy" and never ordered it.

Frustrated by the likelihood of poor performance and fuzzy images, Bushnell

> designed out the need for the computer, because the computers were so slow at that time. . . . So there was this brilliant leap that Nolan made about how he could get rid of just a little bit of logic [and still] do the same thing the computer's going to do, just much, much faster, so he didn't need the computer.[35]

Bushnell and Dabney promptly dumped the stillborn idea of a computer controlling multiple raster-scan displays. They replaced the minicomputer with dedicated circuits based on TV technology that controlled all aspects of the game, from game logic and graphics to player controls. Resetting the project made it possible to finish with a working prototype. After sculpting a futuristic cabinet for the arcade console, Bushnell sold *Computer Space* to Nutting Associates, where the design benefited from the contributions of experienced engineers; Bushnell joined the firm as chief engineer to oversee the final design, manufacturing, and distribution. It was released in August 1971.

Even though Nutting went on to build between 1,500 and 2,300 machines,[36] the historical verdict on *Computer Space* has generally been negative, whether with respect to sales, gameplay, or controls. Video game historians have written that, besides Bushnell's friends, "the rest of the world didn't show any interest in the game at all" and it "failed," that it was "unsuccessful," a "failure," "lacking in mainframe complexity," and a "colossal commercial flop."[37] Bushnell offers only a weak defense, claiming that it "did okay, but it really didn't do as well as it could have" or that it did "very well on college campuses and in places where the educational value was higher. However, there weren't any arcades as such back then."[38]

Computer Space redeemed itself mostly as a negative example for Bushnell and Al Alcorn when they made the next game and as the first step toward the creation of Atari and *Pong*. From his own mistakes and the work with Nutting Associates, Bushnell learned about game console engineering and, especially, the arcade business. On the positive side, Dabney and Bushnell took away $500 in royalties from *Computer Space* to start their new company. On the whole, however, *Computer Space* was the failure that motivated *Pong*'s designers to keep things simple the second time around.

Design Lessons from *Computer Space*

These assessments of *Computer Space* miss its significance for the video game as a technological artifact. It provided more than a learning experience: *Computer Space* established a design philosophy and general technical configuration for arcade consoles,[39] and it reduced the laboratory-based computer game to a format that would launch the video game as a consumer product. When Bushnell noted years later that his engineering friends loved *Computer Space*, even if "the typical guy in the bar" was completely baffled, it is easy to hear echoes of this appreciation in assessments of his technical achievement from engineers, designers, and operators. Most notably, they argued, "The machine is like a historical blueprint of how all arcade games to follow would be made,"[40] and "The brilliance of these machines was that Nolan Bushnell and company took what was computer programming (in *Spacewar*) and translated it into a simpler version of the game (no gravity) using hard-wired logical circuits."[41]

The arcade video game design defined by *Computer Space* was notable on three counts: (1) packaging, both internal and external; (2) optimization; and (3) despite the daunting complexity of its game play, simplification, especially with respect to service requirements. Because Bushnell's lovingly shaped, futuristic cabinet design was visually memorable, good enough to serve as a prop in Hollywood movies such as *Soylent Green* (1973) and *Jaws* (1975), it is easy to forget that this technical configuration of *Computer Space* would remain essentially unchanged for a generation of coin-op arcade consoles. Bushnell divided his machine into component modules. Nutting's sales flyer, probably authored by Bushnell, crowed that there were "only three assemblies in the entire unit": a modified General Electric black-and-white TV set, the front control panel, and the "computer (brain box)." Circuitry, control, and screen were set into the "beautiful space-age cabinet" with a few other parts, such as a power supply and coin acceptor (see figure 8.3).[42]

In *Computer Space*, physical modules in the form of ICs mounted on three printed circuit boards replaced programs and executed game logic in hardware. Bushnell's original concept of a configuration of several raster displays connected to a computer was a failure, so he and Dabney replaced software with electronic components and ICs such as Texas Instruments' 7400 series of TTL circuits. At Ampex, they had been surrounded by engineers (like Al Alcorn) busy at developing products that utilized TTL technology. It was natural to consider how physical logic elements like flip-flops, counters, and registers could provide the synchronization signals needed to display graphical elements and scores, the creation of on-screen symbols, or execution of game logic.[43] For example, a small number of diode arrays connected to logic gates produced the rotating images of rockets seen on the screen; the rocket images were clearly visible even in the pattern of diodes on one of the PC boards (see figure 8.4).

It is tempting to think of these diode arrays as precursors of game ROMs, but this conception reintroduces the notion of program code, exactly what Bushnell and Dabney eliminated from the design. Bushnell's rockets were essentially hardwired bitmaps that could be moved around the screen independently of the background, a crucial innovation that made it possible to produce screen images efficiently. He called these moving images "patches." The design concept would become part of Atari's shared knowledge; even if "nobody could ever understand Nolan's schematic, . . . it was the idea of taking the bit-map in a little area that could be moved around so that it would not be necessary to redraw an entire screen every time an image moved" that every Atari engineer understood.[44] Bushnell's patch solution eventually became a staple of game machines and home computers in the form of "sprites," the term taken from Seymour Papert's briefly popular Logo programming language after a new generation of Texas Instruments graphics chips put it into home computers during the late 1970s. The idea of taking a game design

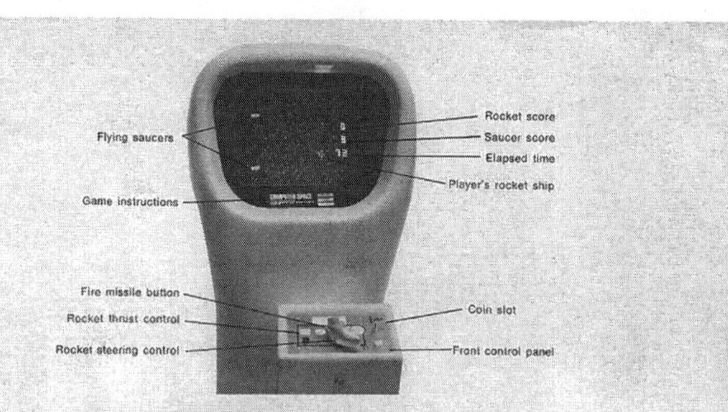

HOW COMPUTER SPACE WORKS AND PRODUCES

BEAUTIFUL SPACE-AGE CABINET attracts players of all age groups. The player is additionally attracted by two flying saucers moving about in formation on the playfield while the unit is in its non-activated state.

EXCITING PLAYER ACTION occurs as coin is inserted and start button is pushed to activate the unit. A rocket ship appears out of nowhere and at the same instant the once friendly flying saucer begins firing missiles at the rocket ship. Now at the controls of the rocket ship, you begin to evade the missiles bearing down on you and maneuver into position to fire your own missiles at the saucers. The thrust motors from your rocket ship, the rocket turning signals, the firing of your missiles and explosions fill the air with the sights and sounds of combat as you battle against the saucers for the highest score. Outscore or hit the saucers with your missiles more times than they hit you for extended play in hyperspace. Attain hyperspace and the playfield turns white and gives you a vision of daylight in outer space. Thrill to the reality of controlling your own rocket ship in gravity-free outer space. Battle the saucers in a duel of wits and coordination!

CHECK THESE UNIQUE FEATURES

★ Coin acceptor, rocket thrust button and rocket steering controls are the only moving parts in the entire unit. No mechanical relays, films or belts.

★ Adjustable play time from 1 minute to 2½ minutes.

★ Standard unit 25¢ play convertible to 2/25¢ by the throw of a switch.

★ Tamper-proof coin meter.

★ Continuous blinking panel lights and back-lit playfield for player attraction.

★ Integrated circuits in low-current solid-state construction insure the ultimate in long life.

★ No repeating sequence; each game is different for longer location life.

COMPONENTS? THERE ARE ONLY THREE ASSEMBLIES IN THE ENTIRE UNIT

Computer (Brain Box) | Front Control Panel | Black and White TV Set

COMPUTER (BRAIN BOX) is sealed and carries a full one-year unconditional guarantee if not tampered with. FRONT CONTROL PANEL houses the only moving parts in the unit—the rocket ship controls and coin acceptor. BLACK AND WHITE TV SET has the life of any new black and white receiver—no modifications have been made to affect its reliability.

Weight 98 lbs.; Dimensions 30" wide x 67" high x 30" deep.

NA NUTTING ASSOCIATES, INC.
500 LOGUE AVE., MOUNTAIN VIEW, CA 94040 (415) 961-9373

Distributed By

11-71-208

Figure 8.3. Nutting Associates' instructions for *Computer Space*. Courtesy of Stanford University Libraries, Department of Special Collections

and making it "more efficient in silicon" persisted in the design of dedicated and programmable game machines such as the Atari 2600 VCS, with the Stella custom graphics chip, even after the introduction of microprocessor control and program storage in ROM[45]; sprites were an important feature of home computers such as the Atari 400/800 (as "player/missile graphics" or

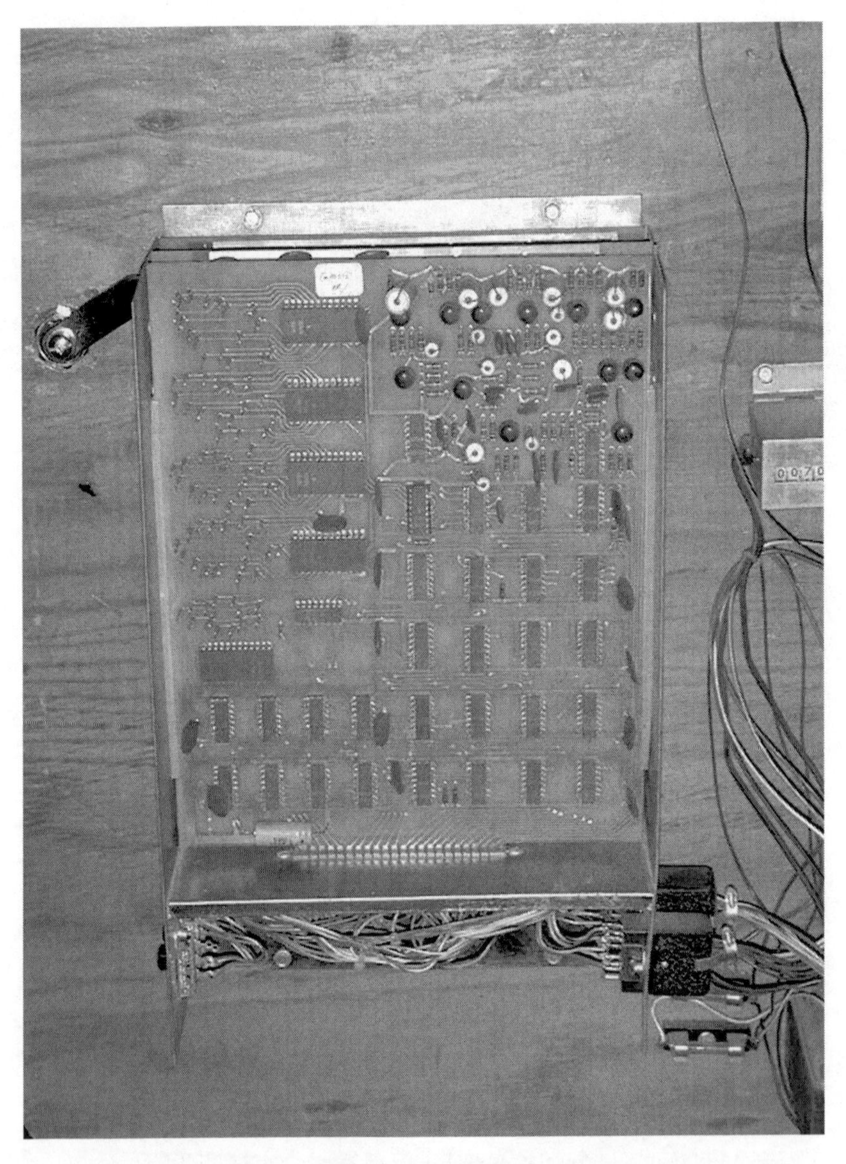

Figure 8.4. Circuit board, possibly from Computer Space, serial number 1. Courtesy of http://www.computerspacefan.com/SN1large.htm

"motion objects") or the Texas Instruments TI99/4A. The cost of expensive microprocessors could not be justified for home machines as they could for arcade consoles through much of the 1970s. Hence, Atari's original coin-op design philosophy was carried on in the design of home machines, reducing the workload on slow central processors by using specialized graphics hardware.

Because the main criticism leveled at *Computer Space* has been the daunting complexity of game controls and game play, Bushnell's efforts to keep the design of the arcade console as simple as possible have been overlooked.[46] His goal was to ensure reliability and ease servicing of delivered units. As he put it in a sales flyer, it was "our object to create a new standard of reliability using the latest technology. We believe that this goal has been met . . . *Computer Space* requires operators to have no more fear of replacing a bad tube than of replacing a bad relay." Because it was built with solid-state circuits, the manufacturers could boast that *Computer Space* had "no mechanical relays, films, or belts" to repair, the only moving parts being the coin acceptor and player controls. Bushnell reminded operators that the display was an ordinary TV set with "no modifications to affect its reliability." It would be "no harder to adjust than any home receiver."

It is worth noting that he played both ends of the stick by designating the internal hardware circuits mounted on PC boards as the "computer." By doing so, he recalled the origins of the game and created a space age aura around it, but on the other hand, he demarcated the "brain box" as a no-touch zone, a black box, by telling operators that it carried an "unconditional guarantee" only "if not tampered with."[47]

Atari and *Pong*

When it became obvious that *Computer Space* would not be a hit in arcades, Bushnell and Dabney severed their relationship with Nutting and founded Syzygy/Atari in June 1972. Before long, they were joined by another talented Ampex engineer, Al Alcorn, who had studied electrical engineering and computer science at the University of California, Berkeley, and since 1968 refined his skills in video and analog engineering at Ampex.[48] Alcorn was particularly skilled at applying his knowledge of transistor logic and ICs to "analog problems." Busy with the two-player version of *Computer Space*, Bushnell assigned Alcorn the task of designing a simple home-consumer game based on Ping-Pong. He inspired the new employee with a story that Atari had a buyer for the game—General Electric, no less. Bushnell failed to mention that he had almost certainly taken the idea from playing a similar game earlier that year on the new Magnavox Odyssey TV game console.[49] Concerned that it was "too big a step for [Alcorn] to go from not knowing what a video game was" to designing a real game, Bushnell's ruse set up a training exercise through which he eased Alcorn into electronic games. So Bushnell came up with "the simplest game I could think of, which was a tennis game." According to Alcorn, he understood the task as simply, "let's just do the most simple game to save time."[50]

In fact, no such contract existed, but Alcorn rose to the challenge and proved his mettle as an engineer. With his previous job experience and training, he was thoroughly familiar with state-of-the-art electronic components such as TTL ICs. The project also demonstrated his mastery of TV electronics. And last but not least, he distilled value from Bushnell's ideas and suggestions, which were as often chaotic as enlightening. Even though he could not decipher the schematics Bushnell showed him of *Computer Space*, Alcorn recalls:

> Nolan had filed a patent on the fundamental trick: . . . how to make a spot appear on a TV screen like Pong without having to do a memory map, a frame buffer, like what you would do today, because there was no memory other than flip flops. And so it was a very, very, very clever trick. I think I perused, glanced at the patent and [learned verbally] from Nolan how it was done; it was really clever. It involved simply making a . . . television sync generator which had counters to count clock pulses to make a horizontal sync, and then counters to count horizontal sync to make vertical sync, and so you'd get the lines set up. If you had another sync generator and you just had it running at the same time, but not synchronous with it, just the same clock and you decided to take the second sync generator output and make a spot where horizontal and vertical sync happen at the same time, that spot would appear randomly, somewhere on that screen, just by happenstance. It was this happy relationship between using the digital TTL circuits, which are absolutely ones and zeros, to do video which in those days was absolutely analog.[51]

Within a few months, Alcorn produced a prototype from a store-bought TV set, a homemade cabinet, about 75 TTL ICs, and some tricks from his bag of analog and television engineering (see figures 8.5 and 8.6). Surprisingly, Alcorn was at first disappointed by his effort because the "criterion then was cost, cost, cost," and he felt that the final chip count was too high. Bushnell and Alcorn named it *Pong* and installed the coin-op prototype in

Figure 8.5. *Pong* prototype front. Courtesy of Allan Alcorn

Figure 8.6. *Pong* prototype rear. Courtesy of Allan Alcorn

Andy Capp's Tavern, a local bar where eager players lined up to stuff quarters into the game.[52]

As I noted earlier, *Pong*'s triumph has been credited to the unrepentant simplicity of its design. Three short sentences on the cabinet's faceplate told

Figure 8.7. Atari's coin-op *Pong* machine. Courtesy of Stanford University Libraries, Department of Special Collections

players everything they needed to know: put a quarter into the machine, a "ball" will be served; move the paddle to hit the ball back and forth. Alcorn felt instructions were unnecessary altogether, but Bushnell insisted on them, so in a semi-sarcastic spirit, Alcorn responded by putting on the faceplate a simple summation that became the motto of the new game: "Avoid missing ball for high score."[53] *Pong* owed much of its success to breaking with the complexity of *Computer Space* and *Spacewar* in a manner suited to bar patrons. Unlike *Computer Space*'s beautiful fiberglass cabinet, the Pong prototype was set in an ugly square box covered with orange paint and wood veneer, with a simple faceplate for control knobs and instructions. In game play and aesthetics, *Pong* and *Computer Space* were polar opposites.

As an engineering design, however, *Pong* followed *Computer Space* in its modularization and optimization of hardware. Compared to Bushnell, Alcorn was better prepared by experience and inclination to build an efficiently engineered arcade console in three respects:

- He built his game with TV technology from the ground up.
- He deftly used digital components to solve the analog problem of mastering TV output, precisely Alcorn's special domain of engineering knowledge.
- The images required for the game were relatively simple. Unlike the oddly shaped objects such as spaceships in *Computer Space* that required ad hoc memory solutions such as diode arrays, the ball, paddle, and other images in *Pong* were all based on simple rectangles that digital TV circuits could easily generate on the fly.

This last point was especially important. Not a single line of software code was involved in the construction of *Pong*. Like *Computer Space*, *Pong*'s game logic and control operations were paced by synchronization signals for the rasterized TV display, but Alcorn understood more intuitively than Bushnell how to work with these signals during every cycle of the TV circuits. Because every image was based on rectangles, he could generate them by gating counters, even the seven segments of the score display. Alcorn was thus able to build *Pong* optimally from a modest number of ICs, and he was more obsessed than Bushnell with reducing the parts count. He eliminated unnecessary parts not only to make the game run more efficiently, but also to reduce the final product cost. Bushnell's original assignment for a simple home console game explains Alcorn's concern that the prototype even had

75 TTL circuits "and would cost way too much for a high volume home machine." His single-minded attention to optimization of the electronic circuitry continued the legacy of *Computer Space* and remained in Atari's engineering culture through the 1970s.[54]

The technology lineage leading from *Spacewar* through *Computer Space* to *Pong* is one way to narrate the complicated historical relationship between the computer and the video game. *Computer Space* and *Pong* were both TV games in the sense that their designers applied techniques of television engineering to make them, and in fact they required a television to operate. Yet, Bushnell's project emerged from the computer space of academic laboratories and large-scale computers, while *Pong* was cut loose from this mooring. In this telling of the story, arcade consoles, the home game foreshadowed in Alcorn's original *Pong* design, and home consoles created during the mid-1970s—including Atari's home version of *Pong* (1975), General Instruments AY-3-8500 "TV game on a chip" (1976), and the microprocessor- and ROM-based Atari 2600 (1977)—solved Bushnell's problem of reducing the computer game to a configuration suitable for delivery as an entertainment product to mass markets.

Atari never gave up on the computer game, however. When Bushnell first assigned Alcorn to the apocryphal GE project, his long-range goal was still to produce games that were "more complex . . . not something simpler" after Alcorn's trial by fire.[55] Atari's misleading advertising encouraged the view that video games were an ambitious coupling of the computer and TV technology (see figure 8.8). The company's early marketing literature characterized games like *Pong* as "video computer games" and claimed to having revolutionized the industry "when we harnessed digital computers and video technology to the amusement game field with *Pong*."[56]

Pong as a Television Game

Narrating the history of Pong as a kind of hyperspace jump from computer space contradicts what has been described as the real story behind video games and, ironically, as the *Pong* story. This story begins with the notion of the video game as "an apparatus that displays games using RASTER VIDEO equipment," not with inspiration from computer games (see http://www.pong-story.com). The inventor of this TV game console was Ralph Baer, a television engineer at the military electronics firm Sanders Associates. Indisputably, in 1966 he designed circuitry to display and control

Figure 8.8. ***Pong*** **advertising depicting Atari as a high-tech company. Courtesy of Stanford University Libraries, Department of Special Collections**

moving dots on a television screen, leading to a simple chase game called *Fox and Hounds* based on his original idea for a home television game. After proving the concept by designing several chase and light-gun games, he received permission from Sanders management to continue his TV Game Project.

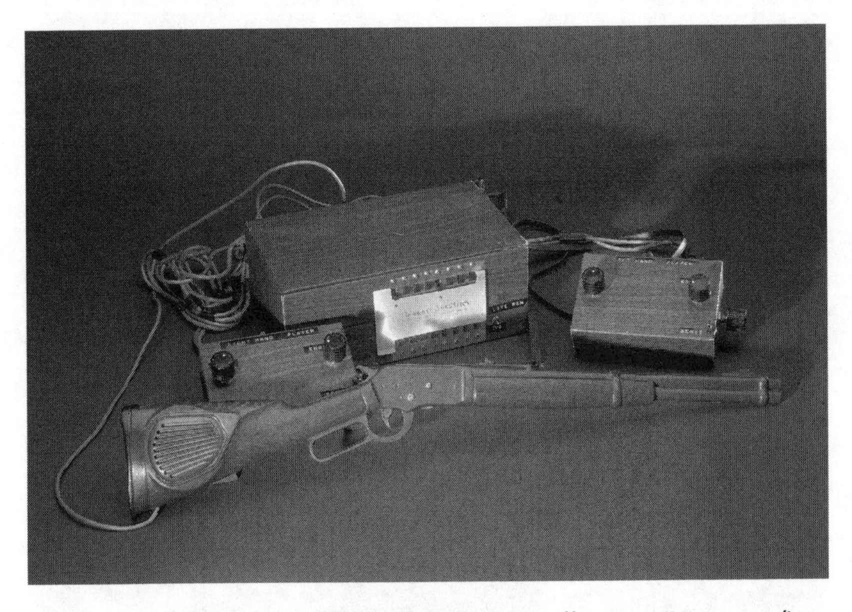

Figure 8.9. The Brown Box with shooting rifle. https://www.ralphbaer.com/how_video_games.htm. Courtesy of Ralph Baer

With two coworkers, Baer improved the game system, leading in 1968 to a seventh version called the Brown Box, a solid-state prototype for a video game console (see figure 8.8). The Brown Box offered handball and light-gun shooting games, in addition to the original chase games and a ping-pong game. Sanders licensed the technology and patent rights to Magnavox, and by mid-1971, a working prototype of what would be marketed as the Magnavox Odyssey console was available for consumer testing. Sanders was awarded several US patents for the technology developed by the TV Game Project in 1972, the same year in which Magnavox began selling the Odyssey—the first commercial TV game console for the home.[57]

Bushnell attended a presentation of the Odyssey in Burlingame, California, on May 24, 1972. Working for Nutting at the time, he viewed and played the Odyssey at the demonstration. When he tasked Alcorn with a ball-and-paddle game, his suggestion must have been influenced by what he had seen from Magnavox.[58] This version of the *Pong* story rebuts the construction of the video game as emerging entirely through Bushnell's engineering of the computer game.

Baer saw the video game as an enhanced, interactive form of television, a key motivation for his project at Sanders Associates. Baer's inspiration thus differed markedly from Bushnell's encounters with computer games. Situating *Pong* in computer space complicates the "keep it simple" version of its success story. A version of this story that includes Baer's seminal role in TV game technology must take note not only of the connection between the Odyssey and *Pong*, but also of the differences between their approaches to the problem of designing video games:

> Mr. Bushnell did indeed play the Magnavox Odyssey's Ping-Pong game hands-on. He clearly needed no instructions on how to play that game. On the other hand, his much more elaborate *Computer Space* game was failing in the marketplace because it was too complicated to play. A light bulb may have gone on in Mr. Bushnell's head the moment he played ping-pong on the Odyssey: "Keep it simple." Complicated games may work for nerds but not for ordinary people.[59]

The failure of *Computer Space* provided the motivation for adopting a design philosophy evident in Baer's "television gaming apparatus." Inspired by computer technology, Bushnell adopted the TV game as the basis for the success of Atari's coin-op console.

After the 1975 Christmas season, a *Business Week* reporter wrote that "at the moment, only two companies are serious factors in consumer electronic games": Atari and Magnavox.[60] Ten years later, the video game industry had crashed, burned, and risen again under the new regime of Nintendo's carefully crafted protection of the technical platform, intellectual property, and content of video games. Nintendo's strategy was zealous control of its technology and products. It staved off competition by using legal, business, and technical means to screen and occasionally block independent software developers from access to its hardware: the Famicom (1983) and its American version, the Nintendo Entertainment System (1985).

Magnavox and Sanders by then were in the habit of tenaciously defending their patents. They had litigated effectively against a virtual who's who of the game industry.[61] A clash was inevitable. As early as 1975, Nintendo had declined Magnavox's offer of a license for TV game technology; a year later in correspondence with Sanders, Nintendo's representative noted a "conflict" with the "concept of Mr. Nolan K. Bushnell," but apparently without knowledge of the licensing agreement that existed between Atari and Magnavox.[62]

More than a decade later, after the introduction of the NES, Nintendo decided to fight the Sanders-Magnavox patents. Conflating the Atari and Baer versions of the *Pong* story, Nintendo's lawyers tried to convince the court that the Sanders patents had been preceded by *Spacewar*. They insisted that the Sanders team must have known about *Spacewar*—a Magnavox patent lawyer had probably seen the game but failed to inform the Patent Office about it, a Sanders engineer had played it at Stanford and later installed a version on a Sanders computer, or they must have read a description of the game in the *Rolling Stone* article or the book *II Cybernetic Frontiers* (Random House, 1974). Nintendo brought in expert witnesses to state that a person "reasonably conversant in the field" would have been able to convert *Spacewar* for a raster-scan display. Nintendo skirted over the difficulties Bushnell encountered in attempting to realize this idea, with its lawyers asserting that the electronic circuitry for a rasterized computer game would have been relatively standard.[63] The dispute was brought before New York Federal District Court Judge Leonard B. Sands, but before the case could be adjudicated, Nintendo realized the futility of its claims against Magnavox and settled out of court in April 1991.[64]

Conclusion

The lesson of *Pong*'s historical journey through computer space is that invention stories are never as simple as a game of *Pong*. Nolan Bushnell's *Computer Space* leads to *Pong* as a product inspired by computer technology but practically realized by TV technology. The result was the coin-op video game console. Baer's TV game apparatus provides a story about who was first to invent the home TV game as a technical, legal, and financial matter.[65]

In later years, Bushnell acknowledged the Odyssey, but he unfairly dismissed its significance as being only an analog game. His historical judgment was more accurate when he admitted, "I feel in some way that I didn't invent the video game—I commercialized it. The real digital game was invented by a few guys who programmed PDP-1s at MIT."[66]

Acknowledgments

I thank Ralph Baer, the inventor of the original video game console, for his comments and suggestions based on an earlier draft of this article, particularly with regard to arguments made in lawsuits related to infringements on the original patents and licenses and how these arguments might have distorted accounts of the early development of the TV console game. I also thank Allan Alcorn, *Pong*'s designer, for reading an early draft of this article and for many conversations about his work and career, including a recent oral history interview.

NOTES

1. I use the more traditional *Spacewar!* on the first reference and then drop the exclamation point thereafter to avoid any awkward sentence punctuation.

2. For example, Henry Lowood, "A Brief Biography of Computer Games," in *Playing Video Games: Motives, Responses and Consequences*, ed. Peter Vorderer and Jennings Bryant (New York: Lawrence Erlbaum Associates, Inc., 2006), 25–41; and Paul Ceruzzi, *A History of Modern Computing* (Cambridge, MA: MIT Press, 1998), 207–210.

3. Brian Sutton-Smith, *Toys as Culture* (New York: Gardner Press, 1986), 64.

4. Geoffrey R. Loftus and Elizabeth F. Loftus, *Mind at Play: The Psychology of Video Games* (New York: Basic Books), 1983, 6–7; and Michael S. Malone, *The Big Score: The Billion-Dollar Story of Silicon Valley* (New York: Doubleday, 1985), 343.

5. Leonard Herman, *Phoenix: The Fall and Rise of Videogames*, 3rd ed. (Springfield, NJ: Rolenta, 2001), 11–15.

6. Two terms used throughout this article require some clarification. "Video game" refers to console games produced for display on televisions and early arcade systems, generally raster-scan displays. Likewise, "video" does not in this article have the general meaning of referring to any signal for any display but rather to the specific analog television signal specifications of the 1960s and 1970s—horizontal and vertical sync, color synchronization, and so forth—and thus pertains to television technology specifically.

7. For a quick summary of the major cases in the business history of video games, see the Patent Arcade Web site, http://www.patentarcade.com/2005/05/feature-video-game-lawsuits.html. See especially *Magnavox Co. v. Activision, Inc.*, WL 9496, 1985 (N.D. Cal.).

8. Ralph H. Baer, *Videogames in the Beginning* (Springfield, NJ: Rolenta Press, 2005), 157–161. For a list of patents, see pp. 197–220.

9. Ken H. Olsen to Peter Elias, "The Story of . . . PDP-1," internal corporate document, Digital Equipment, September 15, 1961, http://research.microsoft.com/~gbell/Digital/timeline/pdp-1story.htm.

10. Steve Russell, quoted in Stewart Brand, "SPACEWAR: Fanatic Life and Symbolic Death Among the Computer Bums," *Rolling Stone*, December 7, 1972, http://www.wheels.org/spacewar/stone/rolling_stone.html.

11. Russell, quoted in "SPACEWAR," http://www.wheels.org/spacewar/stone/rolling_stone.html.

12. J. M. Graetz, "The Origin of Spacewar!" *Creative Computing Video and Arcade Games* 1, no. 1 (Spring 1983): 78–85. This was originally published in *Creative Computing*, August 1981.

13. Dan J. Edwards and J. M. Graetz, "PDP-1 Plays at Spacewar," *Decuscope* 1, no. 1 (April 1962), 2–4.

14. Russell, quoted in "SPACEWAR," http://www.wheels.org/spacewar/stone/rolling_stone.html.

15. Gordon Bell, "Towards a History of (Personal) Computer Workstations (Draft)," *Proc. ACM Conf. History of Personal Workstations* (ACM Press, 1986): 10–11.

16. Russell, quoted in "SPACEWAR," http://www.wheels.org/spacewar/stone/rolling_stone.html.

17. Robert Rivlin, *The Algorithmic Image: Graphic Visions of the Computer Age* (Microsoft Press, 1986) as quoted by the University of Utah, "History of the School of Computing," http://www.cs.utah/dept/history/; Arthur L. Norberg and Judy E. O'Neill, *Transforming Computer Technology: Information Processing for the Pentagon, 1962-1986* (Baltimore: Johns Hopkins University Press, 1986), 122.

18. Don C. Hoefler, "Silicon Valley USA," *Electronic News*, January 11, 1971, 3.

19. See Russell, quoted in "SPACEWAR," http://www.wheels.org/spacewar/stone/rolling_stone.html.

20. Rusel DeMaria and Johnny L. Wilson, *High Score: The Illustrated History of Electronic Games* (New York: McGraw Hill/Osborne, 2002), 16–21.

21. See Herman, *Phoenix*; DeMaria and Wilson, *High Score*; Scott Cohen, *Zap! The Rise and Fall of Atari* (New York: McGraw-Hill, 1984), 15–28; Steve Bloom, "The First Golden Age," *Digital Deli: The Comprehensive, User-Lovable Menu of Computer Lore, Culture, Lifestyles, and Fancy*, Steve Ditlea, ed. (New York: Workman Publishing Co., 1984), 327–332; Van Burnham, *Supercade: A Visual History of the Videogame Age, 1971-1984* (Cambridge, MA: MIT Press, 2001), 64–77; Steven L. Kent, *The Ultimate History of Video Games* (New York: Three Rivers Press, 2001), 28–41; David Sheff, *Game Over: How Nintendo Conquered the World* (Wilton, CT: GamePress, 1999), 133–140; and P. J. Coughlan and D. Freier, "Competitive Dynamics in Home Video Games (A): The Age of Atari," *Harvard Business School Industry and Competitive Strategy Cases*, 9-701-091, June 12, 2001, 1–3.

22. Theodor H. Nelson, *Computer Lib: You Can and Must Understand Computers Now* (Hugo's Book Service, 1983), title page, 2–3.

23. Nelson, *Computer Lib*, 2–3.

24. See Brand, "SPACEWAR," http://www.wheels.org/spacewar/stone/rolling_stone.html.

25. Nelson, *Computer Lib*, 48.

26. Martin Gardner, "The Fantastic Combinations of John Conway's New Solitaire Game 'Life,'" *Scientific American* 223, no. 4 (October 1970): 120–123.

27. According to Al Alcorn, the Atari group did not hear about Nelson until the late 1970s, and "lots of people had ideas but no one . . . built any working machines" (email correspondence, August 2005).

28. J. Mulligan, "Talkin' 'bout My . . . Generation," January 22, 2002, http://www.skotos.net/articles/BTH_17.shtml; Markus Friedl, *Online Game Interactivity Theory* (Boston, MA: Charles River Media, 2003), 4–5.

29. Jamie Parker Pearson, ed., *Digital at Work: Snapshots from the First Thirty-Five Years* (Digital Equipment, 1992), 58, 65.

30. Bill Pitts, "The Galaxy Game," October 29, 1997, Computer History Exhibits, Stanford University, http://www-db.stanford.edu/pub/voy/museum/galaxy.html. Al Alcorn saw the *Galaxy Game* on the Stanford campus with Bushnell while collecting quarters from their *Pong* machine right next to it. Alcorn remembers that "right next to me was Pitts with his, what

do you call it, *Galaxy Game*. And Nolan—we looked at this thing and my goodness, there was a teletype terminal sitting behind it, and he'd be in there modifying code on this thing. There was a vector scan display, from I think Hewlett-Packard or Tektronics, there was a real minicomputer in there." "Oral History of Al Alcorn. Interviewed by Henry Lowood," Computer History Museum, X4596.2008, transcript 2, April 2008, 3.

31. In his book *The Ultimate History of Video Games*, Kent, who interviewed Bushnell, refers to "a new and inexpensive Texas Instruments minicomputer" (p. 31). DeMaria and Wilson's *High Score* reported that Bushnell said, "I originally planned to do it on a Data General 1600," noting that the cost was $4,000 (p. 16). Bushnell probably meant the 16-bit Nova 1200, which cost $3,995 when launched in 1969. Perhaps Kent had in mind Bushnell's use of the TI 7400 series of TTL integrated circuits, such as the 74150 and 74153 multiplexers shown in the design schematics for the board that controlled graphics and motion of in-game rockets, "B-MEMORY 1 or 2 Player, NA 73–103, Computer Space" (January 29, 1973), Computer Space Instructions, http://www.arcadedocs.com/vidmanuals/C/ComputerSpace.pdf. According to Alcorn, Bushnell's original idea was to use the Supernova minicomputer, which came out soon after the Nova ("Oral History," transcript 1, p. 9).

32. Atari did not use vector-generated images until 1979, when it developed the Digital Vector Generator for the coin-operated games *Lunar Lander* (like *Computer Space*, formerly a popular computer game in university labs) and *Asteroids*.

33. Leonard Herman, "The Untold Atari Story," *Edge* no. 200 (April 2009): 94–99.

34. DeMaria and Wilson, *High Score*, 16–21.

35. "Oral History of Al Alcorn. Interviewed by Henry Lowood," Computer History Museum, X4596.2008, transcript, part 1, April 2008, 9–11.

36. See Herman, *Phoenix*, 11–15; Herman, "The Untold Atari Story," 94–99.

37. See Herman, *Phoenix*, 11–15; DeMaria and Wilson, *High Score*, 16–21; Bloom, "The First Golden Age," 327–332; Burnham, *Supercade*, 64–77; and Kent, *The Ultimate History*, 28–41.

38. See DeMaria and Wilson, *High Score*, 16–21; and Kent, *The Ultimate History*, 28–41.

39. Burnham appreciates this point in *Supercade*, noting that "the game established the basic system architecture for nearly every arcade game to follow," 71.

40. A. Maclean, "Computer Space Restoration," http://www.ionpool.net/arcade/archuk/computer_space_restoration.html.

41. L. Kerecman, "Computer Space," Arcade History Database, http://www.arcade-history.com/?n=computer-space&page=detail&id=3388.

42. Nutting Associates, "How Computer Space Works and Produces," flyer, November 1971; N. K. Bushnell, "Computer Space Instructions," typescript, Nutting Associates, February 1972.

43. On this point, I am indebted to Ralph Baer's comments on an earlier draft of this article.

44. Videotaped interview with Atari engineers filmed in August 1997: R. Milner and S. Mayer, "Stella at 20: An Atari 2600 Retrospective," videorecording, CyberPuNKS, 2000. See also T. E. Perry and P. Wallich, "Design Case History: The Atari Video Computer System," *IEEE Spectrum* 20, no. 3 (March 1983), 45–51.

45. Atari's *Tank* (1974) was the first video game to use ROM for storing game graphics.

46. He eliminated some details of gameplay from *Spacewar*, a topic outside this article's scope.

47. Nutting Associates, "How Computer Space Works and Produces," flyer, November 1971; N. K. Bushnell, "Computer Space Instructions," typescript, Nutting Associates, February 1972.

48. Ampex's role as incubator of talented Silicon Valley engineers and entrepreneurs, such as Ray Dolby, Steve Mayer, Steve Bristow, and Lee Felsenstein, deserves study.

49. See Baer, *Videogames in the Beginning*, 157–161; Lowood, "Oral History of Al Alcorn," 9–11.

50. Lowood, "Oral History of Al Alcorn," 9–11.

51. Lowood, "Oral History of Al Alcorn," 2–6; Nolan Bushnell, *Video Image Positioning Control System for Amusement Device*, US patent 3,793,483, February 19, 1974. The patent was actually filed on November 24, 1972, shortly after Alcorn began work on *Pong*.

52. See Loftus and Loftus, *Mind at Play*, 6–7; Herman, *Phoenix*, 11–15; DeMaria and Wilson, *High Score*, 16–21; Cohen, *Zap!* 15–28; Kent, *The Ultimate History of Video Games*, 28–41; Lowood, "Oral History of Al Alcorn," 2–6; Vintage Gaming Network, "Al Alcorn Interview," Vintage Gaming, http://atari.vg-network.com/aainterview.html; D. Owen, "The Second Coming of Nolan Bushnell," *Playboy*, June 1983.

53. Lowood, "Oral History of Al Alcorn," 2–6.

54. See videotaped interview with Atari engineers filmed in August 1997: R. Milner and S. Mayer, "Stella at 20: An Atari 2600 Retrospective," videorecording, CyberPuNKS, 2000; T. E. Perry and P. Wallich, "Design Case History: The Atari Video Computer System," *IEEE Specrum* 20, no. 3 (March 1983), 45–51; Vintage Gaming Network, "Al Alcorn Interview," Vintage Gaming, http://atari.vg-network.com/aainterview.html; D. Owen, "The Second Coming of Nolan Bushnell," *Playboy*, June 1983; Steve Wozniak's reduction of the TTL count for Atari's *Breakout* game provides a famous example; O. W. Linzmayer, *Apple Confidential: The Real Story of Apple Computer, Inc.* (San Francisco: No Starch Press, 1999), 17–20.

55. See Kent, *The Ultimate History*, 28–41.

56. Atari Inc., "Atari Expands Worldwide!" flyer, Arcade Flyer Database, 1972, http://www.arcadeflyers.com/?page=thumbs&db=videodb&id=3303. In the early 1980s, Atari invested heavily in the game machine as home computer in the form of the Atari 400/800.

57. Notably, *Television Gaming Apparatus*, US patent 3,659, 285, filed August 21, 1969. The details of this story can be found in R. Baer's *Videogames*. Baer deserves great credit for the extensive documentation of his activities during the key period of his work at Sanders, both in his book and by donating his significant collection of papers to the Smithsonian Institution's Lemelson Center Archives. See http://invention.smithsonian.org/resources/fa_baer_index.aspx.

58. Baer, *Videogames in the Beginning*, 5–9, 76–82.

59. Baer, *Videogames in the Beginning*, 5–6.

60. "TV's Hot New Star: The Electronic Game," *Business Week—Industrial Edition*, December 29, 1975, 24.

61. For example, *Magnavox Co. v. Chicago Dynamic Industries*, 201 U.S.P.Q. 25 (N.D. Ill. 1977), and *Magnavox Co. v. Mattel, Inc.*, 216 U.S.P.Q. 28 (N.D. Ill. 1982).

62. *Nintendo v. Magnavox*, US District Court, Southern District of New York, document 100. NARA Central Plains Region, duplicate photocopies of selected records at Stanford University.

63. *Nintendo v. Magnavox*, US District Court, Southern District of New York, document 81, 18.

64. *Nintendo v. Magnavox*, "Order of Discontinuance," US District Court, Southern District of New York, document 112.

65. Baer, *Videogames in the Beginning*, 14.

66. D. Becker, "The Return of King Pong," interview, CNET News.com, March 15, 2005, http://news.com.com/The+return+of+King+Pong/2008-1043_3-5616047.html.

9

Game Engines and Game History

Introduction

I have a problem: it's called preaching to the choir. Nobody here lacks faith in the history of games, and I can only hope that my talk will not reduce your enthusiasm. That realization eliminates the need for a homily on the value of historical studies. Instead, I will talk about the possibilities and promises facing the history of games as we plunge forward. What does history offer game studies and what might a history of games give back? These questions guide my thoughts at the end of this conference, which has given so much food for these thoughts.

So, What Is Historiography?

For a historian like myself, the word historiography has a twofold meaning. As the word suggests, it means writing about history and what it means to do history. Second, it encompasses the methods and materials of historical work. Historiography sometimes is used as another way of saying "historical literature," but let's not worry about bibliography today. While preparing for this talk, I checked the *Oxford English Dictionary* to find an example of usage, which led me to a *Wall Street Journal* book review from a few years ago. I quote: "The book is an example of *historiography*, the study of the principles and techniques of history—a discipline that is usually dryness itself."[1] Oh well.

Originally published as "Game Engines and Game History," *Kinephanos*. History of Games International Conference Proceedings special issue (January 2014).

It should come as no surprise to any of you that historians tell stories about the past. I mean, look at the word! HI-STORY. Historiography then might be described as writing about writing about history. Recent historiography has been mightily influenced by Hayden White, author of the much-discussed *Metahistory: The Historical Imagination in Nineteenth-Century Europe*, first published in 1973. White argued that history is less about a particular subject matter or source material than about how historians write about the past. The historian does not simply arrange events in correct chronological order. Such arrangements are merely chronicles. The work of the historian only begins there. Historians instead create narrative discourses out of sequential chronicles by making choices. White puts these choices into the categories of argument, ideology, and emplotment. Without getting into the details of every option open to the historian, the desired result is sense-making through the structure of story elements, use of literary tropes, and emphasis placed on particular ideas.

White says that history is writing a certain kind of way, not writing about a certain kind of thing or using evidence according to a certain kind of method. In his book *Figural Realism: Studies in the Mimesis Effect*, he writes about the "events, persons, structures and processes of the past" that "it is only insofar as they are past or are effectively so treated that such entities can be studied historically; but it is not their pastness that makes them historical. They become historical only in the extent to which they are represented as subjects of a specifically historical kind of writing."[2] The takeaway here is that it opens up the possibility that history can be interpreted as a form of literature, that writing history is, say, like writing a novel—this has been the most controversial implication of White's historiographical writing.

My purpose in bringing Hayden White to your attention is to introduce game studies to this "historical kind of writing." Again, this writing is neither characterized by the object of inquiry nor by source material such as archival records or oral histories. The historical kind of writing is a narrative *interpretation* of something that happened in the past. Let me sharpen the implications by paraphrasing a line written by the late, great John Hughes and spoken by Steve Martin in *Planes, Trains, and Automobiles*, ". . . you know, when you're telling these little stories? Here's a good idea—have a POINT. It makes it SO much more interesting for the reader!" Game history also needs to have a point, and this conference has given us a chance to consider what that point might be.

So I am going to suggest a few points that game studies can make. In order to spice things up, I will lean on my current historical project: the game engine. Like most historical studies, this one is not necessarily universal in its implications, but I believe it is significant. More important for us today, the history of the game engine provides a story through which we can explore the kinds of narratives that historical game studies might deliver. Let me begin by setting the stage.

The Game Engine

Before the 1993 release of its upcoming game *DOOM*, id Software issued a news release. It promised that *DOOM* would "push back the boundaries of what was thought possible" on computers. This press release is a remarkable document. It summarized stunning innovations in technology, gameplay, distribution, and content creation. It also introduced a term, the "*DOOM* engine." This term described the technology under the hood of id's latest game software. The news release promised a new kind of "open game" and sure enough, id's game engine technology became the motor of a new computer game industry.

The "Invention of the Game Engine" was only half the story. John Carmack, the lead programmer at id, did not just create a new kind of software, as if that were not enough. He also conceived and executed a new way of organizing the components of computer games by separating execution of core functionality by the game engine from the creative assets that filled the play space and content of a specific game title. Jason Gregory in his book on game engines writes, "*DOOM* was architected with a relatively well-defined separation between its core software components (such as the three-dimensional graphics rendering system) and the arts assets, game worlds, and rules of play that comprised the player's gaming experience."[3]

Before I circle back to historiography, let's look more closely at the chain of events that produced the game engine and the decision to package assets separately.

Google's Ngram Viewer allows us to analyze historical usage of the term "game engine" in texts available in the Google Books database. This analysis confirms id's bravado. The term first appeared in print during the year after *DOOM*'s launch. It occurs frequently during 1994 in fact. We find it in André

LaMothe's *Teach Yourself Game Programming in 21 Days* or *Tricks of the Game Programming Gurus* by LaMothe and John Ratcliff, articles in *PC Magazine* and the inaugural issue of *Game Developer*, and—no surprise—*The Official DOOM Survivor's Strategies and Secrets*, by Jonathan Mendoza. Google Books identifies citations before 1994, but these have turned out to be unrelated to game development or simply false hits. Fun fact: I bet you didn't know that the term "game engine" first appeared in Richard Burn's *The Justice of the Peace and the Parish Officer* in 1836! Well, it didn't really. The erroneous result was caused by the appearance of the word "engine" in an annotation in the right margin one line before the word "game" was printed on the left margin of the text itself. A more serious candidate for earliest published use of the term is Douglas A. Young's *Object-Oriented Programming with C++*, published in 1992, but the Engine it describes is an object in a demonstration program of a tic-tac-toe game. It is the prime mover in the game and controls the computer's intelligence, but the term is specific to this one program, and does not refer to a class of software. Interesting perhaps, but not relevant. Google's tool also suggests that the new term "game engine" kept pace with the increasingly prevalent "game software" while "game program" declined as a description of the code underlying computer games. These analytics suggest that "game engine" was a neologism of the early 1990s and support the conclusion that the invention of this game technology was a discrete historical event of the early 1990s.

Where in fact did the term "game engine" come from? The answer to this question begins with the state of PC gaming circa 1990 and its invigoration by id Software. It is fair to say that as the 1990s were beginning the PC was not where the action was. Video game consoles dominated, while the popular 8-bit and 16-bit home computers of the 1980s were on the last legs of a phenomenal run. The strengths of the PC as a platform for game design had not yet proven their worth. By the time id released *DOOM* on one of the University of Wisconsin's FTP servers in late 1993, the decade of the PC game was about to begin. *DOOM* was the technological tour-de-force that heralded a "technical revolution" in the words of id's news release, a preview issued nearly a year before the game itself. *DOOM* showcased novel game technology and design: a superior graphics engine that took advantage of 256-color VGA graphics, peer-to-peer networking for multiplayer gaming, and the mode of competitive play that id's John Romero named "death match." It established the first-person shooter and the PC became its cutting-edge platform, even though *DOOM* had been developed on NeXT machines and cross-compiled for DOS execution. Last but not least, *DOOM* introduced Carmack's separation of the game engine from "assets" accessible to players and thereby revealed a new paradigm for game design on the PC platform.

The release of *DOOM* was a significant moment for the chronicle of events, whether we are focused on game technology or the history of digital games more generally. Before we can build a historical narrative that takes this event into account, we obviously have to know more about the historical contexts for it. Let's begin with the dramatis personae. Many of you know id's story, so I will focus on elements related to the history of the game engine and its impact. In 1990, Carmack and Romero were the key figures in a team at Softdisk tasked with producing content for a bimonthly game disk magazine called *Gamer's Edge*. They had come to the realization that advanced machines running DOS were the future and they would need to "come up with new game ideas that . . . suit the hardware."[4] Following the release of MS-DOS 3.3 in 1987 and the maturing of the hardware architecture based on Intel's x86 microprocessor family, the PC was poised to become an interesting platform for next-generation computer games.

Romero and Carmack cut their teeth as young programmers on the Apple II. They had perfected their coding skills by learning and working alone. As they thought about future projects at Softdisk, they also realized that they would need to make software as a team. The plan was to divide and conquer. Carmack would focus on graphics and architecture, Romero on tools and design. Their first project was Slordax, a *Xevious*-like vertical scroller. Over the course of his career Carmack has displayed a knack for figuring out fundamental programming innovations while working on specific game projects. As Slordax was taking shape, he easily showed how to produce smooth vertical scrolling on the PC, for example. Romero was modestly impressed, but he "wasn't blown away yet." He challenged Carmack to solve a more difficult problem: NES-like *horizontal* scrolling.

Some of you know the story, I'm sure. Carmack answered the challenge, coding furiously while Romero egged him on. Then one morning Romero found a floppy diskette on his desk. With help from Tom Hall, Carmack had passed the previous night creating a frame-for-frame homage to *Super Mario Brothers* that ran on the PC with smooth horizontal scrolling. They had used some graphics created for Romero's next Dangerous Dave game to create the demo, so they called it "Dangerous Dave in Copyright Infringement," acknowledging that it was a blatant copy of Nintendo's game. *Now* Romero *was* blown away. "I was like, Totally did it. And I guess the extra great thing is that he used Mario as the example, which is what we were trying to do, right, to make a Mario game on the PC."

The story I just recounted both connects and separates Nintendo and PC games as creative spaces. This theme is worth pursuing. For now, let's stick with the development of id's game technology. Horizontal scrolling was the team's ticket out of Softdisk. They realized as Romero recalled in his oral history that, "We can totally make some unbelievable games with this stuff. We need to get out of here." The separation from Softdisk turned out to be gradual. Their new company, id Software, continued to produce games for their old company. While this was happening Romero came into contact with Scott Miller of Apogee Software, a successful shareware publisher. After showing Miller the Dangerous Dave demo, id began to produce games for him too. These overlapping commitments produced a brutal production schedule, even without taking into account independent projects and technology development. Id met this schedule, in part, by producing games in series.

The first series of Commander Keen games, "Invasion of the Vorticons" was published between late 1990 and the middle of 1991. It featured Carmack's smooth horizontal scrolling. He used other projects to make progress with 3D rendering, which seemed like a promising technology for a second Keen trilogy. Carmack and Romero began to call the shared codebase for these three games the "Keen engine." The engine then was a single piece of software that produced common functionality for multiple games. The idea of licensing such an engine as a standalone product to other companies emerged quickly. Id briefly tested the idea by offering a "summer seminar" in 1991 to potential customers for the Keen engine. They demonstrated the design of a *Pac-Man*–like game during the workshop and waited for orders to pour in. They received one . . . from Apogee. Romero recalls that, "so they get the engine, which means they get all the source code to use it." Apogee made one game with id's engine, *Biomenace*, then made their own engine after gaining access to id's code. "And then," as Romero put it, "they didn't license any more tech from us."

While the trial balloon of the licensing concept was a failure, the game engine stuck as a way of designating a reusable platform for efficiently developing several games. Romero later recalled that, "I don't remember, at that point, hearing of an engine, like you know, Ultima's engine, because I guess a lot of games were written from scratch." In other words, game programs had been put together for one game at a time. But why call this piece of game software a game *engine*? Carmack and Romero were both automobile enthusiasts and, as Romero explained, the engine "is the heart of the car, this is the heart of the game; it's the thing that powers it . . . and it kind of feels like it's the engine and all the art and stuff is the body of the car." We can now make a more precise entry in our chronicle: id Software invented the game engine around 1991 and revealed the concept no later than the *DOOM* press release in early 1993.

So that was a short excerpt from the Chronicles of the Game Engine. We know roughly where our timeline begins and can put a few events in order, such as the creation of id, the invention context, and the enunciation of the game engine as a game technology.

DOOM's game engine is a significant event in the history of game software. Jason Gregory recalls in his introduction to *Game Engine Architecture* that when he bought his first system, the Mattel Intellivision, in 1979, "the term 'game engine' did not exist." Of course, we knew this already. He also observes in the same sentence that games of that era were "considered by adults to be nothing more than toys" and these games were created individually for specific platforms. Not so today. Gregory marvels that "games are a multi-billion dollar mainstream industry rivaling Hollywood in size and popularity. And the software that drives these now-ubiquitous three-dimensional worlds—*game engines* like id Software's *Quake* and *DOOM* engines, Epic Games *Unreal Engine 3* and Valve's *Source* engine—have become fully featured reusable software development kits that can be licensed and used to build almost any game imaginable."[5] In other words, the development of engine technology traces the growth and maturation of the game industry.

I have encountered the game engine in various historical projects. In one of them, I came across a John Carmack interview from several years ago. He remarked that *DOOM* was a "really significant inflection point for things, because all of a sudden the world was vivid enough that normal people could look at a game and understand what computer gamers were excited about."[6] Contrast this remark to Gregory's observation about adults having no clue what was going on with digital games during the early home console years. *DOOM* and the game engine technology that powered it marked the beginning of the modern computer game, not only as a technical achievement, but as the springboard for a whole host of changes in perception and play: along with games like this we got networked player communities, modifiable content, fascination with the sights and sounds of games, and concerns about hyper-realistic depictions of violence and gore.

Most of us would agree with Carmack about the inflection point in computer gaming circa 1993 and its aftereffects. A specific example close to my heart is the history of machinima, and this topic represents another of my encounters with the game engine. Before there was machinima, there were *Quake* movies, which Carmack's separation of the game engine from assets made possible. Demo files were a particular kind of asset file in both *DOOM* and *Quake*. A few players figured out how to change these files and produce player-created movies that the game engine could then play back. Id had not anticipated movie-making, but enabled it as an affordance of their game technology.

Recently, thinking about how game engines interact with assets as input has even informed my work on game software preservation in the second Preserving Virtual Worlds project. The separation of game engine from assets in *DOOM* suggested a possible solution to the problem of auditing software in digital repositories. It turns out that the version-specific ability of a game engine to play back demo files constitutes what media preservationists call a significant property of computer games. Put another way, we can check the integrity of game software stored in a digital repository by seeing if it can run a historical data file such as a *DOOM* demo. The game asset provides a key for verifying the game software, and the game engine pays back the service by playing back historical documentation such as replays. Our case study for this work was *DOOM*.

These touch-points lead me back to *DOOM* as a historical moment. Gregory's memory of video games before the modern game industry and technology is probably a fairly typical one. So let's turn now from the chronicle to the historical narrative. We have seen that the game engine concept was in place in 1991, two years *before DOOM* was released. The Keen Engine served efficiency of serial game development and raised the possibility of licensing to external parties in order to create new games. During the two years between 1991 and 1993, Carmack worked feverishly on new technology and games and turned his attention especially to the problem of 3D rendering. *DOOM* showed what he had accomplished during that period.

Lev Manovich has described the impact of *DOOM* as nothing less than creating "a new cultural economy" for software production. He had in mind the full implications of its software model of separated engine and assets. Manovich described this new economy as, "Producers define the basic structure of an object and release a few examples, as well as tools to allow consumers to build their own versions, to be shared with other consumers."[7] Carmack and Romero opened up access to their games in a fashion that might be construed in other media as giving up creative control. And yet, id's move was not a concession; they embraced its implications as the company focused increasingly on technology as a foundation for game development. They encouraged the player community and worked with third-party developers who modified their games or made new ones on top of id's engine. Carmack's support for an open software model can be explained in part by his background as a teenage hacker. Now he had created a robust model of content creation that would allow players to do what he had wanted to do: change games and share the changes with other players. Carmack's attention shifted as a result to improving the technology, rather than working on game design.

Users—players that is—played a significant role in shifting id's focus to the game engine as a content creation platform. When id released *Wolfenstein 3D* in 1992, the efforts of dedicated players to hack the game and insert characters like Barney the Dinosaur and Beavis and Butthead made an impression on Carmack and Romero. Michael Adcock's "Barneystein 3D" patch and others like it documented the eagerness of players to change content, even though the game did not offer an easy way to do this. Romero has recalled in his oral history that *Wolfenstein 3D* demonstrated that players wanted "to modify our game really bad." He and Carmack concluded about their next game that, "We should make this game totally open, you know, for people to make it really easy to modify because that would be cool." Carmack's solution then was a response to a perceived demand. Assets such as maps, textures, and demo movies could now be altered by players without having to hack the engine. Stability of the engine was important for distribution and sharing of new content. Moreover, access to assets encouraged the development of software tools to make new content, which then generated more new modifications, maps and design ideas, and so on. Id's corporate history boasts to this day that after *DOOM* was released, "The mod community took off, giving the game seemingly eternal life on the Internet."[8]

Manovich points to the implications of the changes introduced in *DOOM* in terms of support for content modifications and reuse of the engine, but he does not say very much about the motivations. David Kushner, who wrote a history of id Software, says that Carmack's separation of engine and assets resulted in a "radical idea not only for games, but really for any type of media. . . . It was an ideological gesture that empowered players and, in turn, loosened the grip of game makers."[9] At the same time, as Carmack and Romero had predicted, it was also good business. Eric Raymond took up this theme in *The Magic Cauldron* by bolstering his case for the business value of open-source software by analyzing id's decision to release the *DOOM* source code. The media artist and museum curator Randall Packer also noticed that a cultural shift had occurred only a few years after the release of *DOOM* and *Quake*. He observed that games had become the exception among interactive arts and entertainment media because game developers did not view the "letting go of authorial control" as a problem. He meant id, of course.

My narrative suggests that the logic of game engines and open design meant that id would become a different kind of game company. The PC, as a relatively open and capable platform during the 1990s, was also conducive to this logic. In line with Carmack's focus, id's game was now technology. A few years after *DOOM* he reflected that technology created the company's value and that there was not much added by game design over what "a few reasonably clued-in players would produce at this point."[10] Id's technology was expressed primarily through the game engine while the provisions for modifying assets opened up possibilities for player-generated content. That was the formula. This combination of strategic innovations was strengthened and emphasized in id's next blockbuster: *Quake*, released in 1996. It proved to be an especially fertile environment for player creativity. Expected areas of engagement such as mods were matched by unexpected activities such as *Quake* movies. During the five years from *Keen* to *Quake*, id Software had worked out a game engine concept capable of prodding the rapid pace of innovation in PC gaming during the 1990s.

Narrative

It seems that Jason Gregory's casual observation about the connection between game engines and the history of game development can be expanded into a meaningful historical narrative. As we learned from Hayden White, historical writing incorporates arguments, ideologies, and emplotments to build such a narrative. Assembling the story elements I have discussed today provides material for building a history of the game engine on an argument that is familiar in technology studies: it is called technological determinism. A historical narrative does more than answer questions, it compels us to ask more of them. So here is one: Does the history of the Game Engine support the case for technological determinism? I will conclude this brief history of the game engine with a few thoughts in response to this question.

First, let's refine the determinist argument a bit. In *The Nature of Technology*, Brian Arthur argues that modern technologies are inherently modular. This modularity is an important theme for his theory of technological change. Arthur argues that, "a novel technology emerges always from a cumulation of previous components and functionalities already in place."[11] This cumulation is not achieved by replacing one technology with another, but by reorganizing and improving previous technologies. Usually, this process is driven by a novel need or phenomenon. Arthur suggests that this kind of change depends on the modular nature of contemporary technologies, such as the systems and components of a computer, automobile, or jetliner. He observes that, "We are shifting from technologies that produced fixed physical outputs to technologies whose main character is that they can be combined and configured endlessly for fresh purposes." Arthur adds elsewhere that "the modules of technology over time become standardized units."[12] This historical process is evolutionary rather than revolutionary in character because coherent modules endure as components of other technologies in which they are embedded.

Arthur's way of thinking about technology resonates with the game engine concept, which relies on modularity for a "swift reconfiguration" of game software to "suit different purposes."[13] The game engine is modular not just in the general way proposed by Arthur. Modularity was an aspect of the architecture of computer games such as *DOOM* and *Quake*. Like a jet engine, the game engine is a component of a finished product, the computer game, which is completed by adding other assets (modules, if you will) created by other digital technologies. We should perhaps not be at all surprised that a quintessentially "modern" technology was the foundation for the modern game industry, Q.E.D. Recall Carmack's opinion that id derived its value as a company from *technology*, not game design. A historical narrative consistent with these points would argue that the success of *DOOM* and *Quake* encouraged developers and players to exploit id's new game technology and modular content model. Technological change fueled the dynamic growth of the PC game industry through the 1990s. Ripples became waves as innovations in "middle" technologies such as graphics and sound cards in hardware or AI and physics engines in software kept pace with the expanding game industry, and the intense development of these component technologies sparked secondary innovations as well. For example, it has been argued that the game industry created demand for 3-d graphics hardware, which in turn provided a favorable environment for improvements in computer-aided design.

So, the determinist shoe fits, right?

Well, it was fun to try it on, but before we buy that shoe, let's keep shopping. A reading of communications historian Susan Douglas provides different plot ideas. In her Da Vinci Medal address in 2009, Douglas relied on a "poetic structure" or trope specifically identified in Hayden White's historiographic writing, although she did not refer to him. This structure is irony. White presented historical tropes as a way of categorizing how historians relate language and thought, that is, how they use narrative to relate ideas about history. Influenced by literary scholars such as Northrop Frye, White argued that historical narratives are characterized by a "tropological structure." This means that a particular trope prevails in every piece of historical writing. Tropes also correspond to emplotments. White identified four tropological structures: metaphor, metonymy, synecdoche, and irony, which are tied respectively to romance, tragedy, comedy, and satire. According to White, this breakdown provides us "with a much more refined classification of the kinds of historical discourse than that based on the conventional distinction between linear and cyclical representations of historical processes."[14] The payoff for us is recognizing that there are deep formal similarities between literature and historical writing.

So what? Shouldn't we only be concerned with history's truth value? Does it matter whether a historian is giving us a tragedy or a comedy? While there has been a huge amount of debate about what some have perceived as White's reduction of history to literary fiction, I would characterize White's objective as consciousness-raising. The historian is as dependent as the writer of fiction on language and the structure of narrative form. White says about irony, for example that, "A mode of representation such as irony is a content of the discourse in which it is used, not merely a form—as anyone who has had ironic remarks directed at them will know all too well. When I speak to or about someone or something in an ironic mode, I am doing more than clothing my observations in a witty style. I am saying about them something more and other than I seem to be asserting on the literal level of my speech.

So it is with historical discourse cast in a predominantly ironic mode . . ."[15] Thus irony would be used by a historian who stresses a contrast or disjuncture in historical elements (people, events, movements, etc.) that were thought to be similar or closely affiliated.

Douglas tells us about the "irony of technology." Like many historians weaned on social construction of technology, Douglas rejected long ago the technological determinism a graduate student trained after the 1970s would recognize, say, in the writings of Marshall McLuhan. When she later rethought the relationship between technology and social context, she found "a new attention to what are now called technological affordances" that tell us what "certain technologies privilege and permit that others don't." We might say that as often happens as polarizing debates wind down, the middle position began to look attractive. Douglas concluded that a reasonable take on communications media would be that technologies define a suite of affordances, yet these affordances do not determine actual historical use. The process of "producing often ironic and unintended consequences" is indeed as the social constructionists would have it, a process of negotiation. While the affordances are not to be denied, their impact is critically shaped by business imperatives, use, legal constraints, and other messy historical complications.[16]

Douglas's contemplation of technology as ironic can be applied to the game engine. The id developers originally considered the game engine as a way to solve the problem of producing computer games more efficiently. Build one engine to produce a series of three *Commander Keen* games. This is efficiency on the order of Peter Jackson's filming of three *Lord of the Rings* films more or less at the same time. Id's 1991 summer seminar revealed a more general idea, but still an idea associated with efficiency. The licensed game engine could become a platform upon which diverse games would be constructed. In either case the engine developer provides core functionality, that is, a set of affordances, and game developers decide how to deliver new game mechanics and assets using the engine. Games built this way are the product of a modular design concept built around the game engine, but the use of this technology is shaped by constraints ranging from business practices and commitments to user needs and objectives. The history of television is an often discussed problem that exhibits technological irony. McLuhan, for example, had high hopes for TV and the "global embrace" that electronic media technologies would deliver. Instead, we have reality series and the fragmentation of news and ideas, as a variety of nontechnical factors have emphasized different affordances than those seen by McLuhan. In the case of the game engine, we might ask if expectations about open software and player-generated content made possible by game technology have been realized as they were imagined. Ideas about game production, design, business models, and player creativity have played upon the expectations made possible by the affordances of the game engine; these ideas complicate the deterministic model while raising new historical questions.

More historical work is needed to produce a better understanding not only of the rise of game engine technology, but also of the business and creative decisions that conditioned the use of technology to make new games. I wish to suggest that a binary question—technologically determined or not?—often blurs as ironic and other narrative structures add nuance and messiness to our histories.

Conclusion: A Future for Game History

Musing about the history of games generally and the history of the game engine in particular has led me to call upon my home field, the history of technology, several times already. In concluding, I am going to this well one last time.

In an essay called "The History of Computing in the History of Technology," published in 1988, Michael Mahoney reflected on the former subject's puzzling lack of impact on the latter until then. The importance of computing and informatics was widely recognized and there was a natural relationship between these subjects. So why had the history of computing lagged as a subdiscipline of the history of technology? His analysis of writing in the history of computing showed that the field had been dominated by three kinds of contributions. See if you recognize them in game history. The first was "insider" history. "While it is first-hand and expert, it is also guided by the current state of knowledge and bound by the professional culture," Mahoney argued.[17] The second group of contributors consisted of journalists, whose work he described as long on immediacy but short on perspective. The third segment of the literature consisted of "social impact statements" and related writings, composed in the service of futurism or policy or studies of social impact; he considered these studies to be more polemical than historical. Finally, Mahoney identified "a small body of professionally historical work."[18]

Mahoney next considered what he thought of as the big questions in the history of technology. He suggested that as the history of computing matured as a discipline, it would be able to contribute answers to these questions, just as these questions would guide and inform work in history of computing. Here are a few examples of the kinds of questions he had in mind: "How has the relationship between science and technology changed and developed over time and place?" "Is technology the creator of demand or a response to it?" or "How do new technologies establish themselves in society, and how does society adapt to them?"[19] Big questions, in other words.

So where am I going with this? I would like to suggest that the history of games is in a similar situation to the history of computing, only twenty-five years later and with a less clear notion of its natural parent discipline. I draw two points from this comparison, and I would like to add one more from the example of the Game Engine and its history. First, the history of games is in its infancy. As with the history of computing in 1988, most of game history has been written to answer questions that arise from firsthand familiarity, journalism, or implications for policy or business affairs. None of this is unwelcome, of course. However, let me remind you what Mahoney said about the history of computing written by computer scientists. He wrote that, "while it is first-hand and expert, it is also guided by the current state of knowledge and bound by the professional culture."[20] He meant that decisions or results that such an author might take as a given, an outside viewer such as a historian might consider as a choice.

Second, the history of games will develop in more interesting ways if it finds connections to big questions. From my own selfish perspective, we could certainly do worse than reflect on some of the questions that have shaped work in the history of technology. Paraphrasing Mahoney, for our purposes we could ask, "How has game design evolved, both as an intellectual activity and as a social role?" or "Are games following a society's momentum or do they redirect it by external impulse?" "What are the patterns by which games are transferred from one culture to another?" And so on. Only by asking such questions will the history of games find connections to other areas of historical research and cultural studies, which in turn will invigorate our own work with fresh perspectives and interpretive frameworks.

Finally, my last point derives from the history of the game engine. Consider the contrasting view of this topic that we acquire by backing away from technological determinism and considering Douglas's notion of technological irony. This contrast encourages us to look more closely at the messy interplay of intentions, users, and the marketplace. My point is not that either of these points of view is the one and true answer, but rather that we are still so far from creating a critical mass of divergent ideas and perspectives, our last three days at this conference notwithstanding. One last quotation from Mahoney about the history of computing: "What is truly revolutionary about the computer will become clear only when computing acquires a proper history, one that ties it to other technologies and thus uncovers the precedents that make its innovations significant."[21] Asking big questions of the history of games and answering those questions in creative and diverse ways strikes me as the right strategy for reaching a similar goal: not creating a separate enterprise called the history of games so much as finding connections with related fields surrounding this subject, from history of technology to intellectual history. In other words, it will not be enough to create more data points about the history of games. We will also need good questions and big ideas to help us make sense of that history and, ultimately, to have a point.

NOTES

1. John Steele Gordon. "Review of Seth Shulman, *The Telephone Gambit*," *Wall Street Journal*, January 16, 2008, D10.

2. Hayden White, *Figural Realism: Studies in the Mimesis Effect* (Baltimore: Johns Hopkins University Press, 2000), 2.

3. Jason Gregory, *Game Engine Architecture* (Wellesley, MA: A. K. Peters, 2009), 11.

4. Quotations below from unpublished oral history interviews with John Romero were conducted by the author at the Computer History Museum. Quotations without specific citations are from these interviews. Publication is forthcoming.

5. Gregory, *Game Engine*, 3.

6. John Carmack, "*DOOM* 3: The Legacy." Transcript of video retrieved June 2004 from the New *DOOM* Web site, http://www.newdoom.com/interviews.php?i=d3video.

7. Lev Manovich, *The Language of New Media* (Cambridge, MA: MIT Press, 2001), 245.

8. id Software, "Id Software Backgrounder," retrieved February 2004 from id Software Web site, http://www.idsoftware.com/business/home/history/.

9. David Kushner, *Masters of DOOM: How Two Guys Created an Empire and Transformed Pop Culture* (New York: Random House, 2003), 71–72.

10. John Carmack, "Re: Definitions of terms," Discussion post to Slashdot, January 2, 2002, Retrieved January 2004, http://slashdot.org/comments.pl?sid=25551&cid=2775698.

11. Brian Arthur, *The Nature of Technology: What It Is and How It Evolves* (New York: Free Press, 2009), 124.

12. Arthur, *The Nature of Technology*, 25, 38.

13. Arthur, *The Nature of Technology*, 36.

14. White, *Figural Realism*, 11.

15. White, *Figural Realism*, 12.

16. Susan J. Douglas, "Some Thoughts on the Question 'How Do New Things Happen?'" *Technology and Culture* 51 (April 2010): 300.

17. Michael Mahoney, "The History of Computing in the History of Technology," *IEEE Annals of the History of Computing* 10, no. 2 (1998): 114.

18. Mahoney, "The History of Computing in the History of Technology," 115.

19. Mahoney, "The History of Computing in the History of Technology," 115–116.

20. Mahoney, "The History of Computing in the History of Technology," 114.

21. Mahoney, "The History of Computing in the History of Technology," 123.

10

Putting a Stamp on Games

Wargames, Players, and PBM

Playing games by mail (PBM) may strike some as an all too obviously *historical* topic. This is mail without the "e-," replete with cursive writing, stamps, and stacks of envelopes. What could possibly be more dated, dead, and gone? I hope to make the case for the history of PBM (play-by-mail) in a different sense. This history of PBM tells us something useful even today about player communities and alternative modes of play, reminding us that organizing play involves creative work. I will focus on the early history of PBM systems for playing a group of board games usually called wargames, and sometimes known as conflict or historical simulations. Wargames played a central role in the nascent board game industry of the 1960s. Wargame designers enunciated key concepts of game design, including "game design" and "game development," during a golden period that stretched from the early 1960s through the late 1970s. Wargames and wargame designers also influenced other game forms, such as role-playing games (beginning with *Dungeons and Dragons*) and other forms of simulation and strategy games.

Let me begin by briefly summarizing a few points about what was new in wargame design and the business practices constructed around it. Charles S. Roberts designed the first commercial wargame in the US, *Tactics*, in 1952. After selling this game via mail order, he was convinced he might be able to build a successful company around game design and publication. He founded the Avalon Game Company, later renamed the Avalon Hill Game Company (henceforth: Avalon Hill), in 1958. Roberts exemplified what we think of

Originally published as "Putting a Stamp on Games: Wargames, Players, and PBM," *Journey Planet* no. 26 (December 2015): 14–22.

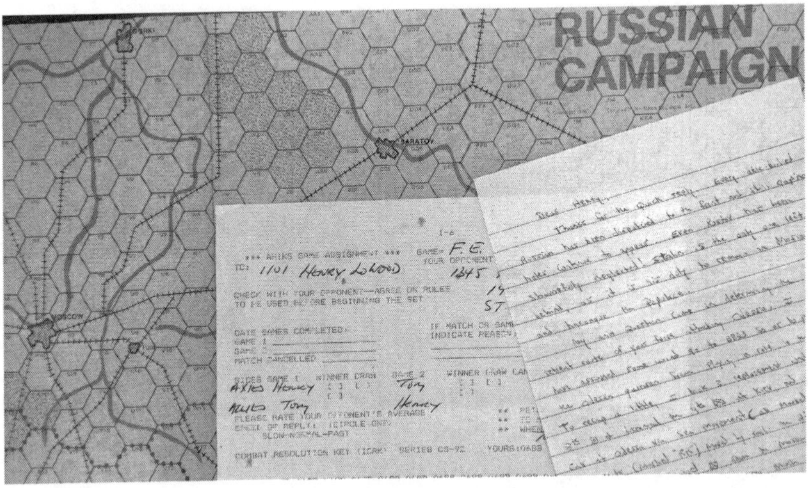

Figure 10.1. Correspondence, forms, and maps from *The Russian Campaign*

today as a game designer, and Avalon Hill was the first company to make a serious go of marketing wargames to adults.

Roberts designed ten games for Avalon Hill between 1958 and the end of 1963. These games included the first historical wargame, *Gettysburg* (1958), followed by other titles that established the concept of the game as a study of historical battles or campaigns. He also produced the design vocabulary, rules conventions, and physical component systems that defined wargames even after Roberts retired from the scene. Three aspects of wargames assembled and promoted by Avalon Hill through the 1960s were especially important. First is the assemblage: "When a game is designed, generally the first things that go into it are the map, the values on the playing pieces (combat strength and movement allowance), the Terrain Effects Chart and the Combat Results Table."[1] Second, this assemblage determined procedures of play. Players read the rules to learn about the units represented by the playing pieces ("counters"), how to move them on the map during a turn using the Terrain Effects Chart, and how to resolve combat between opposing units (and other aspects of the games as simulations) by calculating battle odds, rolling dice, and consulting the Combat Results Table (CRT). Generally, these games were turn-based and designed to be played by two players: I move and we resolve combats, you move and we resolve combats, and so on.

Figure 10.2. PBM archive 1965. Stanford University

Third, Avalon Hill's marketing efforts concentrated on the creation and development of a community of hobbyist wargamers who would buy and play these games. These efforts accelerated after Roberts turned Avalon Hill over to its creditors and left the company in January 1964. The community-building strategy was led by Thomas N. "Tom" Shaw, whom Roberts had brought to Avalon Hill in 1960. Rex Martin, who worked with him at Avalon Hill and later wrote a PhD dissertation on the wargame industry, described Shaw as the "godfather" of the wargaming "subculture."[2] Shaw presided over the company's strategy for expanding the hobby. The communication platform for executing this strategy would be *The Avalon Hill General* (henceforth *The General*), a magazine that premiered in May 1964, just a few months after Roberts's departure.

PBM turns our attention to game design in a sense not often encountered in game studies, particularly those devoted to digital games. That sense has less to do with designing games than providing the means to play them. Mail-based methods for playing with remotely located opponents extended Avalon Hill's games of the 1960s; they did not replace these games. Later PBM games were more frequently constructed from the ground up with mail-based communication as a core component. By contrast, playing Avalon Hill wargames by mail opened up a format for competitive play that had not been anticipated by game designers.

So let's start with a player who has just bought a few of these newfangled Avalon Hill wargames in the mid-1960s, probably by mail-order. He or she (mostly he) has probably browsed through the rules and components. What next? Play the game, of course. Yet, many potential players in this new, niche hobby did not know anyone nearby who was interested in this kind of game. Solitaire was an option, of course, and it was frequently the method chosen. Another option—one he might have read about in *The General*—was to play against remote opponents via the postal system.

The *General* launch signaled that by 1964 Avalon Hill considered it good business to nurture the community of players devoted to its games. They used the bimonthly publication not just to market products and excite readers about future projects, but also to highlight the existence of other players, wherever they might be. For example, regional correspondents reported on events and club activities, and the names and addresses of subscribers could be found on the back page of the first few issues. "Opponents Wanted" ads appeared right from the beginning, although the first issue only included one such listing. The hobby was new, and its players were spread out across towns, cities, and army bases around the country (and the world); identifying remote *General* subscribers would have a role to play in pulling the player community together. Certainly, PBM systems that opened up a new format of competitive play also encouraged players who did not share the same kitchen table—or even zip code—to get to know each other. And vice versa, players had to find each other, in order to play by mail. More specifically, remote play required, or at least encouraged, the development of systems for matchmaking, evaluating the trustworthiness of other players, ranking players, and the like. In other words, PBM was not just a system for sharing and resolving game moves, it was also part of a system for building a community of wargamers.

Before all of that could happen, however, players needed to figure out how to play by mail.

Wargames of the 1960s and 1970s were generally turn-based affairs, and movement followed by combat resolution were the essential activities of a turn. The heart of Roberts's game procedures was the interaction of rules with charts, tables, and map overlays. These procedures governed movement and combat. Indeed, two of Roberts's major contributions to game design were the key elements of the movement and combat mechanics, respectively: the hexagonal grid map overlay that regulated the movement and combat of

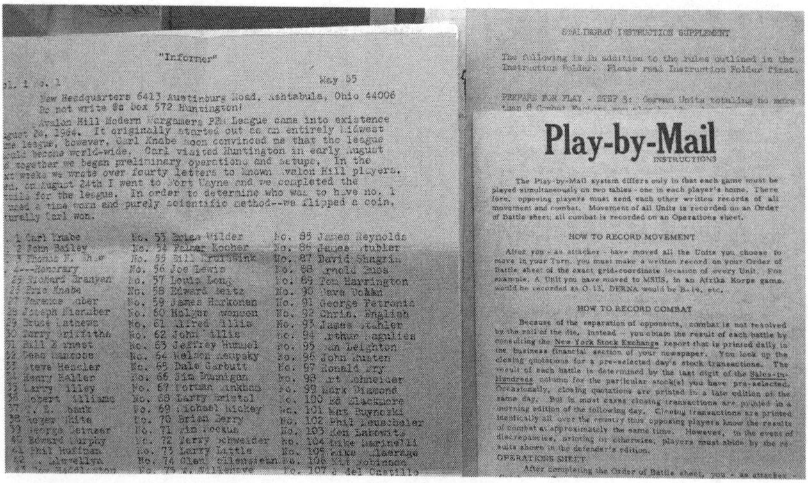

Figure 10.3. PBM archive 1965. Stanford University

units on the gameboard and the Combat Results Table (CRT). The CRT produced simulation out of the rules of play. In the rules manual for Avalon Hill's *Tactics II*, Roberts advised players that "an examination of the Combat Results Table is imperative. As this game attempts to be as realistic as possible, the Combat Results Table reflects the fact that an attacker must have a strength advantage in order to be reasonably sure of success . . . Please examine the Table very carefully."[3] The CRT became a problem for PBM, because resolving battles against this table depended upon exact specification of unit locations, mutually agreed-upon odds calculations, a dice roll, looking up results, and applying the results to counters on the map. Conducting this process through player actions separated by time and space required a system for accurately recording the positions of units, producing random numbers, and communicating results. The PBM problem boiled down to two essential needs: a system for record-keeping and a fair process for generating die rolls.

Let's focus on die rolls. They raised trust issues. Playing across the table from each other, two players easily agreed on the result, the occasional "cocked" die notwithstanding. PBM upended this stability, and that was a problem. Die rolls also reinforced the questionable simulation value of early historical games such as *Gettysburg*, *Afrika Korps*, and *Stalingrad*. The simulation value of the die roll, or more precisely the randomness that dice represent, was to bring the element of uncertainty to combat results. Rex Martin notes that while various aspects of historical battles were simulated with the aid of die rolls—weather, supplies, etc.—combat resolution was the point "when players seem most conscious of the contingency of events." He explains the preference for die rolls as a favorable confluence of publishers' and players' benefits: "Although several tools exist for determining uncertain outcomes, the use of one or more dice is the method of choice in most wargames, inexpensive for game publishers and mystical for game players."[4]

How does one produce a series of D6 (six-sided die) rolls openly and fairly by postal mail? Game designers, clubs, and players answered this question by providing a slew of PBM die rolls. The various methods were reduced to whether or not a game needed to be modified in order to use them. The first method was probably no method at all: rely on your opponent to roll real dice and report the results. This method was called the Honors System. One blogger has attributed the system to Tom Shaw himself, adding that it was a "hare-brained idea," because "the vast majority of gamers" are "an intensely suspicious lot." He concluded that, therefore, the system was almost universally rejected.[5] The Honors System failed to address the issue of trust. Yet, some players continued to use it.

In 1964, Avalon Hill issued its first PBM kit in the form of a four-page booklet called "Play-by-Mail Instruction." It provided instructions about how to keep "written records of all movement and combat" for seven Avalon Hill game titles: *D-Day*, *Stalingrad*, *Waterloo*, *Afrika Korps*, *Tactics II*, *Gettysburg*, and *Battle of the Bulge*.[6] Two aspects of the system of record-keeping became fixtures of PBM. First, the *Instruction* told players how to identify individual hexagons on the map grid by imposing a strict row-and-column system of annotation according to a specified orientation of the map. It would be several years before Avalon Hill's primary competitor, SPI, introduced the practice of numbering individual hexagons on a printed game map; with that enhancement, the issue of locating units was settled. Second, Avalon Hill printed "order of battle" sheets for recording unit (counter) map placement on a turn-by-turn basis, as well as listing the combats to be resolved at the end of the movement phase of a turn. Having acquired or made forms for recording moves and combat, opponents would send these paper records back and forth by mail, essentially a turn-by-turn archive for a game in progress. So far, so good.

But what about those die rolls? The first issue of *The General*, published in May 1964, included a contest—the magazine's first—that challenged en-

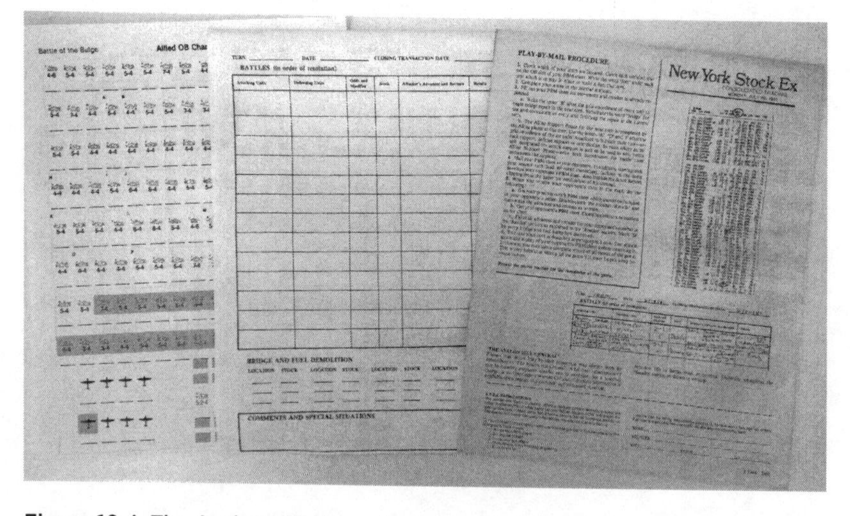

Figure 10.4. The Avalon Hill Game Company PBM forms

trants to provide the best solution for a series of combat situations in the game *Afrika Korps*. The contest required that "combat will be resolved exactly as in the *Afrika Korps* Play-by-Mail kit."[7] This may have been the first public announcement of Avalon Hill's own system, other than its inclusion in copies of that game. It specified two features of the PBM kit. First, "we will obtain the result of the combat by consulting the New York Stock Exchange report for closing transactions of Monday, June 1, 1964." Since contest entries had to be postmarked on or before the 30th of May, these stock results would be unknown to players submitting entry forms. The "die rolls" would be obtained from the "last digit of the Sales-in-Hundreds column for each stock you have on the Operations Sheet." Because the sales figures were unknown when the entries were submitted, these digits functioned as random numbers. The second feature of the PBM method altered the game. Since the last digit of the sales figures could be any number from 0 to 9, the method simulated the roll of a 10-sided die, not the 6-sided die that determined results on the standard CRT that shipped with the game. So Avalon Hill changed the CRT along with other chance tables (such as weather tables, for example. Players seem to have been troubled by different results resulting from different tables, leading to a "modest controversy."[8] As an example, consider the results for an attack at 3:1 odds. Rolling dice against the standard D6 table gave a 1/3 chance each for the "defender eliminated," "exchange," and "defender back two

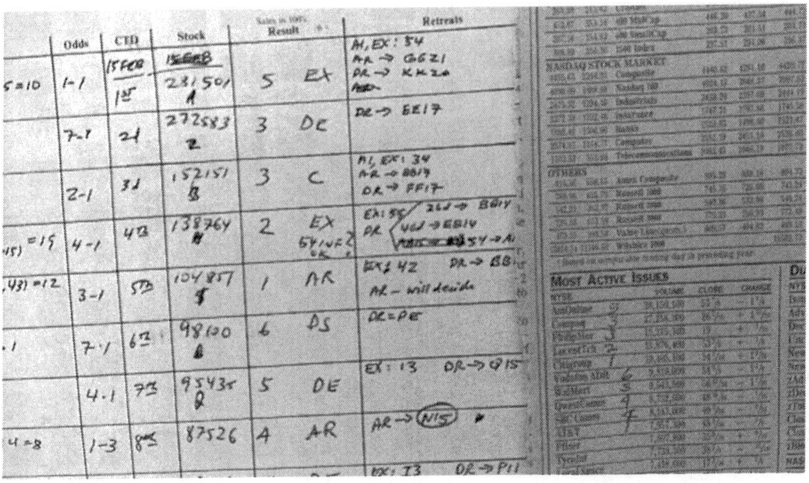

Figure 10.5. Combat resolution stock sales

hexagons" results. Against the PBM D10 table using stock sales the respective chances were 4/10, 3/10, and 3/10. As in this example, players concluded that the revised CRT for postal play produced a game that was more favorable to an attacking side than the original design. Many players were not satisfied by a PBM system that was not fully compatible with the original game.

Players proposed alternatives to the Honors System and modified CRTs regularly in magazines and newsletters. The eventually dominant methods for PBM incorporated a simple insight: postal play should be based on six-sided die rolls in a fashion that mitigated the trust issue. PBM could then supplement game systems without the need for an approved change such as Avalon Hill's revised CRT. Players could proceed with the standard, published version of the game they had (presumably) already purchased. Furthermore, eliminated authorized game changes created an opening for supplementary systems that were not created by Avalon Hill. Independent PBM clubs recognized this opportunity and began to issue alternative systems by the mid-1960s. Without venturing too far into the weeds of specific details, the summary version is that two systems were dominant. Avalon Hill distributed revised PBM kits after 1974 or thereabouts that endorsed a system still based on stock sales, but without requiring a special CRT. This involved stipulating a (future) "closing-transaction date" or CTD as before, then dividing the sales figures by six and using the remainder (0 = 6) as the D6 result. The other popular system was provided by a PBM organization founded in 1966 called the Avalon Hill Intercontinental Kriegspiel Society, better known as AHIKS. Its system was called the ICRK (pronounced "irk"), which stood for "Individual Combat Resolution Key." It was a printed form (handmade at first, computer generated after 1975) with hundreds of randomly distributed die roll results, such as the numbers 1 through 6 for a six-sided die. These numbers were arranged in a table, with columns and rows labeled by letters and numbers, respectively. The forms were distributed by a neutral third-party, called the Match Services Officer (MSO), during the matchmaking process. The MSO was responsible for matchmaking, distributing forms for recording of movement and combat, and generating ICRK forms for every match.

The existence clubs like AHIKS and their active attention to PBM hardly constituted a challenge to Avalon Hill; rather, it was a direct outcome of the company's community-building strategy. *The General* published profiles of these new clubs. The clubs, in turn, generally preferred Avalon Hill's games as the focus of their efforts to promote tournaments, PBM, and other modes of competitive play. The founding of AHIKS, for example, was reported in the September 1966 issue of *The General*, with a short article titled "A Message from the AHIKS." The anonymous author or authors asserted the uniqueness of the society, for reasons that included, "Banished are the piles of collected stock market clippings, gone are questions of their authenticity, forgotten is the wait for 'that certain date' to resolve a battle, stored away is the die-throwing cup."[9] In any case, the ICRK system became the primary service that AHIKS offered to members, along with a newsletter called *The Kommandeur* that it published in various print and electronic formats for nearly fifty years. The Society also offered matchmaking and ranking services and ICRK. Its format for PBM—supplemented by email and pseudo-random number generators—continues to this day.

The 1966 "Message from the AHIKS" also emphasized sociability among adult players spread around an international network of members. Sociability, of course, was an essential aspect of PBM. The AHIKS Member's Guide distributed in 1981 advised members that, "you should write your opponent immediately (don't wait for him to write you: letters crossing in the mail is better than no correspondence at all) and decide on rules, edition of the game being played, scenarios, special procedures, etc. Make and keep a copy of all such agreements. Also, it is helpful if you briefly introduce yourself in a friendly way to your opponent. Now, you are ready to play."[10] Based on the evidence of letters surviving from PBM games from the 1960s through the 1990s, these brief introductions often led to long-running exchanges of letters between players, often peppered with personal news, hobby gossip, questions about rules, move corrections, and virtually anything about a game that could not be summarized by unit locations, combat results, and die rolls.

The early history of PBM exemplifies Martin's conclusion that "taken together, *The General*, play-by-mail, the clubs, conventions and newsletters made Roberts' unintentional hobby permanent and profitable."[11] Martin's notion of wargaming as a subculture differs somewhat from the usual scholarly uses of this term over the past half century or so; it is closer to current use of the term "community." In this context, PBM is a chapter in a story about community-building within a small (or perhaps "niche") culture rather than resistance to a dominant culture. For example, when Avalon Hill dropped the modified CRT, it responded to player preferences, and the popularity of the resulting system reinforced the cohesion of Avalon Hill's efforts with those of clubs and player groups.

After the publication of Avalon Hill's original PBM kit in 1964, postal play seems to have caught on quickly, judging from the number of letters and articles about PBM in *The General*, as well as the increasing number of advertisements for opponents. In a letter to the editor published in the second issue of *The General*, a reader enthused, "Let me be one of many, I'm sure, who will congratulate you on your idea of 'play-by-mail' kits(s) [*sic*]; how about making these kits for your other games?" PBM became an important part of the wargame hobby from the mid-1960s forward. Avalon Hill communicated with players through *The General*, and its robust mail-order system meant that these players could also buy its games from just about anywhere. Just as mail delivery was a platform for publication and mail order, it became a platform for play. The availability of PBM solved the practical problem of access to remote players, as long as players were willing to put up with the accounting and record-keeping moments in managing a game played by mail. As compensation, much pleasure derived from this mode of play came from social interaction mediated by correspondence. Asynchronous, distant communication by postal mail perhaps implies less rather than more direct contact between players than a game played face-to-face. Nonetheless, correspondence was a social step up from the alternative method of playing solitaire. Separation in time and space had its ways of enhancing the enjoyment of games that depended upon mail delivery. The less hurried pace was

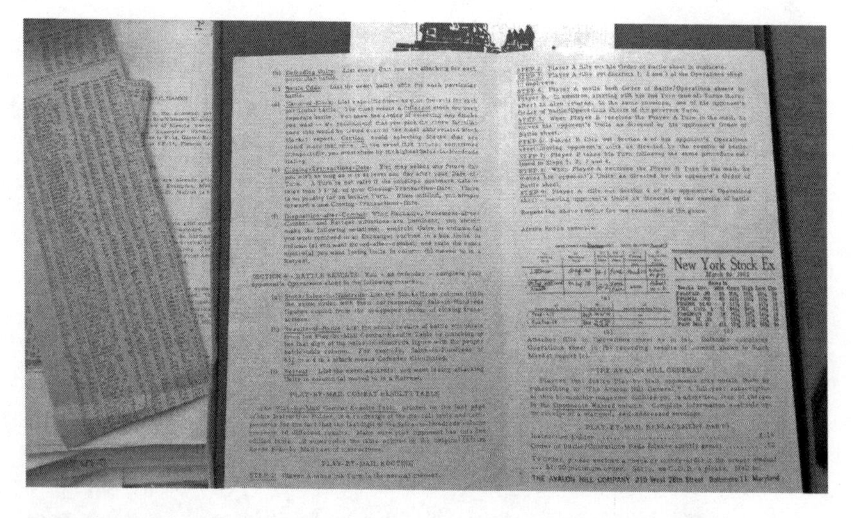

Figure 10.6. The Avalon Hill Game Company PBM instructions, 1965

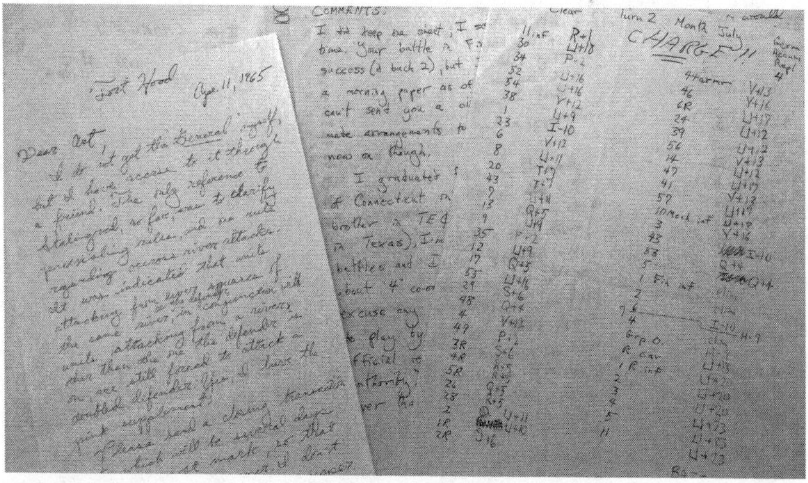

Figure 10.7. PBM archive 1965. Stanford University

accompanied by the leisure to dwell on a particular move for days rather than minutes; die rolls that meant waiting for the CTD also required a trip to the local library to find the specified stock listing; and there was that sliver of excitement that came with checking the mail in anticipation of the delivery of an opponent's move. The correspondence itself offered handwritten, typed, or eventually even electronically delivered gossip, game discussion, personal news, banter, and friendly trash talk in a register that differed from the in-the-moment performance of a live FTF game.

Let's not understand postal play by comparison to digital games as "networked multiplayer, but with stamps." The point of looking at the history of PBM is not to find an essential similarity across time and differing styles of play. Unlike later digital games, PBM systems extended playing options for manual wargames designed to be played face-to-face. Rather than being integrated into the core rules systems of early wargames, they were supplementary systems. Avalon Hill's experiment with modified CRTs notwithstanding, the need for these systems provided an opportunity for players and independent clubs to add something to the games that they wanted to play. Their solutions to the problems of PBM—from matchmaking to ranking players based on the results of games—augmented the efforts of companies like Avalon Hill and SPI. Put simply, the impact of PBM on the expansion of a new hobby and its community of players was far more significant than its impact on game design.

Postscript: PBM Archives

The various written records and forms associated with a PBM match constitute an archive of a played game. This documentation differs from other ways of understanding played games as events or performances. Unlike written summaries of the events that games represent, such as fictional stories written to summarize role-playing narratives or the "after action reports" of wargamers, these documents do not recount, fictionalize, or provide order. They are not summarizing narratives, but primary records constructed on a turn-by-turn basis. Unlike computer game replays that provide exact recordings of game performance based on uninterrupted and complete data capture, PBM records present only limited data points, more like baseball box scores than surveillance videos. Unlike these other kinds of game records, PBM correspondence is also rich in contextual information, providing insights about topics ranging from personal details and the emotions of players to the organizational matrix that supported postal play.

PBM archives, for want of a better phrase, consist primarily of paper records at least through the 1990s. The records that I have seen divide roughly equally between printed forms and handwritten or typed correspondence. Few of these records have been preserved. I have had access to two collections of PBM documentation while carrying out my own research on its history.

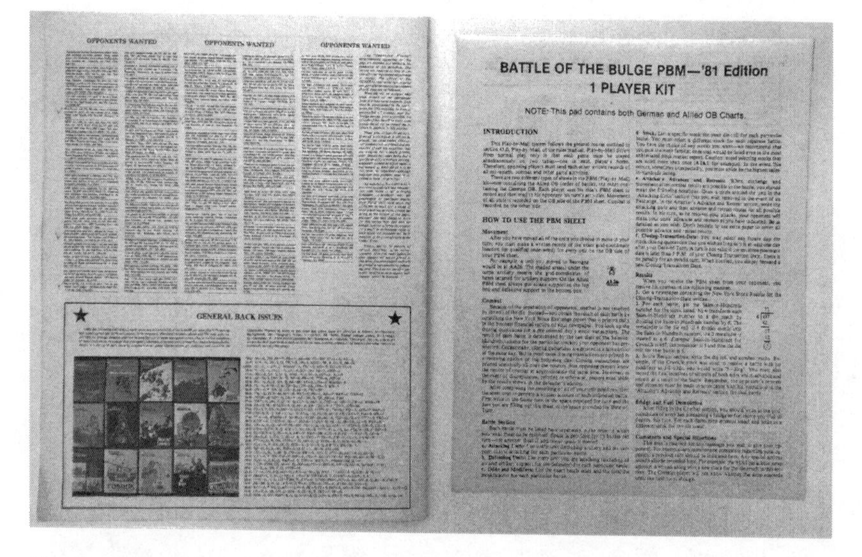

Figure 10.8. Avalon Hill PBM kit and Opponents Wanted ads, 1990s

Not only were they useful, but both collections reminded me that my engagement with this subject has been shaped by my life with games in ways other than being a historian: specifically, as a curator of library collections and as a player.

The first collection of documents was a library acquisition for the Stanford University Libraries, where I am the Curator for History of Science and Technology, as well as Film and Media Collections. Since the late 1990s, this work has included the collection of materials related to the history of digital games and simulations. More recently, I have begun to add selected historical documentation relating to the history of game design and play more generally, particularly in the areas of conflict and historical simulations, including wargames. In March 2015, I learned of a collection that had become available through an eBay auction with the intriguing title, "Rare big box Avalon Hill *Stalingrad* + PBM History & early letters from clubs!" This particular version of the historical wargame, *Stalingrad*, is indeed less common than the boxed versions one usually encounters, but that is not what piqued my interest. Rather, it was the inclusion of documentation. Auction photographs included with the auction listing showed that this documentation included handwritten letters from games played during the mid-1960s. (PBM games often lasted for months, sometimes even for a year or more.) The "item specifics" spelling out the contents of the documents included this remarkable statement:

> This is a unique and historic lot. Let me explain.
>
> For starters this is a rare big box version of Avalon Hill's classic wargame Stalingrad. . . .
>
> But what is even more unique and exciting about this lot is what it comes with inside. The entire original game is present of course, but inside the box are a stack of original early 1960s play-by-mail letters and correspondence. The letters are all post marked and dated and include turn by turn records and analysis for multiple games played with others around the country all via the postal system. In addition many of the play-by-mail games were played against one of the most historic and venerable wargaming clubs of that era, the MIT Strategic Game Society! The MIT Strategic Game Society still exists to this day although they now focus largely on Euro games and not on competitive wargaming as they did in the early 1960s. It's also interesting to learn that originally the MITSGS was called the MIT Wargaming Society. The President of the club writes in his PBM letter to the owner of this game in 1966 that just a few days before they had officially changed their name to MITSGS. We now have an absolute date for the birth of the MITSGS, it was Feb 12 1966. Neat stuff.
>
> . . .
>
> An unobservant person might consider tossing out all this extra stuff in the box but once you realize what it is, it's really, really fascinating and terribly rare. I've never seen another record of . . . PBM exploits preserved with the original game move by move from this period of early AH history. Just awesome!!"

The documentation just described give the description of the item as a "rare big box" new meaning. This big box did not just contain a game, it contained an archive! I was able to review the contents more carefully after the acquisition arrived at Stanford. The documentation comprised correspondence mostly around two or three PBM games involving the player who presumably owned the game, numerous completed PBM forms, clipped stock sales tables, the Avalon Hill PBM kit from 1964, and a few documents from clubs, such as the above-mentioned MIT Strategic Game Society and the first issue of *The Informer* (May 1965), a publication of the Avalon Hill Modern Wargamers PBM League, founded in August 1964. In short, these documents provide a relatively complete picture of the kinds of materials that a first-generation PBM wargamer would have accumulated. They include record-

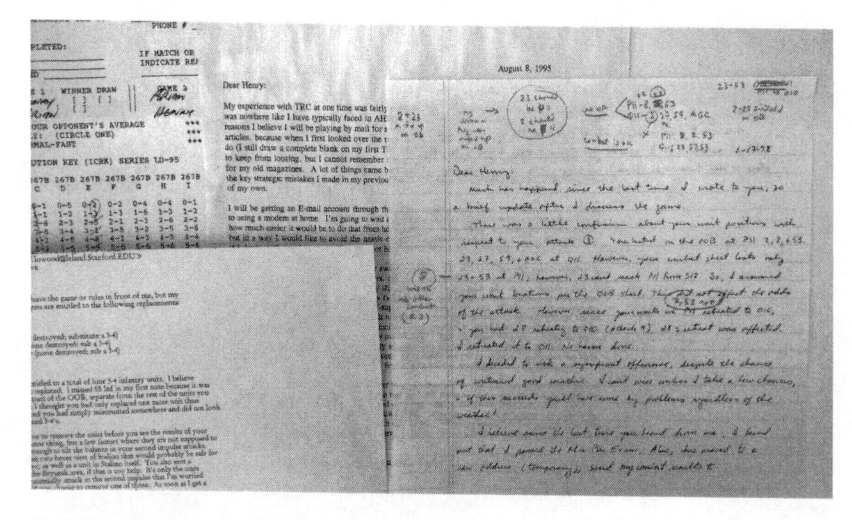

PHONE # _

PLETED:

IF MATCH OR
INDICATE RE

WINNER DRAW

GAME 2

OUR OPPONENT'S AVERAGE
(CIRCLE ONE)
MAL-FAST

UTION KEY (ICRK) SERIES LD-95

Dear Henry:

My experience with TRC at one time was fairly
was nowhere like I have typically faced in AH
reasons I believe I will be playing by mail for a
articles, because when I first looked over the
do (I still draw a complete blank on my first T
to keep from loosing, but I cannot remember
for my old magazines. A lot of things came b
the key strategic mistakes I made in my previou
of my own.

I will be getting an E-mail account through th
to using a modem at home. I'm going to wait
how much easier it would be to do that from ho

lowood@leland.Stanford.EDU>

have the game or rules in front of me, but my
you are entitled to the following replacements

August 8, 1995

Figure 10.9. PBM archive *The Russian Campaign*, 1995

keeping for turns, club communications, corrections of moves, comments on results, questions about game rules, and more. The correspondence, in particular, documents the intermingling of social correspondence and game moves, as well as implications of the pace of postal play for these interactions; for example, one handwritten letter opens, "Dear Art, I'm in the hospital, that's why I haven't gotten in touch with you sooner. It all happened very suddenly. While I had finished your move, I hadn't had a chance to get it in the mail." Other letters testified to the banter and tone of PBM among mail friends: "Art, Well nobody's perfect, and you are no exception. You made a few mistakes on your move."

The second PBM archive is my personal collection of correspondence and order-of-battle sheets from games played mostly from the late 1980s through the late 1990s. The mix of materials is similar to those in the *Stalingrad* collection. Yet, the historical context is quite different, with implications for the records, as well. For example, printouts of email correspondence are mixed in with handwritten; stock clippings with ICRK forms. The Avalon Hill order-of-battle forms date from the 1980s have a professionally produced appearance, while the ICRK sheets of that era were computer printouts.

The sameness of the mix of social interaction, gameplay recordkeeping and organizational support evidenced in the records from the 1960s and 1990s contrasts with the different impressions of the historical context for wargaming, one characterized by novelty and experimentation, and the other by mature products and routines and changing modes of communication and gameplay. The thirty years that separate these two collections of PBM records encompassed the rise and decline of wargaming as a hobby, at least in the United States. It is not at all surprising that historical records from two rather different decades for this hobby deliver different impressions about PBM. Indeed, access to such documents is a powerful reminder of the potential value of primary sources for historical game studies.

NOTES

1. James F. Dunnigan, *The Complete Wargames Handbook* (New York: Morrow, 1980), 34–35.
2. Rex A. Martin, "Cardboard Warriors: The Rise and Fall of an American Wargaming Subculture, 1958–1998" (PhD diss., Pennsylvania State University, 2001), 190.
3. *Tactics II*, 10.
4. Martin, "Cardboard Warriors," 154–156.
5. [JCB III], "The Curious Saga of the D10 "Postal" Combat Results Table," *Map and Counters* (blog), November 17, 2011.
6. [The Avalon Hill Game Company], *Play-by-Mail Instruction* (1964).
7. "Contest No. 1," *Avalon Hill General* 1, no. 1 (May 1, 1964): 5.
8. [JCB III], "The Curious Saga of the D10 'Postal' Combat Results Table," *Map and Counters* (blog), November 17, 2011.
9. AHIKS, "Message from the AHIKS," *Avalon Hill General* 3, no. 3 (September 1966): 6.
10. AHIKS, *AHIKS Member's Guide* (1981).
11. Martin, "Cardboard Warriors," 234.

11

Game Counter

> One of the least discussed and least understood aspects of conflict simulation design is, ironically, that which is most obvious: the graphics and physical systems that make a game a reality in the hands and eyes of the gamer. In fact, the better the graphic design, the more likely it will *not* be noticed.
>
> —Redmond A. Simonsen (1977)

The US Department of Defense defines a wargame as "a simulation, by whatever means, of a military operation involving two or more opposing forces, using rules, data, and procedures designed to depict an actual or assumed real life situation."[1] During the mid-1970s, commercial game designers (who abhorred the word "hobbyist") produced many of the ideas shaping the development of military simulations. In the United States, military traditions of wargaming, with historical roots reaching back to the Prussian *Kriegsspiel* of the early 19th century, had been driven down by the perceived failures of political-military gaming since the 1950s, including the Vietnam War. By contrast, sophisticated wargame designs were thriving in the commercial sector, beginning with the founding of the Avalon Hill Game Company by Charles S. Roberts in 1958. Roberts's own *Tactics* (1952) and *Tactics II* (1958) and subsequent Avalon Hill titles such as *D-Day* (1961) popularized wargame conventions such as hexagonal grids on maps to regulate movement (borrowed from RAND Corporation, one of the military think tanks),

Originally published as "Game Counter," in *The Object Reader*, ed. Fiona Candlin and Raiford Guins (Abingdon, UK: Routledge, 2009), 466–469.

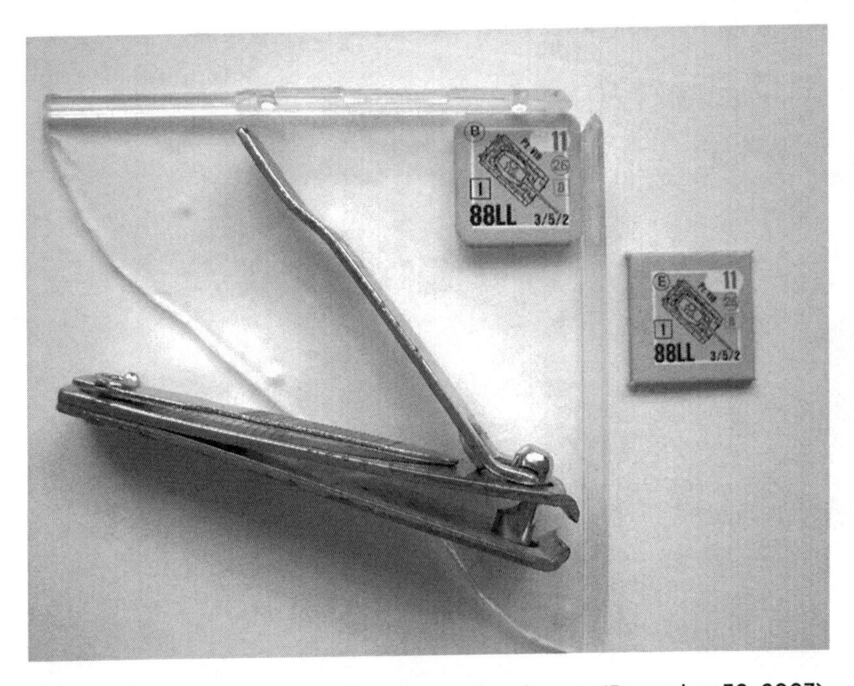

Figure 11.1. Clipping a counter. Boardgamegeek forums (December 30, 2007). Courtesy of Ernst Knauth (Warbear)

combat result tables, and the use of printed cardboard counters to represent military units and display their individual characteristics. These innovations shifted the mechanics of board game design from abstract strategy (as in chess) or chance (as in *Monopoly*) to representations of historical reality defined by complex systems of rules and data, that is, simulation. Board games became the hotspot of game design in the 1970s, and wargames were at their leading edge.

While Avalon Hill introduced the modern conception of historical wargames as simulations, further refinement and popularization of this genre was the work of Simulations Publications Inc. (SPI), led by James F. Dunnigan and a group of game designers that included Redmond Simonsen, Al Nofi, and others. While a student at Columbia University, Dunnigan designed his first game, *Jutland*, for Avalon Hill in 1966. In 1969, he became the publisher of *Strategy and Tactics* magazine, which had been founded two years earlier. Early issues analyzed data and rules in existing games, but before long *Strategy and Tactics* published game modules, add-ons, and eventually complete, original games in every issue. Just before taking over the magazine, Dunnigan had founded SPI, which took over publication of *Strategy and Tactics* as well as publishing boxed wargames. His company also became the leading publisher of boxed commercial wargames, for which Dunnigan coined the term "conflict simulations," and disseminated information on military systems and history in the magazine. By the late 1970s, Dunnigan and his collaborators were working closely with the US Army and other service arms to re-invigorate military wargaming, planting the seeds of a deeper collaboration among military and commercial designers that would eventually lead to the development of high-end computer simulations for military training.

What distinguished issues of *Strategy and Tactics* physically from other magazines was not simply that there was a new game in every issue, but the inclusion of a printed map and cardboard counters that spilled out of the magazine. Removing the map and counters naturally took precedence over the articles. Deconstruction for once preceded reading, as these objects were first eagerly removed from each new issue, followed by unfolding the map and—for those who chose to play the game—carefully "punching" (cutting away) counters from their cardboard sheets. Whether from a magazine issue or a boxed game, playing a conflict simulation began with physical construction of the game. The essential building blocks were the counters, typically half-inch square playing pieces of cardboard-and-ink. They were called "unit counters," because they represented historical or hypothetical military units or perhaps individual leaders or politicians. (A few others were merely "markers" that helped players to keep track of the game state.) For example, the rules for Avalon Hill's *Bitter Woods* (1998) tell players that "the cardboard pieces or unit counters represent individual combat units that fought during the battle" (in this case, the Battle of the Bulge).

But how do players know which units are represented or what the counter is telling them? A typical unit counter features a symbol taken from standard military schemes such NATO's "Military Symbols for Land Based Systems"; if the counter depicts an individual person or vehicle, it usually shows an image or silhouette instead. This symbol identifies the unit type (say, cavalry or infantry or king). From that point, the information becomes more abstract. Larger numbers usually placed below the symbol quantify unit capabilities, such as fighting strength and movement rate; smaller numbers usually to the side of the symbol stand for any number of additional characteristics, from morale to weapon range. Additional numbers and letters, usually at the top of the counter, reveal the historical identity of a unit, and colors or color bands indicate at a glance nationality or affiliation with other units on the board. Reading information on a typical counter depends heavily on conventions of representation that are virtually undecipherable to anyone but a wargamer. As complicated as the grid of information on this tiny cardboard square might appear to be, by the mid-1970s it had become so conventional that players immediately scanned and interpreted counters upon opening a new game, then launched into heated discussions of the game system's accuracy as a detailed simulation. It is through the understanding of these conventions that game counters link simulation and subculture.

The consumption of media by readers, viewers, and players typically leads to the evaluation of books, movies, or games in terms of content rather than packaging, to experience of using rather than system of constructing. And so it has been with board games. Redmond Simonsen, who led SPI's efforts to systematize development and production of historical simulations in the late 1960s and early 1970s, nevertheless sought frequently to unveil the production process for players. In one essay written for a book on wargame design, Simonsen described the role of graphic design in "simulation games" and "how a game is produced" as *terra incognita*. He meant that effective production and graphic design should in fact be invisible. What better way to illustrate a wargame design maxim than with a military metaphor, which Simonsen called "the signal-to-noise ratio in graphics." As Simonsen put it, "the player is an unspecialized demolitions man defusing a complex bomb and receiving instructions on how to do so via a radio. The game is the bomb, the game designer is on the other end of the radio and the artwork *is* the radio."[2] His explication of artwork and physical system design in wargames owed as much to Claude Shannon as it did to industrial product design, for Simonsen viewed good design of counters, maps, and rules as a communicative act.

Thinking of counters and other board game artifacts as code words and radios of course makes it much easier to ruminate about games as a medium, but it also raises the question of how the materiality of these games works for them. If the game design is easily "read" in terms of the rules system and components, and the act of moving a piece or reading a results table is entirely contingent upon the strategies and tactics of play, the physical components of board games such as historical simulations become purely informational; their physicality becomes in some sense "immaterial." We can easily appreciate the rush to "computerize" simulation that began in the late 1970s.

As this rush was about to begin (but without reference to it), Simonsen used the term "physical system" in his essay to explain how graphics design functions in conflict simulation board games; his idea of a system portrayed physical components as aids to the player's "work" of digesting and using information needed to play a game and encompassed rules, tables regulating movement on the map, and data presented on unit counters.

Dunnigan, the rockstar designer of SPI's board games, realized that many wargamers ignored the possibilities offered by agonistic play; they were only interested in the system itself, not the interactivity of gameplay. He pointed out often that studying the components of historical simulations was a way of quite literally reading game systems as authored accounts of historical events. Gamers who did this collected board games without ever punching the counters, let alone playing them. Dunnigan recognized, however, that "this does not mean they are not used." Such a "player" engaged the simulation in a different way, "unfolded maps, looked at counters, read rules, or maybe set up an opening move for solitaire play to discern orders of battle, the lay of the land, and the historical commander's options." These moments of single-player simulation always occurred in a player's "head with the aid of game components."[3]

Today computer simulations and video games have usurped the terms simulation and game. The dominant game cultures form around digital games. Therefore, it makes sense to contrast the material board game to its computer-based counterpart. Consider "real-time strategy" games such as the *Warcraft* series. In these games, information is interface; players master interface as the syntax of tactics. This interface mastery calls upon a physicality (reflexes, fast hand movements) absent in "physical," paper-based board games. By contrast, how one moves the playing piece of a board game is indeed immaterial; whether a player can move his cardboard counter quickly is of no consequence. But perhaps it is even more important to credit the ways in which the physical components of a board game convey modes of simulation disentangled from the expectations of experience, interactivity, and performance of digital simulations and games. Rulebooks and graphics systems can be read in order to understand a complex model of historical reality as a system rather than an interface, counters scanned on the sheet or sorted as a kind of historical manifest, counter values translated into abstract representations of historical or predicted performance, or maps carefully laid out on the table as windows through which a flow of events can be perceived and understood. Surprisingly perhaps, the material objects of paper-and-cardboard conflict simulations open up an abstract information space that calls upon the (often solitary) player to contemplate them by understanding deep issues of scale, quantification, and modeling in simulation that are often hidden from view in digital games. The board game counter teaches us that we can hold a simulated world in our hands.

NOTES

1. Joint Chiefs of Staff, *Publication 1, Department of Defense Dictionary of Military and Associated Term* (Washington, DC: Government Printing Office, 1987), 393.

2. Redmond Simonsen, "Image and System: Graphics and Physical Systems Design," in Staff of *Strategy and Tactics* magazine, *Wargame Design: The History, Production and Use of Conflict Simulation Games* (New York: Hippocrene, 1977).

3. See James F. Dunnigan, *The Complete Wargames Handbook* (New York: Morrow, 1980).

12

War Engines

Wargames as Systems from the Tabletop to the Computer

This chapter will be a lengthy but incomplete response to the question, "What can we learn from the history of wargames about games as systems?" I will focus on the history of commercial or hobbyist wargames in the United States, mostly board games produced from the early 1950s into the 1970s, but also a few computer games. Wargames—also sometimes called historical, conflict, or military simulations—generally emphasize themes rather than mechanics, and simulation over gameplay. The themes can be historical or they can be speculative; they involve military conflicts that have occurred between forces that existed, might have existed, actually do exist, or might exist in a future or alternative world. The thematic emphasis of wargames has been contrasted with more abstract forms of strategy games, such as "Eurogames."[1] One might conclude from the specific emphasis on conflict simulation that wargames have relatively lightly influenced game design generally, yet they paced the evolution of modern board game design through the 1970s, particularly in the United States. Hobbyist wargame designers such as Charles S. Roberts and James Dunnigan were responsible for an eruption of creative ideas in the United States, and their work stimulated debates about game development during the 1960s and 1970s. Along with role-playing games, wargames also played a featured role in the transition from physical to computer-based games during the 1980s, and in the increasing use of simulations by the military from the 1990s forward.[2]

Originally published as "War Engines: Wargames as Systems from the Tabletop to the Computer," in *Zones of Control*, ed. Matthew Kirschenbaum and Patrick Harrigan (Cambridge, MA: MIT Press, 2016): 83–105.

The Avalon Hill Way

Your July column mentions the regular hexagon tessellation as "so familiar to bees and users of bathrooms." Perhaps you are not aware of another common use of this tessellation, that of compartmentalizing game maps, particularly what are called wargames or military simulations.

—John E. Koontz to Martin Gardner[3]

Charles S. Roberts completed a four-year stint in the army in 1952, returning to a commission in the Maryland National Guard. While waiting for a commission in the regular army, one that never came, he pondered how to prepare for his military career while a civilian. Roberts concluded that he would "gain some more general applications for the study of the principles of war" by designing a game.[4] Most accounts of the history of commercial wargames in the United States begin with that game, *Tactics*.

Roberts's idea of a wargame thus filled a void, rather than continuing a tradition. It is perhaps surprising, but no evidence has emerged to suggest that Roberts was aware of the history of earlier wargames, such as the Prussian *Kriegsspiel*, while designing *Tactics* in 1952. He simply reasoned that games could be useful in a way that books were not, because "to be conversant with the Principles of War is to a soldier what the Bible is to a clergyman," yet books can be easily read, while "wars are somewhat harder to come by." However, "since there were no such wargames available, I had to design my own."[5] He did not envision wargames as simulations for training purposes, however; he stressed principles rather than experiences. Decades after *Tactics*, he still believed that "when you're talking about a rifle platoon, that's what it all comes down to, and there isn't a single game that will tell you how it's done. You can't get that flavor in a game."[6] After considering that others might be interested in his game, he began in 1954 to sell *Tactics* by mail order. He sold perhaps two thousand copies, roughly breaking even financially.[7] Encouraged by these sales, he founded the Avalon Game Company, later renamed Avalon Hill Game Company (henceforth Avalon Hill) in 1958.

Roberts published three games under the Avalon Hill imprint in its first year: *Tactics II*, *Gettysburg*, and *Dispatcher*. He also worked on a fourth title, *Game/Train*, which was not published. Avalon Hill would continue to publish a variety of "adult" or family and business games; *Dispatcher* was only the first of Avalon Hill's railroad simulations, close to Roberts's heart as a historian.[8]

Having worked in the advertising industry, Roberts produced clever and inviting packaging and marketing. A *Tactics II* insert was typical: "Always, to play an Avalon Hill game is an exhilarating challenge . . . to give one, a subtle compliment." The components were modest and even crude by today's standards for printed board games, but the company's print advertising depicted these games as sophisticated and demanding. While Roberts put games on many American kitchen tables as wargaming's first entrepreneur, he was also its first game designer, at least in the modern era, and he was prolific. He produced ten games for Avalon Hill between 1958 and the end of 1963, when he sold the company for financial reasons. Counting *Tactics* and *Afrika Korps* (1964), ten of the twelve games he designed were wargames; they defined the essential elements of this kind of game, both the set of components and the terminology to describe game mechanics and structure.

What was new about these games? A superficial accounting of *Tactics/Tactics II* includes a game board, counters, dice, printed play aids and rule book, nothing particularly remarkable. Board games with visually similar configurations had been published in the United States since the mid-nineteenth century. *Monopoly*, first published by Parker Brothers in 1935, utilized similar components.[9] Roberts's distinctive talent was inventing game procedures through the interaction of rules with charts, tables, and map overlays. These combinations defined the wargame. Two innovations deeply associated with Roberts's games are the Combat Results Table (CRT) and the hexagonal map grid to regulate the movement of units. Roberts introduced the CRT in the 1952 version of *Tactics*. The table's columns each represented an odds calculation corresponding to an attacker's strength divided by that of adjacent defending units; each row corresponded to "the figure thrown on the cubit," i.e., a six-sided die. After working out odds, rolling the die, and consulting the table, the result determined by the intersection of row and column was applied, usually with one or both players losing or retreating units involved in combat. The CRT integrated military simulation and procedures of play. In the rules manual for Avalon Hill's *Tactics II*, Roberts advised players that "an examination of the Combat Results Table is imperative. As this game attempts to be as realistic as possible, the Combat Results Table reflects the fact that an attacker must have a strength advantage in order to be reasonably sure of success. . . . Please examine the Table very carefully."[10] Some features of Roberts's system of combat resolution failed to catch on (e.g., indicating the side affected by a result in terms of the odds

Figure 12.1. Avalon Hill's *Gettysburg* (1958). Wikimedia public domain collection, https://commons.wikimedia.org/wiki/File:Gettysburg_Board_Game_1958.jpg

factor rather than specifying "attacker" or "defender"); nevertheless, his mechanism for combat resolution and use of an odds-based table to randomize results became a staple feature of wargames. Principles embedded in the table also persisted, such as the rule-of-thumb that an army on the offensive must achieve a 3:1 ratio of combat strengths in order to ensure a positive result.

Roberts's "invention" of the CRT indirectly led him to his second big design idea, the hexagonal map grid introduced in the 1961 edition of *Gettysburg*. The invention also illuminates Roberts's contacts with "professional" game creators, specifically the RAND Corporation think tank, a leading center for game theory and game-based research. Dunnigan tells us that there are "mechanical elements common to most" wargames that drive the specific "tactics" of playing them; "The chief among these is the hexagon grid itself."[11] Before 1961, game boards for *Tactics/Tactics II*, the first *Gettysburg* (1958), and *U-Boat* (1959) consisted of maps on which terrain features were drawn to conform to a square grid overlay. As Roberts explained in the *Tactics II* rules, "the mapboard is marked off in one-half inch squares. Each counter, representing a division or HQs [Headquarter units] fits a square. All movement is based on these squares. Units may move vertically, horizontally or diagonally."[12] The obvious problem with this scheme is diagonal movement. The hexagonal map grid for representing terrain and movement was the elegant solution introduced in Avalon Hill's 1961 titles *D-Day*, *Chancellorsville*, and *Civil War*, as well as the new version of *Gettysburg*.

The documentary evidence concerning Roberts's brief encounters with RAND wargamers suggests only a tenuous link between hobbyist and professional games. Lou Zocchi, whose association with Avalon Hill began in 1959, recalled that Roberts was invited to visit RAND in 1960 because the researchers there were "so impressed by *Gettysburg*."[13] After Roberts told them he had come up with the CRT on his own, he was offered and declined a position at the think tank. While visiting RAND, he "noticed that they were using a hex-pattern overlay on their maps, which diminished diagonal movement distortion," so he quietly adopted the same overlay scheme for Avalon Hill's new games in 1961.[14] According to Stephen Patrick, the resemblance of the *Tactics* CRT to a "more complex one" in use at RAND was already a cause for consternation at the think tank "in the early fifties." Roberts informed RAND's researchers that it had taken him but fifteen minutes to develop the concept, but their inquiry piqued his interest in their work, according to Patrick. Years later when he noticed photographs depicting a RAND game, his eye picked out the hexagonal grid on their game map. He immediately recognized this as the way to regulate movement in his games.[15] The differences between these two accounts are significant, but both connect RAND to Roberts's CRT and from there to his discovery of hexagonal map overlays. Yet there is little independent evidence to suggest that dice-driven results tables played a significant role in any RAND games during the 1950s. Roberts's CRT seems unlikely to have caused much consternation. While it is possible that Roberts took the general idea of a results table from another game such as a version of the historical *Kriegsspiel*[16] or even from RAND, one wonders why if this were the case, the details of his homespun CRT were so distinctive. As for the hexagonal map, the Gordian knot is easily cut: the photograph mentioned by Patrick probably is the one in a photo essay by Leonard MacCombe published in *Life* in May 1959.[17] A photograph over the caption "Playing War Games" depicts two teams of RAND researchers playing out an air battle.[18] Maps on the gaming table exhibit a clearly visible hexagonal overlay. It is difficult to imagine Roberts not staring intently at any photograph with that

Figure 12.2. Typical wargame components: *Advanced Squad Leader*

image and caption and running with the idea. In any case, Avalon Hill's key design innovations seems not to have derived from any personal collaboration with the professional simulation community represented by RAND's mathematicians and social scientists.

Roberts delivered the core components of the wargame. When Dunnigan in his handbook of wargame design specified the designer's first priorities, his list might have been an inventory of *Tactics II*: "When a game is designed, generally the first things that go into it are the map, the values on the playing pieces (combat strength and movement allowance), the Terrain Effects Chart and the Combat Results Table."[19] In other words, Roberts's assemblage defined the field and mechanics of play. Avalon Hill also promoted wargames in two important ways that did not directly involve design. First, it created the wargame industry and built a community of players to consume its products. Prior to Avalon Hill, the dearth of hobbyist board games in the United States contrasted with the professional military's occasional embrace of the *kriegsspiel* or the many enthusiasts of military miniatures, a hobby centered in Britain. Avalon Hill filled this vacuum by developing business practices for production, distribution, and sales. After Roberts's departure from the company in 1963, Avalon Hill under the leadership of Tom Shaw hired and trained staff in areas ranging from game design and research to printing and sales, as well as working closely with players and freelance designers.[20] By the end of the 1960s Avalon Hill's bench of available designers was still short, but it had lengthened to include Shaw, Lawrence Pinsky, Sid Sackson, Lindsley Schutz, and, as we shall see, James Dunnigan. As for players, it invested in communication with and among them through its house publication, *The Avalon Hill General* (henceforth *The General*), the premiere issue of which appeared in May 1964. *The General* made players aware of other players. Its regional correspondents reported on events and club activities, and readers found "Opponents Wanted" notices and services such as a Q&A column, news about the company, and previews of upcoming games. Some authors and correspondents appearing in the early issues later achieved star status in the game industry or related endeavors: Lou Zocchi (the Southwest editor), Jerry Pournelle, Al Nofi, George Phillies, James Dunnigan, Dave Arneson, and Gary Gygax. Game designer Greg Costikyan concluded years later that, "just as the letter columns of the science fiction pulps were instrumental in forming science fiction fandom, so the classified ads in *The General* were critical in the creation of the wargaming hobby."[21] Through its business practices and publications, Avalon Hill exposed playing customers to the emerging practice of game design.

A second way in which Avalon Hill shaped expectations about game design was a shift after Roberts's departure in 1963 toward an emphasis in its marketing and magazine on historical research. During the Roberts years and beyond, Avalon Hill wargames often portrayed hypothetical or abstract conflicts (*Tactics*, *Tactics II*, *Blitzkrieg*, *Kriegspiel*, *Nieuchess*). Historical titles like *Gettysburg* and *D-Day* were always prominent in its lineup, but they offered at best rough approximations of the events depicted. In 1964, readers of *The General* may have been skeptical when they read that the company had always "specialized in designing all-skill, realistic games based on actual battles out of the historical past." The article with this claim introduced a new game on the Battle of Midway by describing a research process that included "hundreds of hours pouring [*sic*] over data" gathered from the Library of Congress, National Archives, and other libraries. The designers sought out the help of retired Rear Admiral C. Wade McClusky, who played an important role in the battle. He was impressed by "the design staff's devotion to authenticity" and agreed to provide technical advice for *Midway*, released later that year. He also joined the company's new Board of Technical Advisors, followed in early 1965 by General Anthony C. McAuliffe, who had been consulted during development of Avalon Hill's *Battle of the Bulge* (1965). *The General* reported that he "checked over all the game parts" and even discovered an error in the game's order of battle, "to our embarrassment." This was not a *mea culpa* for sloppy game development, but evidence that historical authenticity had assumed a place of prominence in quality control at Avalon Hill.[22]

From the Monograph to the War Engine

Dunnigan began it all in 1969 with *Tactical Game 3 / PanzerBlitz* . . .

—Rodger MacGowan[23]

Avalon Hill's titles through the mid-1960s were *monographic* games. The *Oxford English Dictionary* defines a monograph as "a detailed written study of a single specialized topic." A monograph is not a general work "in which the topic is dealt with as part of a wider subject." Whether Avalon Hill produced games that were abstract studies of military operations, like *Tactics*, or covered historical conflicts, like *Gettysburg* or *Waterloo*, every game stood alone, covering a single conflict situation with a bespoke system, components, and rules. They were fixed on a single topic. Of course, rules and components were occasionally reused in another title. Yet, even when specific rules or tables crossed over from one game to another, say, from *D-Day* to *Afrika Korps*,[24] every game had its own system. It was a corollary perhaps of their monographic nature that Avalon Hill's early games were operational or strategic in scope, because the subject matter was generally a coherent battle or historical campaign. This coherence fit Avalon Hill's emphasis on playable systems and its newfound respect for historical research based on archival documents and expert approval. An alternative model of game systems based on modularity and reusability of core components to produce a variety of scenarios, and with a more analytical approach to historical simulation, emerged in 1970 with the publication of Avalon Hill's first tactical game, *PanzerBlitz*, designed by James Dunnigan.

Unlike Roberts, Dunnigan had played wargames when he returned from the army in 1964. While stationed in Korea, soldiers in his artillery battalion introduced him to Avalon Hill's games. Back in the United States, he participated in the wargame community; he read and contributed to *The General*, wrote for a new magazine called *Strategy and Tactics*, and edited his own historical 'zine, *Kampf*. He studied games, read strategy articles, and analyzed military history. Unlike Roberts, who started by designing a game but rarely played them, Dunnigan took apart games he played to learn how to design his own. His eclectic, skeptical, and analytical attitude toward design was summarized by the title given to his designer notes for *Tactical Game 3*: "The Game Is a Game."

In 1966, Dunnigan met Shaw (then in charge of Avalon Hill), who had noticed one of his contributions to *The General*, an article in which he "blew away any pretensions to historical accuracy" in the 1965 *Battle of the Bulge* game. Shaw soon asked him to make a game for Avalon Hill.[25] This game, *Jutland*, was Dunnigan's first and was published in 1967. It took the game as monograph one step further; *Jutland* was a scholarly treatise in the form of a game. Dunnigan's commitment to historical simulation trumped easy gameplay. He jettisoned traditional components such as a game board and cardboard counters. Instead, he used maps and miniatures-like printed ship pieces to simulate the historical campaign at two scales—one for ship detection and the other for combat. The editors of *The General* fielded an "avalanche of inquiries" about this unfamiliar system, justifying it as flowing from "greater emphasis . . . on re-capturing historical accuracy than for any other game."[26] Dunnigan added that when asked to design the game in 1966, "I had never thought of designing a game. My interest had always been in history." Indeed, production delays were due to zealous historical research. In line with Avalon Hill's marketing emphasis on such research, Dunnigan joked about being dragged out of the Butler Library on the Columbia campus to finish the game. Between the lines, this story revealed a tense relationship with the Avalon Hill way of making games: "I had to prove a point to Avalon Hill, the delay was caused by my researching the historical data from every conceivable source imaginable, cross-indexing this information over and over again." After tangling about commitments to historical accuracy in a *General* interview about the game, Dunnigan turned the tables on Avalon Hill by prodding his interlocutor (presumably Shaw) to state the company line: "From this point on, Avalon Hill's philosophy will be to place historical accuracy uppermost in the future design of games."[27] This discussion of *Jutland*'s design principles in the wargame hobby's principal forum doubled down on Avalon Hill's commitment to research as the foundation of conflict simulations.

The relationship between Dunnigan and Avalon Hill would continue as a mixture of collaboration, disagreement, and eventually competition. He created games for the company during the late 1960s. His second effort was *1914*, which Avalon Hill itself later described as a "sales success" but a "lousy game" and "too good a simulation" to be "fun to play."[28] By the time those words were published in 1980, Dunnigan's Simulations Publications Inc. (SPI) had become Avalon Hill's main competitor. When *1914* was published in 1968, he was its promising design talent, having created the company's only wargames between *Guadalcanal* (1966) and *Anzio* (1969). For Dunnigan, designing two games for Avalon Hill was instructive. He "carefully" observed their design process and concluded that there must be a "more effective way to publish games."[29] He then decided in 1969 to jump into the game industry with both feet. He gathered a group of wargamers and writers with whom he had worked on various projects. This group included Al Nofi, a likeminded wargamer/historian, and "graphic design ace" Redmond A. Simonsen, who worked out SPI's systems for production and physical component design—he coined the term "game developer" to distinguish his role from game design.[30] In July Dunnigan acquired the magazine *Strategy and Tactics* (*S&T*) from its founder, Christopher Wagner. Dunnigan, Nofi, Simonsen, and others in the SPI orbit had contributed to Wagner's *S&T*. With these pieces in place, Dunnigan began work on the two "basic concepts" behind SPI: (1) backing "games published by gamers" who controlled "all of the game development, production and marketing decisions" and (2) "publishing more games."[31]

The seventeen issues of Wagner's *S&T* had provided an alternative to *The General*; it was partly a "journal of American wargaming," partly a fanzine, and at times critical of Avalon Hill fare. Yet Wagner recruited Dunnigan as a contributor because of the editor's enthusiasm for *Jutland*, an Avalon Hill game.[32] As publisher of *S&T*, Dunnigan worked through a growing rift between his and Avalon Hill's design philosophies.[33] He later claimed that by the middle of 1969, "after doing *1914* for Avalon Hill in 1967 they lost interest in my work for a while." His games were considered "too complicated." By 1969, he concluded that Avalon Hill had been "wandering in the wilderness" since Roberts's departure. The decision to take over *S&T* and start SPI gave Dunnigan his platform for producing more games "by gamers." His team needed to make "good, playable, realistic and authentic games" more efficiently. The first Dunnigan issue of *S&T*, number 18, was published in September 1969. It included a complete "mini-game," *Crete*, about the German airborne invasion of the island in World War II. Providing a "game in a magazine" became the defining feature of *Strategy and Tactics*,[34] with the implication that its publication schedule put intense pressure on Dunnigan and his team to create games at an unprecedented pace. An advertisement in this issue

Figure 12.3. A game in a magazine: *Strategy & Tactics*. Courtesy of Stanford University

promised readers "tired of the 'one game a year' routine" no fewer than ten new games, ranging from revisions of Avalon Hill titles to new subjects such as ancient warfare and a facsimile reprint of an American version of the *kriegsspiel* originally published in the late nineteenth century. Rodger MacGowan, who later made his own mark on game magazine publishing and game production, remembered the promise years later: "This was unheard of. We had all become accustomed to one new release per year from AH and an occasional 'independent' title. . . . This ad marked the beginning of a flood of wargames."[35] By comparison, Avalon Hill produced four wargames between 1966 and 1969, two by Dunnigan. Efficiency in game design became a critical goal for SPI if it was to fulfill its promise to produce so many new games.

Beginning in December 1969, *S&T* began to issue a bimonthly *Supplement* as a forum for its readers. An advertisement in the first issue revealed one of SPI's methods for filling an expanded pipeline of games under development: Test Series Games. It opened with a provocative question, "Tired of the 'one game a year' routine[?] Well, you don't have to depend on Avalon Hill any longer. We have a new idea. Why not cut costs to the bone and just publish games?" The Test Series concept combined mail-order distribution, standardization of components, and user feedback in the selection and development of game titles. The prospectus for the series promised "at least" six new games per year and described thirteen under development, with another four in progress outside the series. Some of these titles had already been previewed as part of the first shot fired across Avalon Hill's bow represented by the advertisement in *Strategy and Tactics* 18. Their appearance in the *Supplement* represented Dunnigan's "solution" for the problem of producing games in volume.[36] It is interesting to note the number of titles that were connected to previously published or forthcoming Avalon Hill titles. Criticism, competition, and collaboration between the two companies could coexist. Most of the Test Series games were battle or campaign studies, such as *1918*, *Tannenberg*, *Normandy*, or a game titled *1914 Revision*, a "cleaned up" version of Dunnigan's Avalon Hill title; these games held to monographic game design. Three Test Series titles stood out as exceptions to the monograph model: *Tactical Game 3*, *Deployment*, and *Strategy I*. These games were systems of components and rules, rather than monographic studies. *Deployment*, for example, was a "unique departure in wargames." It would provide a "wide selection of counters" to "combine different types" of military forces from the eighteenth and early nineteenth centuries in various battle scenarios. Its description promised "limitless variations." Likewise, the Test Series prospectus portrayed *Strategy I* as "based on a 'module' system of rules and components." It was designed for players "who seek variety or . . . wish to design their own" games. The term was not used, but these games introduced the concept of *universal simulators* for whipping up conflict games set in any period from antiquity to World War III. They were kits for exploring vast conflict simulation possibility spaces. Dunnigan's designers would provide a few scenarios, yet such systems invited players to roll their own.

The third outlier on the original Test Series list, *Tactical Game 3*, would have the greatest impact on wargame design when published by Avalon Hill in 1970 as *PanzerBlitz*. The *S&T Supplement* described it as "Tactical Game 3 (Russia, 1944)—A new departure in games. A platoon and company level game whose main objective originally was to compare different weapons and tactical systems. Out of it all came a game that both miniature and board game enthusiasts can enjoy." After a few details about the historical setting and weapons, the synopsis emphasized its "radical new approach to historical gaming" and that it would be "the first in a series of similar games." *Tactical Game 3* was quite unlike a monographic game, but what was it? It did not portray a historical battle or campaign, but instead gave players the means to compare weapons and systems. It was not operational or strategic, but tactical, "the first game to go below the battalion level." Not a monograph, it was a system for serial production of multiple games.[37]

The *Tactical Game 3* in the Test Series list was almost certainly a playtest kit distributed to a small number of testers. It followed earlier Dunnigan projects *Highway 61* and *State Farm 69*, tactical games inspired by miniatures rules and based on his historical research about mobile warfare on the Eastern Front during World War II. The new game reflected his interest in extending historical study to a more experimental, "analytical" format. For Dunnigan, "analytic history differed from the more common narrative history in that it, like the games, took a more numbers oriented and 'systems' approach."[38] Willingness to experiment led him to borrow rules from miniatures systems for his board game simulation, but without the "complex and tedious procedures often required in miniatures games."[39] The first version of *Tactical Game 3* consisted of rules, a rough map of the area near "State Farm 90" and two sheets of paper unit counters, some with hand-drawn silhouettes of vehicles. They provided the necessary information to play scenarios, also called "mini-games." The nearly finished version appeared in issue 22 of *Strategy and Tactics* in July 1970.[40] The development process was nearing completion, so this was essentially a preview of *PanzerBlitz*, released a few months later as Avalon Hill's fall game for that year. Dunnigan published it as a companion to his magazine's featured game, another Tactical Game project called *The Renaissance of Infantry*. There were differences between the *S&T* and Avalon Hill versions, such as the sequence of play and combat values assigned to specific unit counters. Dunnigan addressed these changes in his Designer Notes, along with an analytical article by Steve List about the design, feedback, and revision processes during playtesting, so that readers learned about the development process.[41] This version of the game did not include all of the components of *PanzerBlitz* and lacked a complete map, but it was possible to begin digesting the new game system, study a selection of the counters, and play through one mini-game. Unintentionally perhaps, the rough presentation and ad hoc nature of *Tactical Game 3* in the magazine underscored the openness of the system.[42]

Dunnigan concluded his Designer Notes for *Tactical Game 3* by expressing his wish that "*Panzer* [i.e., *PanzerBlitz*] will usher in a new era of quality in the design and presentation of historical games."[43] He might have written: quality *and quantity*. In contrast to monographic games, *PanzerBlitz* introduced the game system as a generator for multiple mini-games. Wargamers came to call these mini-games "scenarios," possibly borrowing from the term's currency among RAND's Cold War gamers to describe synopses of imagined or hypothetical political crises or military situations.[44] Henceforth, I will call this combination of system + scenarios a "War Engine."

Games based on this design concept paced wargame development through the 1970s and 1980s, from *PanzerBlitz* to *Advanced Squad Leader*. In the *PanzerBlitz* rules, a "Note to 'Veteran' Players of Previous Avalon Battle Games" warned that it might look like other wargames, but "many of the concepts, techniques and details of play are totally unlike other Avalon Hill games." Among these differences one might have noticed that credits for game design, graphics design, and playtesting named Dunnigan, Simonsen, and the staff of *Strategy and Tactics*. A more substantial one was the inclusion of a dozen situation cards, each describing a "scenario" in terms of specified arrangements of three "geomorphic" maps, unit counters used by both sides, victory conditions, the number of turns to be played, and the historical situation that would be simulated.

There was also "Situation 13," called "Making Your Own Situations." It gave players advice about how to use historical research to put together their own scenarios, noting that "you must work under the same restrictions when designing new situations" as the designers of *PanzerBlitz*. The suggestion that players use the *PanzerBlitz* engine to make their own games derived from Dunnigan's ideas about games and analytical history. In his Designer's Notes for *Tactical Game 3*, he had provocatively asserted, "How does this sound? 'Most game players are really trying to be game designers.' It's a thought I've been playing with these past few years, the idea that the 'game' itself is really not what people are interested in, at least not in the long run."[45] The point of a historical game was to test out alternative scenarios, to experiment with variables that might produce different outcomes; it was an interactive medium. Dunnigan felt that "this ability also implies that the game itself can be changed"—thus the title for his *Tactical Game 3* designer's notes, "The Game Is a Game." His War Engine connected the generation of scenarios to empowerment of the player through its flexibility and accessibility. Players took up the challenge. One bibliography of articles about *PanzerBlitz* lists roughly 275 published articles, letters, variants, and replays appearing in more than two dozen magazines and other outlets over a period of about three decades. It would also become the best-selling wargame of the twentieth century, with more than 320,000 copies sold between 1971 and 1998.[46]

While *PanzerBlitz* was an Avalon Hill title, the War Engine also delivered games to the SPI catalog. Simonsen's ideas about games as a communication medium defined the roles of artwork and physical components and produced a standard format for rules presentation.[47] The relatively consistent presentation, especially for the magazine games, helped players to deal with the increased flow of games. For example, they learned to expect a sequence of rules, procedures, and cases, followed by specific details ("chrome," as Dunnigan called it) such as optional rules, design notes, and scenarios, often with their own special rules. SPI defined both the game as system and a process for system design.[48] The company usually described its products as "conflict simulations"; scenarios accordingly gave Dunnigan's analytical historians cases for study using SPI's systems' rules as simulation engines. Thus, the introduction to *Grenadier* (1971) described "a historical simulation of company/squadron/battery level combat in European warfare of the Eighteenth century and Napoleonic Wars," not a battle monograph. The player then recreated specific engagements "by means of scenarios," each of which "is a complete game-simulation in itself, and simulates reality by use of the game equipment." The scenarios covered battles ranging historically from Blenheim (1704) to Palo Alto (1846), yet rarely required more than a brief paragraph of special rules to supplement *Grenadier*'s core system of rules, components, and generic map.

The impact of the War Engine was twofold. First, it was an alternative to the monographic game. Second, SPI's transparency about its process highlighted innovative design practice and thus encouraged the publication of different kinds of games and simulations—and many of them. *Strategy and Tactics*, which issued between 1,000 and 1,500 copies in 1970, reached a peak circulation in 1980 of 37,000. Dunnigan estimated that fewer than 100,000 wargames were sold in 1969, mostly by Avalon Hill; in 1980, sales reached 2.2 million copies.[49] Another implication of SPI's emphasis on systems and scenarios was that it opened up another method for producing multiple games from a single system: inviting the player to design them. The War Engine exemplified by *Tactical Game 3/PanzerBlitz* introduced a flexible, modular approach to game design. Dunnigan concluded that by the late 1970s, players had lost their "awe" of game designers and publishers. Many decided to "do it themselves," which led to a proliferation of both games and companies that published them. In order to understand this impact of SPI's game systems, it will be helpful to compare the War Engine's "system + scenario" template for game design to the options available for modifying Avalon Hill's monographic games.[50]

Several titles in SPI's original Test Series revised, extended, or applied game designs previously published by Avalon Hill. An example was *Anzio Beachhead*, published in the twentieth issue of *S&T* in 1970. Dave Williams, the designer, provided a small game on the Anzio landings of World War II that could be played as a supplement to his *Anzio* game published by Avalon Hill in the previous year. More ambitiously, Dunnigan's system of game design met the Avalon Hill monograph head on in the *Blitzkrieg Module System*, which was published as a bonus game for *S&T* 19, published in November 1969. Like Roberts's *Tactics* and *Tactics II*, Avalon Hill's *Blitzkrieg* (1965) was an abstract, fictional game about modern warfare. Designed by Lawrence Pinsky, it was a big, complex game; other than a few optional rules and modes like the fast-play "tournament version," it was also a finished game. Players intrigued by the unprecedented array of military options in the game noticed the potential for experimentation, and a few articles proposing optional rules and other variants appeared in *The General* along with dozens of strategy articles. SPI's *Module System* revisited *Blitzkrieg* not to improve Avalon Hill's game, but to morph it into a different, open system. Dunnigan and Simonsen described it as an "alternative design" or "starting point." (In this regard perhaps it was a playtest for *Strategy I*, a Test Series game.) They put together eighteen modules in the *Module System*: "They cover every aspect of the game and, if all are used, create an entirely new game."[51] These modules were constructed so that players could use some or all of them, also picking and choosing physical components from *Blitzkrieg* or others provided in *S&T*. They were not scenarios, but rather blocks of rules, with titles like "fluid impulse" (or "rigid impulse"), "railroads," "air forces" and "weather." Players were thus given access to the engine itself, mixing rules in any combination they chose with the rules of the original game. As the title implied, the *Blitzkrieg Module System* tested whether game systems could be broken down into parts, reused, and recombined. One reviewer praised SPI's modularized *Blitzkrieg* as a "major improvement" in game design over the original. He approved of its flexibility that allowed players to simulate "areas of interest" to them by adding modules to the "basic skeleton" of rules. Naturally, the reviewer suggested modifications to those modules, and players continued the discussion about changes to the *Module System* and *Strategy I*.[52] Dunnigan later explained these contributions as a natural feature of "mushware," which he defined as "what people do with complex procedures in their brain, without benefit of a computer." Players of "manual games" gladly jumped the intellectual hurdle of understanding complex rules systems and their interaction with scenario statements and physical components. Then they thought about how to modify them. His conclusion was that players "exposed to manual wargames became, whether they wanted to or not, wargame designers."[53]

PanzerBlitz was published by Avalon Hill, so players may have taken its modularity as an invitation not just to make new scenarios, but also to revise Avalon Hill's other games just as the *Blitzkrieg Module System* had done. Monographic games were not particularly conducive to extension or revision, however. Yet players had since its first year filled *The General* with suggestions for variants and changes, ranging from new units and mapboards to rules modifications.[54] One correspondent perceptively argued that Roberts's *Tactics* was an attractive target for changes, since it presented an abstract theme: "It's flexible because you needn't worry about historical accuracy."[55] These contributions to *The General* documented a potential for players improving and designing games. Dunnigan recognized the potential and activated it. It is hardly surprising that his transparent efforts to modularize game systems would encourage players to make or add to games. SPI's design team showed them in-house projects—the *1914 Revision* in the Test Series or *the Blitzkrieg Module System* in *S&T*—that taught them how to modify Avalon Hill titles.

Despite this encouragement, the fact remained that not every game was a modular system. How does one revise a published monographic game? John Edwards's *The Russian Campaign* provides the most successful example. Edwards was Australian; he became an avid player after picking up a couple of Avalon Hill games during a visit to the United States in 1968. As a new player, initially, he understood the monographic nature of these games: "Changing the rules then seemed akin to sacrilege. . . . I regarded the rules as gospel, and would never have dreamed of making my own modifications." Eventually he did turn a critical eye toward the historical accuracy and playability of these games and began to fiddle with rules, orders of battles, and combat factors. "Slowly but surely I was developing my own version" of Avalon Hill's *Stalingrad*.[56] He corresponded with the company about his project, but it decided not to publish a new game based on his ideas. Instead, he published his "suggestions" for *Stalingrad* as "Stalingrad: Australian Style" in SPI's *S&T Supplement*. It turns out that the best way to update a monographic game is to make a new one. While praising Avalon Hill's game in the article, he noted that "many wargamers" had proposed ways to improve it, and so he did.[57] Meanwhile, his contact with Avalon Hill led to an exclusive agreement to import and distribute its games in Australia. This business arrangement taught him that he could reduce costs by self-publishing his game at home rather than dealing with import duties and the like. He then founded a game design company, Jedko. Its first game was *The African Campaign* (1973), inspired by Avalon Hill's *Afrika Korps*. His *Stalingrad* successor, *The Russian Campaign* (*TRC*), followed in 1974.

Like computer game developers years later, Avalon Hill decided that it made more sense to join than fight players who modified their games. After redeveloping Edwards's improved treatments of subjects covered by its older monographic titles, it "introduced these playability-emphasis games to an enthusiastic American audience." Beginning in 1976 with *TRC*, Avalon Hill published Jedko games such as *War at Sea* and *Fortress Europa* that had been difficult to obtain in the United States. As it admitted in *The General*, Avalon Hill's developers found that although *TRC* "covered much the same ground" as its own *Stalingrad*, Jedko's games were "too good to ignore." Avalon Hill improved the game's graphical presentation, cleaned up the rules, and added a few "scenarios" and player aids; the result was a game that might "cover the same ground" as *Stalingrad*, but was "an entirely new game."[58] Following Edwards's lead, Avalon Hill had replaced an old monograph with a new one. The new game begat various revisions and new editions over the years, producing a chain of connected games stretching from *Stalingrad* (1963) to Jedko's *TRC* (1974) to the Avalon Hill versions of the Jedko game (1976, 1977, 1978, and reprints), then an updated *Russian Campaign II* (1986) by Edwards for Jedko, and on to a fourth edition published by L2 Design Group (2003), followed by a *Southern Expansion Kit*, also by L2 (2004). In the company "Timeline" published in 1980, Avalon Hill acknowledged that *TRC* had driven its own *Stalingrad* "from the retail shelves," while reminding readers that an Avalon Hill developer, Richard Hamblen, edited the third edition rules with "meaningful changes." The experience of working with Jedko opened Avalon Hill's eyes to the benefits of outside developers. It also opened its wallet. When the editors of *The General* explained that *TRC* had been developed as "the initial product of another company," they added that this product was "small potatoes" compared to Avalon Hill's acquisition of Games Research Inc., the publishers of *Diplomacy* and the impending conclusion of negotiations to acquire 3M Games, the makers of a diverse line of games. In the

Figure 12.4. The War Engine at work: *Advanced Squad Leader*. Courtesy of Will Merydith

following year, it concluded the purchase of 3M and *Sports Illustrated*'s game titles.[59] The company timeline consequently dubbed 1977 the "Year of the Acquisition." Its business moves brought more than twenty-five new game titles into the Avalon Hill catalog, diversifying its contents substantially.

The example of *TRC* as Avalon Hill's model for the revision of monographic games contrasts sharply with the impact of modular systems introduced by SPI. During the early 1970s, the monograph and the War Engine followed separate paths. Six years after the release of *PanzerBlitz*, the author of an exhaustively detailed "hex by hex" analysis of its maps was baffled that Avalon Hill had not published new map boards for the game system. This inattention to an obvious extension of the game's scenario possibilities was a missed opportunity, because "*PanzerBlitz* could have sold mapboards the way Barbie sold midget bikinis."[60] That they did not—while publishing the critical remark in their house publication—suggests that Avalon Hill was not ready to jump on the War Engine, despite publishing its most influential game. Indeed, the puzzled *PanzerBlitz* expert's remarks appeared in the same issue in which the company announced its acquisitions strategy as a primary method for adding to the product line. Avalon Hill did not build its own War Engines until the mid-1970s, publishing tactical systems like *Tobruk* (1975), *Squad Leader* (1977), and *Advanced Squad Leader* (1985), the "complete game system" that a magazine advertisement would dub the "crowning achievement" of the company's thirty-year history.[61]

There was always a connection between Dunnigan's conception of the player as designer—or at least as rules fiddler—and the launching of SPI's various design projects and innovations. These projects proved his point by showing that games could be configurable and extensible systems, and they did so in the name of raising efficiency in game design in order to publish more games. The handshakes between design efficiency and productivity on the one hand, and flexibility and modularity on the other, led to the War Engine model of game design. Building games as combinations of systems and scenarios made it possible for publishers such as SPI to produce games serially, while also encouraging players to add their own scenarios on top of the same systems of rules and physical components. And yet, I do not mean to tell a Whiggish story about how SPI's War Engine replaced Avalon Hill's monographic model. By the mid-1970s, both companies were producing games of both kinds. Dunnigan's company also produced monographic battle studies and found other ways to design efficiently. It produced "Quadrigames" like *Napoleon's Last Battles*, each one a set of four games or "folios" based on a core set of rules and linked thematically by a historical campaign. It also recycled well-received rules concepts by using them in new games. For example, Dunnigan's *Panzergruppe Guderian*, originally published in *S&T* and then by Avalon Hill as a boxed game, simulated the Battle of Smolensk in 1941. It introduced several novel aspects in the turn sequence and rules, such as an additional movement phase for armored units and a mechanism for randomly assigning combat strengths to untested Soviet units. Several subsequent games applied these design ideas in games covering other battles of World War II, so that one spoke of the "Guderian System" as a feature of this or that title. These methods for producing series of games from systems or even from specific design concepts gave SPI more tools for realizing Dunnigan's emphasis on efficiency and productivity. Competitors did not miss the point. As we have seen, Avalon Hill acquired publication rights, including those for numerous SPI titles (*Panzergruppe Guderian*, *Frederick the Great*, and others), and eventually began to produce its own War Engines. New companies might build an entire brand or product line around a single Engine. Game Designer's Workshop, founded in 1973, did just that with the *Europa* series, a massive project to deliver a series of games at one geographic scale and governed by a unified rules set. Assuming the project could ever be completed, GDW held out the possibility of combining multiple games to play out the entirety of World War II, provided one could find a table large enough for the maps. Looking back at the early days of the wargame industry, Dunnigan was satisfied that his model for design and production had spawned many game companies, "each following the SPI system to one degree or another."[62]

From Paper to PC

> Most of those gamers were not aware that simply playing manual games turned them into game designers, although over time most of them realized it.
>
> —James Dunnigan, *Wargames Handbook*[63]

> There is not a hell of a lot of difference between what the best designer in the world produces, and what quite a few reasonably clued in players would produce at this point.
>
> —John Carmack, 2002[64]

Much recent work on the history of wargames has focused on military simulation, especially the rise and impact of the so-called Military-Entertainment Complex.[65] This chapter has been concerned with a different kind of wargaming. The professional simulation community—whether R&D think tanks like RAND or military labs—had little direct contact with Charles Roberts or SPI during the late 1960s and early 1970s. Still, it was perhaps inevitable that as with military simulation, computation would have an impact on the industry they built. Dunnigan has described the 1980s as the "conversion era" for wargames, by which he meant a general shift of energy and ideas from "manual" games to computer games. Specific game design concepts and even games also moved over to the computer. As Dunnigan put it, "By the mid-1980s, many manual [paper] wargames had been directly transferred to computers." In the early conversions, adapting manual games essentially "meant displaying a hex grid on the computer screen."[66] The first steps in this direction were hybrid games, in which software provided an opponent for a manual game. Chris Crawford introduced this approach with a game originally called *Wargy I*, first programmed in FORTRAN for an IBM 1130 computer, then published in a version renamed *Tanktics* for the Commodore PET microcomputer in 1978.

Crawford remembers it as "the ONLY commercially available wargame when I first released it." A revised version for multiple systems was released by Avalon Hill in 1982, which had launched a line of computer wargames. David Myers has observed that "the main appeal of *Tanktics*—and most other computer wargames of the period—was that a game could be played without having to solicit another (human) player." By the time Avalon Hill picked up the title, Crawford had raised that bar by programming *Eastern Front 1941* for the Atari 400/800 series of 8-bit home computers. Based on an earlier project rejected by Atari due to lack of interest in wargames, it was released through the Atari Program Exchange (henceforth APX) and may have been the best-selling software title for these machines.[67]

Crawford's wargames followed a development path that resembled the movement from monograph to War Engine led by SPI. He began with monographs. *Tanktics* was a study of a single battle, and the first APX version of *Eastern Front* (1941) was a monographic study of a campaign. However, the second, cartridge version of *Eastern Front* approached the War Engine concept through the provision of a separately sold "Scenario Editor." The APX catalog for fall 1983 urged players to use this Scenario Editor to "establish your own criteria for the battles of the Eastern Front." It enabled players to "control more than a dozen factors that could alter the outcome" of their games. The utility of this tool was reinforced by other items for sale in this catalog. These included a set of three "challenging new scenarios" by Ted Farmer for the game designed to extend the war past the 1941 campaign and Stephen Hall's MAPMAKER as something "of special interest to players of the hugely popular" *Eastern Front*. MAPMAKER was described as the software Crawford had used to create the scrolling map for his game; the APX made it available to anyone "thinking of creating a strategy game."[68] Together, a scenario editor and mapmaking utility certainly encouraged ambitious players to create new scenarios. Yet the system was not modular or extensible in the sense of SPI's early manual projects, such as the *Blitzkrieg Module* system or even *PanzerBlitz*, because the rules system (program code) was not accessible in the same way.

Manual games were not just physical games; they were also games with rules *manuals* that players read and studied. Supporting player-generated scenarios was one thing, but without providing access to the underlying game system, players could not analyze how games worked. In this sense, *Eastern Front* exemplified Dunnigan's criticism that "computer wargames

Figure 12.5. Chris Crawford's *Eastern Front* (1941)

plunged the games' inner workings into darkness."[69] This limitation was shared by computer wargames of the 1980s that provided scenario generators or built "universal" simulators covering different historical periods. Noteworthy efforts included the Strategic Study Group's (SSG) *Battlefront* series (1986–), Strategic Simulations Inc.'s (SSI) *Wargame Construction Set* (1986), versions of Rainbird's *Universal Military Simulator* (1987–) and more recent scenario-generating games such as HPS Simulations' *Point of Attack 2* (2004), which was issued in both a military and a "civilian" release.

Computer War Engines such as those developed by SSG, SSI, and Rainbird in the mid-1980s explored and sought to extend the system + scenarios formula for wargames. In so doing they were implicitly following Crawford's prediction that "wargames on personal computers will not be just like boardgames." His vision of the "future of computer wargaming" was that game designers would strive for optimization of the "strengths of the computer" and avoidance of its weaknesses.[70] Crawford's work on *Eastern Front*, for example, had stalled initially. He restarted the project after realizing the potential that scrolling map graphics available on Atari home computers offered for a "monster game, a game with everything."[71] The designers of the *Wargame Construction Set* (*WCS*) and *UMS: The Universal Military Simulator* also focused on the potential computers offered as a vast simulation space, not just in terms of scale but also in terms of flexibility of representation. The influence of a way of thinking about computation with roots in Turing's Universal Machine resonated in this version of the War Engine and could be read—if faintly—in the title of Firebird's game. The game manual for *WCS*—yes, computer games had them too—promised that players could "explore a wide range of conflict situations in military history, fantasy, and science fiction." Buyers of *UMS* read on its box cover that "from ancient battles of Classic History to the bloodiest Science Fiction fantasies, UMS lets you re-create them all with its unique 3D graphics system." Designers considered how the computer as a platform for wargames "with everything" might move them beyond the limitations imposed by cardboard counters, physical maps, and game manuals.

While the possibility of a War Engine "with everything" seemed to promise unlimited scenario generation powers, the standard practice of describing the process as "editing" hinted at limitations such as those encountered with APX's "scenario editor" for *Eastern Front*. *WCS* likewise offered players access to the game's "EDITOR," which made possible map and unit editing.

One reviewer, the author Orson Scott Card, considered the *Wargame Construction Set* as part of a family of computer "building" games of the early 1980s that included *Lode Runner*, *Pinball Construction Set*, and *Adventure Construction Set*. He described SSI's offering as "simple, elegant, infinitely variable" and focused in his short review on what it offered for game *design*, rather than how it was played.[72] His review underlined the ambitious vision of the limitless simulation opened up by scenario editors and construction sets. The constraints imposed by database editing as scenario generation can be illustrated by a comparison of *PanzerBlitz* and *WCS*. In the manual game, the first important set of player-created scenarios appeared in *The General* in 1974 under the apt title, "Beyond Situation 13."[73] These scenarios—called "Situation 13 variants" by the author—appeared in the magazine in the format of *PanzerBlitz* situation cards numbered from 14 through 25. The scenarios each included a historical description, "map configuration" for arranging the game's geomorphic maps, setup instructions with unit manifests for both the Soviet and German forces, and a turn track. Significantly, they also included scenario-specific rules and "victory conditions." For example, Situation 18, under the title "Combined Russian Offensive: Somewhere in Russia (1944)" included a rule that "Soviet forces may not cross from one board to another (e.g., from board 1 to board 2, etc.). They may, however, fire across from one board to another."[74] Game components were fixed by the original game, but rules were not. The scenario designer used the situation cards to provide instructions about how to deploy the existing components, but also had the option of extending or even modifying the rules by writing them in a free textual form. This was more than editing; the scenario designer (optionally) authored new rules. In the computer game, the process of creating a scenario involved using the EDITOR to create a map by selecting and placing "terrain icons," editing map colors, setting the number and type of units in a scenario, and defining the characteristics of military units, from the firepower and movement rates of friendly, player-controlled units to the selection of "aggression levels" for those under the control of the game software. The editing features provided flexibility and variability through combinations of decisions about database elements, expressed by actions such as selecting, placing, and assigning a color value to a map tile or giving a unit a numerical "firepower" rating between 1 and 99. Roger Damon, who developed *WCS*, informed players in the instructions that came with the game that "the central concept of this construction set is to first define the variables and then make them as easy to manipulate as possible. The designer's focus can turn then to wrenching a game out of the available data."[75] The *WCS* scenario designer edited a database, without the option of writing new rules.

A second, related take on wargame systems leads us to another kind of martial game technology for computers: the game engine. Recounting the historical development of game engine technology and the first-person shooter genre would take us far afield, and I have already done so elsewhere.[76] The question I will briefly consider here is whether the War Engine as a modular system based on systems and scenarios ought to be considered as an influence on the game engine concept—that is, a software architecture that divides engine from content. Both systems were driven into existence by daunting production schedules, at SPI and id Software respectively. Both enabled the reuse of core systems with new creative assets, such as scenario cards or game maps. These are indeed meaningful similarities. SPI and id were not working with the same technologies, but they were addressing related issues: efficiency in production, design innovation, and player creativity. It is certainly tempting to trace a recurring form of game design manifested in both the War Engine and Game Engine. This interpretive strategy might be a stretch, however. It is true enough that both SPI and id had efficiency of design in mind when they produced systems that could be reused and combined with thematic data to produce different game experiences, such as scenarios or mods. Dunnigan insists in his *Wargames Handbook* however, that manual and computer games are essentially different: "With board wargames, you could not ignore the details of how the game did what it did. With computer wargames, you could, and most gladly did."[77] In his view, a game like id's *DOOM* was a success due to superior technology. This technology, such as the game's compelling graphics and first-person perspective, were products of the engine, the secrets of which were locked away. In the terminology of the War Engine, what was left—and all that was left—to the player as designer was scenario editing, not simulation design. This was exactly why Dunnigan viewed the decline of board games as bad news. Dunnigan's and John Carmack's compatible visions aside, the transition from manual to PC games complicated access to engines and, in turn, shortened the player's field for new design content.

The designers of computer games easily repeated SPI's move away from the wargame as an authored monograph, but whether War Engines like *Eastern Front* or Game Engines like *DOOM*, the role of the player as designer was changed by an important feature: other than their programmers, nobody had direct access to the system/engine.[78] Bringing board game and computer game designs into the same conversation about wargame systems and player-created scenarios is nonetheless a worthy exercise; doing so helps us to understand the persistent challenge of designing game systems and the relevance of "manual" wargame design for digital games. In his introduction to a collection of essays on "analog game design," former SPI game developer Greg Costikyan insisted that "digital and non-digital games are not different in essential nature." As a historian, this claim encourages me to pay more attention to tabletop games not just for their own sake, but as part of the history of digital games. One more justification from Costikyan: "As many game studies programs have discovered, tabletop games are particularly useful in the study of game design, because their systems are exposed to the player, not hidden in code."[79] I have tried to show that the Wargame Engine was an important and useful concept in game system design. If so, we have another reason to revisit the history of games on the tabletop.

NOTES

1. See Stewart Woods, *Eurogames: The Design, Culture and Play of Modern European Board Games* (Jefferson, NC: McFarland, 2012); and Greg Costikyan, "Board Game Aesthetics," in *Tabletop: Analog Game Design*, ed. Greg Costikyan and Drew Davidson (Pittsburgh, PA: ETC Press, 2011), 179–184.

2. On military simulation and its connection to commercial games and the entertainment industry, see Lenoir and Lowood (2005) and Perla (1990). James Dunnigan has observed that the US military had "largely abandoned" wargames by the 1950s; commercial wargames "attracted the troops' attention in the early 1970s and led to a renaissance of military wargaming" (Dunnigan 2000a, loc. 196).

3. John E. Koontz to Martin Gardner, June 11, 1975. Box 33, folder 8, Martin Gardner Papers, SC647, Stanford University Libraries.

4. See John J. Vanore, "Interview: Charles S. Roberts—Founder of the Avalon Hill Game Company and Founding Father of Board Wargaming," *Fire and Movement* 56 (1988): 17; and Charles S. Roberts, "Charles S. Roberts: In His Own Words," 1983, http://www.alanemrich.com/CSR_pages/Articles/CSRspeaks.htm.

5. Roberts, "Charles S. Roberts: In His Own Words," 1983.

6. John J. Vanore, "Interview: Charles S. Roberts—Founder of the Avalon Hill Game Company and Founding Father of Board Wargaming," *Fire and Movement* 56 (1988): 18.

7. See Roberts, "Charles S. Roberts: In His Own Words," 1983.

8. Frederick N. Rasmussen, "Charles S. Roberts, Train Line Expert, Dies at 80," *Baltimore Sun*, August 28, 2010.

9. For more examples from the 1880s, see Margaret Hofer, *The Games We Played: The Golden Age of Board and Table Games* (Princeton, NJ: Princeton Architectural Press, 2003).

10. Charles Roberts, *Tactics II*, Avalon Hill Company, 1958, 10.

11. James F. Dunnigan, *Wargames Handbook, Third Edition: How to Play and Design Commercial and Professional Wargames* (Lincoln, NE: Writers Club Press, 2000); see also Garry D. Brewer and Martin Shubik, *The War Game: A Critique of Military Problem Solving* (Cambridge, MA: Harvard University Press, 1979), 59–66; Thomas B. Allen, *War Games: The Secret World of the Creators, Players, and Policy Makers Rehearsing World War III Today* (New York: McGraw-Hill, 1987), 141–147.

12. Charles Roberts, *Tactics II*, Avalon Hill Company, 1958, 7.

13. He became one of Avalon Hill's early playtesters and in 1964 became the first Southwest US "editor" of the *Avalon Hill General*, the company's newsletter.

14. Lou Zocchi, "Gettysburg," in *Hobby Games: The 100 Best*, ed. James Lowder (Seattle: Green Ronin, 2007), Kindle e-book, 2650–2655.

15. Stephen B. Patrick, "The History of Wargaming," in *Strategy and Tactics Staff Study Nr. 2: Wargame Design* (New York: Hippocrene, 1977), 11–12; Peter P. Perla, *The Art of Wargaming: A Guide for Professionals and Hobbyists* (Annapolis, MD: Naval Institute Press, 1990), 115–116.

16. Jon Peterson, *Playing at the World: A History of Simulating Wars, People and Fantastic Adventure, from Chess to Role-Playing Games* (San Diego, CA: Unreason Press, 2012), 289.

17. Peterson, *Playing at the World*, 106.

18. This photograph appears as Figure 30.1 in Elizabeth Losh's chapter in Pat Harrigan and Matthew G. Kirschenbaum, eds., *Zones of Control: Perspectives on Wargaming* (Cambridge, MA: MIT Press, 2016).

19. James F. Dunnigan, *The Complete Wargames Handbook* (New York: Morrow, 1980), 34–35.

20. It is important here to note the company's close relationship with its printer, Monarch, whose owners eventually acquired Avalon Hill after Roberts's departure.

21. Greg Costikyan, "A Farewell to Hexes," Internet Archive copy, 1996, http://web.archive.org/web/20040212100739/http:/www.costik.com/spisins.html.

22. "Midway—Newest Battle Game!" *Avalon Hill General* 1, no. 3 (1964): 1–2; "General McAuliffe Added to Advisory Staff," *Avalon Hill General* 1, no. 6. (1965): 1–2.

23. Rodger MacGowan, "20 Years Later and 10 Years After *Squad Leader*," *Fire and Movement* 53 (1987).

24. Rex Martin, "Cardboard Warriors: The Rise and Fall of an American Wargaming Subculture, 1958–1998" (PhD diss., Pennsylvania State University, 2001), 230.

25. James F. Dunnigan, *Wargames Handbook, Third Edition: How to Play and Design Commercial and Professional Wargames* (Lincoln, NE: Writers Club Press, 2000); James F. Dunnigan, "Transition: S&T Change Publishers," *Strategy and Tactics. Book IV: Nrs. 16–18* (N.p.: Simulations Publications, Inc., n.d.), inside covers.

26. "Cover Story," *Avalon Hill General* 4, no. 1 (1967b): 2.

27. "The Avalon Hill Philosophy—Part 3," *Avalon Hill General* 4, no. 1 (1967a): 2–4.

28. Avalon Hill General. *Index and Company History, 1952–1980. Volume 1–Volume 16* (Baltimore: Avalon Hill, 1980), 9.

29. Dunnigan, *The Complete Wargames Handbook*, 2994.

30. Dunnigan, *The Complete Wargames Handbook*, 3009; Redmond A. Simonsen, "Physical Systems Design in Conflict Simulations," *Moves* 7 (1973): 22–24.

31. Dunnigan, *The Complete Wargames Handbook*, 3004; Dunnigan, "Transition: S&T Change Publishers," n.d.

32. Christopher Wagner, "Background on S&T Nrs. 16 & 17," *Strategy and Tactics. Book IV: Nrs. 16–18* (N.p.: Simulations Publications, Inc., n.d.), inside front cover; Dunnigan, "Transition: S&T Change Publishers," n.d.

33. Dunnigan's operations appeared under a few different organizational names, with SPI the eventual wargame publishing flagship. These included Infinity Corporation, Poultron Press for magazine publishing, and Operations Design Corporation for game design.

34. As of 2014, *S&T* had published nearly three hundred wargames under a series of publishers.

35. Rodger B. MacGowan, "20 Years Later and 10 Years After Squad Leader," *Fire and Movement* 53 (1987): 34.

36. "Test Series Games," *S&T Supplement* (December 1969–January 1970): 24.

37. James F. Dunnigan, "Designers Notes: The Game Is a Game," *Strategy and Tactics* 22 (1970): XS3.

38. Dunnigan, *The Complete Wargames Handbook*, 3268; Dunnigan, "Designer Notes," 1970.

39. *S&T* 23 returned the favor with the publication of *T-34*, designed by Arnold Hendricks, a version of *Tactical Game* 3 for miniatures.

40. SPI published a reprint of the playtest kit several years later, with improved counter art, perhaps for collectors.

41. Steve List, "Game Design: Down Highway 61, through State Farm 69, around Tactical Game 3, and into *Panzerblitz*," *Strategy and Tactics* 22 (1970); Dunnigan, "Designer Notes," 1970.

42. Dunnigan, "Designer Notes," 1970; Alan R. Arvold, "A Comprehensive Index to *Panzerblitz*," n.d., http://grognard.com/info1/pbartrev.html; Michael Dorosh, "Tactical Game 3," *The Tactical Wargamer*, 2008, http://www.tacticalwargamer.com/boardgames/panzerblitz/tacgame3.htm; "Avalon Hill Philosophy—Part 24, "Why *Panzerblitz*?" *Avalon Hill General* 7, no. 4 (1970).

43. Dunnigan, "Designer Notes," 1970.

44. See for example Herman Kahn, *Thinking about the Unthinkable* (New York: Horizon Press, 1964), 150.

45. Dunnigan, "Designer Notes," 1970.

46. Arvold, "A Comprehensive Index to Panzerblitz," n.d..; Dunnigan, *The Complete Wargames Handbook*, 3056.

47. See Henry Lowood, "Game Counter," in *The Object Reader*, ed. Fiona Candlin and Raiford Guins (Abingdon, UK: Routledge, 2009), 466–469.

48. Redmond A. Simonsen, "Physical Systems Design in Conflict Simulations," *Moves* 7 (1973): 22–24; Redmond A. Simonsen, "Image and System: Graphics and Physical Systems Design," in *Wargame Design: The History, Production and Use of Conflict Simulation Games*, ed. Staff of *Strategy and Tactics* (New York: Hippocrene, 1977).

49. Dunnigan, *The Complete Wargames Handbook*, 3142.

50. SPI also produced some monographic games, as did most game companies.

51. James F. Dunnigan and Redmond Simonsen, "The Blitzkrieg Module System," *Strategy & Tactics* 19 (1969): 17.

52. Russel Reddoch, "Comments on Module *Blitzkrieg*," *S&T Supplement* 2 (1970): 9–11; Richard Bauer, "Thoughts on *Strategy I*—Part 1." *S&T Supplement* 3 (1970a): 7–11; Richard Bauer, "More Thoughts on *Strategy I*." *S&T Supplement* 4 (1970b): 18–22.

53. Dunnigan, *Wargames Handbook, Third Edition*, 161.

54. See for example Jon Perica, "Putting More Realism into Tactics II," *General* 1, no. 4 (1964): 7–12; Victor Madeja, "Midway, D-Day, Tactics II, Stalingrad Re-worked," *Avalon Hill General* 1, no. 5 (1965): 3.

55. Eric R. Shimer, "Meanwhile—Back at Tactics II," *Avalon Hill General* 1, no. 5 (1965): 11.

56. John Edwards, "Interview: John Edwards," *Avalon Hill General* 15, no. 1 (1978): 16–17; Avalon Hill General, *Index and Company History, 1952–1980. Volume 1-Volume 16* (Baltimore: Avalon Hill, 1980), 12.

57. John Edwards, "Stalingrad: Australian Style," *S&T Supplement* 3 (1970): 12–18.

58. "Avalon Hill Philosophy, Part 53," *Avalon Hill General* 12, no. 5 (1976): 2.

59. Avalon Hill General, *Index and Company History*, 12.

60. Larry McAneny, "*Panzerblitz*: Hex by Hex," *Avalon Hill General* 12, no. 5 (1976): 3.

61. "The Ultimate Wargame." *Avalon Hill General Special Issue*: 59 (1988).

62. Dunnigan, *Wargames Handbook, Third Edition*, 3066.

63. Dunnigan, *Wargames Handbook, Third Edition*.

64. John Carmack, "Re: Definitions of Terms," discussion post to Slashdot, January 2, 2002. http://slashdot.org/comments.pl?sid=25551&cid=2775698.

65. See for example Timothy Lenoir and Henry Lowood, "Theaters of War: The Military-Entertainment Complex," in *Collection, Laboratory, Theater: Scenes of Knowledge in the 17th Century*, ed. Helmar Schramm, Ludger Schwarte, and Jan Lazardzig (Berlin: de Gruyter, 2005), 427–456; Roger Stahl, *Militainment, Inc.: War, Media, and Popular Culture* (New York: Routledge, 2010); Patrick Crogan, *Gameplay Mode: War, Simulation and Technoculture* (Minneapolis: University of Minnesota Press, 2011).

66. Dunnigan, *Wargames Handbook, Third Edition*, 3530.

67. Chris Crawford, *The Art of Computer Game Design* (Berkeley: McGraw-Hill, 1982a), E-version cited, 1997, 35; David Myers, "Chris Crawford and Computer Game Aesthetics," *Journal of Popular Culture* 24, no. 2 (1990): 17–32.

68. *APX/Atari Program Exchange Product Catalog: Fall Edition 1983*, Internet Archive scan, 24, https://archive.org/details/Atari_Program_Exchange_catalog_Fall_1983.

69. Dunnigan, *Wargames Handbook, Third Edition*, 178.

70. Chris Crawford, "The Future of Computer Wargaming," *Computer Gaming World* 1, no. 1 (1981): 3.

71. Chris Crawford, "*Eastern Front*: A Narrative History," *Creating Computing* 8, no. 8 (1982b): 102.

72. Orson Scott Card, "Gameplay," *Compute (Greensboro)* 104 (January 1989): 12.

73. Robert D. Harmon, "Beyond Situation 13," *General* 11, no. 4 (1974): 7–12.

74. Harmon, "Beyond," 9.

75. *Wargame Construction Set*, 23.

76. Henry Lowood, "Game Engines and Game History," *Kinephanos*. History of Games International Conference Proceedings, 2014, http://www.kinephanos.ca/2014/game-engines-and-game-history/.

77. Dunnigan, *Wargames Handbook, Third Edition*, 69.

78. Of course, open-source software leaves the engine's hood up, so to speak, but for practical purposes only to skilled programmers who understand what they are seeing. The manual game's "engine" (the manual itself) is open for all to read.

79. Greg Costikyan, "Board Game Aesthetics," in *Tabletop: Analog Game Design*, ed. Greg Costiky and Drew Davidson (Pittsburgh, PA: ETC, 2011), 13–14.

III FURTHER DIRECTIONS

Sports Games and e-Sports

Author's Introduction

In recent years, e-sports have become the conversation starter for digital games. More specifically, the concept of professional competition, of making money by playing video games, seems especially to fascinate people with only casual knowledge of the games played, as it does investors, advertisers, and professional sports teams looking to expand their fan base.

"Beyond the Game," one of two previously unpublished essays in this book, appears in this volume as originally written in 2007, with only a few minor changes. It was originally an invited contribution to a compilation of original essays with the working title, "Play and Politics: Games, Civic Engagement, and Social Activism." That volume never saw the light of day, and so the essay languished in the proverbial desk drawer—or more accurately, a Windows file folder. Over a decade later in August 2020, when it was my turn to present to a writing group in the Stanford Libraries, I decided to ask the group for their thoughts about whether this essay could be revived and revised for publication. It occurred to me that the growth of interest in e-sports might provide the oxygen for resuscitating this article. Surprisingly, the consensus of this group of readers was that it would be more interesting to publish this essay without revision. They recommended doing just that, supplemented perhaps by a contextual introduction or notes, to serve as a historical document about the reception of e-sports in its earlier days.

That explains, perhaps even justifies, the hubris of publishing an essay written in 2007. What about that context? I would identify three aspects of e-sports and cyberathleticism (a term used just as frequently around that time) that prompted "Beyond the Game" and remain relevant in critical discussion of the past, present, and future of e-sports today: (1) connections between sports and game studies; (2) the focus on players; and (3) the tension between competitive ideals and commercial partnerships. These three themes all made appearances in "Beyond the Game," and simply on that basis alone, I would argue that they were apparent at an early stage in the development of e-sports before the expansion of opportunities characterized by streaming and sponsorships in the 2010s.[1]

The possibility of a relationship between critical sports and game studies reflects similarities or analogies between sports and games. An essential similarity might be constructed out of the commonplace observation that sports are game forms by most definitions. Katie Salen and Eric Zimmerman, for example, refer to professional sports as a "category of game."[2] However, structural similarities or design questions are not what has piqued my interest in exploring what sports tell us about games. As is usually the case with me, it was not a theoretical question, but an event—or rather, two events—that focused my attention.

The first event was the appearance of a well-known competitive player, Dennis Fong (a.k.a. "Thresh"), in a History of Digital Games course that I taught at Stanford in 2001. Two moments during this visit left an impression. The first was that Dennis spoke for several minutes about the importance of a background in competitive sports—he was an excellent high school tennis player, if I recall correctly—and athletic ability for a competitive gamer. The second was the line of students that formed at the speaker's lectern after class to ask for his autograph! These moments planted the seed of an interest in cyberathletic competition both as connected to sports and as a potential rival for the attention of spectators and fans.

The second event was my participation as a referee for the *Warcraft 3* tournament in the World Cyber Games (WCG) held in October 2004 in San Francisco. This was the first time that these "video game Olympics" were held outside South Korea. As the senior (oldest) referee at WCG 2004, I was asked to be available to media covering the event, and in this role I was able to gauge the reactions of both enthusiast and relatively mainstream journalists to the then current state of e-sports. The WCG experience led specifically to an article on "beautiful play" in digital games,[3] as well as two workshops co-organized with T. L. Taylor, the preeminent academic expert on e-sports, one at Stanford and other at the IT University of Copenhagen. The WCG featured opening and closing ceremonies (I carried the referees' flag in the latter), which is what probably started my thinking about the relationship between the

Olympics and their video game namesake, which became the starting point for writing "Beyond the Game."

For me, the obvious place to start to construct useful analogies is with players. Competitive games, like competitive sports (as opposed to mere athletic prowess) call on players to perform creatively, in response to an opponent, and in real time. They do so within a framework of authored, constraining rules, the domain of game design. My interests tend to run with the players. It may seem that in "Beyond the Game" I am more concerned with organizers and sponsors, but for me two quotations encompass my take on the convergence of the Olympic Ideal and e-sports. The first from the Baron de Coubertin (1892), who stated that "rowers, runners, and fencers" would be the "free trade of the future," by which he meant to underline his conviction that in the modern world, athletics would be "democratic and international."[4] This remark was echoed by Atlanta mayor Andrew Young about a century later when he defended the "Coca-Cola Olympics" held in his city by insisting that "the commercialization of sport is the democratization of sport."[5] The line that runs through these positions is that those who would promote access to global spectacles of sport, with its claimed ideals and benefits, must accept international exchanges not just of ideas, but also of the commercial benefits and promotional activities that support these events.

The problem that the World Cyber Games exemplified for e-sports is that leaning into both the rhetorical ideals and commercialization of the Olympic Games inevitably leads to tensions. The uneasy coexistence of two "global" visions: Coubertin's elevation of the Olympics to a demonstration of the ideal of the competitive spirit, which the WCG picked up in its own way for these video game Olympics, and the "democratization" of e-sports through commercial exploitation of these same competitions via the reliance on marketing and sponsorship. Such tensions are nothing new for the Olympics, but they resonate on a specific wavelength for game studies generated by the historical work of Johan Huizinga. In this 2007 essay I perhaps overemphasized the "magic circle" concept in *Homo Ludens* as the focus of this tension; today I would put the main issue as consideration of whether commercialization erodes the "play element" that Huizinga considered essential for the cultural development of civilization. In other words, perhaps commercialization of international sports and e-sports contributes to the breakdown of "noble play" and the "observance of play-rules" that was the crucial indicator of civilization's health for Huizinga.[6] From this point of view, my conclusion in 2007 that the WCG stood for a globalism rooted "as deeply in commerce as culture" seems to have held up pretty well for the development of e-sports since then.

I continue to be fascinated by connections between digital games and sports. The "beautiful play" article already mentioned above is one example of taking a concept ("the beautiful game") from a sport (football/soccer) and exploring its applicability and relevance to competitive digital games. The obvious next step is to take a closer look at sports games. Besides personal interest in the topic, another compelling reason to look at these games is that they have been woefully neglected in historical game studies, and not just digital games, but also board games, card games, and other kinds of games with sports as a theme.[7] A good place to begin the work of extending game history to sports games is one of the most successful, profitable, and long-lasting games on the world's most popular sport, the *FIFA Football* series that has been produced by the EA Sports division of Electronic Arts since 1993. In an essay in a book entitled *EA Sports FIFA: Feeling the Game* (2022), coedited with Raiford Guins and Carlin Wing, my interest in linking topics has brought me to the question of how players learn how the game they are playing (*FIFA*) simulates the game they are, well, also playing (football) in a different mode. While a study of a sports game and its players, it also might be thought of as software "users" engaging with software; sports game studies, meet technology studies. That is not the only way in which this project harmonizes my interests. Tackling neglected topics like this one—and I had promised myself to avoid sports puns here—offers opportunities for both the historian and the curator in me. On the one hand, writing, and on the other, gathering physical and digital materials or producing interviews that can be preserved in the Stanford Libraries and our digital repository.

NOTES

1. Documented in T. L. Taylor's *Raising the Stakes: E-Sports and the Professionalization of Computer Gaming* (Cambridge, MA: MIT Press, 2012) and *Watch Me Play: Twitch and the Rise of Game Live Streaming* (Princeton, NJ: Princeton University Press, 2018).

2. Katie Salen and Eric Zimmerman, *Rules of Play: Game Design Fundamentals* (Cambridge, MA: MIT Press, 2004), 276.

3. "Joga Bonito: Beautiful Play, Sports and Digital Games," in *Sports Videogames*, ed. Mia Consalvo, Konstantin Mitgutsch, and Abe Stein (New York: Routledge, 2013), 67–86.

4. From "Physical Exercises in the Modern World: Lecture Given at the Sorbonne," in *Pierre de Coubertin, 1863–1937: Olympism, Selected Writings*, ed. Norbert Müller (Lausanne: International Olympic Committee, 2000), 297.

5. *Le Mouvement olympique = The Olympic Movement* (Lausanne: International Olympic Committee, 1993), 34.

6. Johan Huizinga, *Homo Ludens: A Study of the Play Element in Culture* (Boston: Beacon, 1950), 4–5. Orig. ed. *Homo ludens: proeve eener bepaling van het spel-element der cultuur* (Haarlem: Willink, 1938).

7. A notable exception, though not focused on historical questions per se, is Mia Consalvo, Konstantin Mitgutsch, and Abe Stein, *Sports Videogames* (New York: Routledge, 2013), which included the aforementioned "beautiful play" essay.

13

"Beyond the Game"

The Olympic Ideal and Competitive e-Sports

On March 28, 2007, heavyweight game developers Sega and Nintendo made a "historical announcement." Their iconic characters, Sonic the Hedgehog and Mario, respectively, would set aside more than fifteen years of intense rivalry to appear together for the first time in a digital game. Sega and Nintendo surely expected *Mario & Sonic at the Olympic Games* to be a blockbuster hit, scheduling its release in time for the Christmas season. The announcement struck a loftier note inspired by the modern Olympic Ideal, that of bringing the world together through competition. Like recent Olympic Games and other international sport competitions, however, the theme of amity through agon was joined to an equally powerful message of globalization through commercialization. Sega president and COO Hisao Oguchi announced that its license would enable Sega to "further enhance our global entertainment business and drive sales in critical Asian markets like China that offer an extremely high growth potential."[1] In announcing this business and intellectual property partnership, the partners spoke in terms derived from the venerated Olympics themselves. Oguchi looked forward with excitement at the prospect of the "world's greatest games' characters" meeting to "compete in the world's greatest sporting event"; star designer Shigeru Miyamoto of Nintendo portrayed the Olympics as the "perfect opportunity" for Mario and Sonic to meet at last as "respectful rivals." Raymond Goldsmith, chairman and CEO of International Sports Multimedia (ISM) explained that "the Olympic Games represent the true spirit of competition and passion."[2] Indeed, these digital games were connected to the "real life" Olympics at

This essay was written in 2007.

many levels. Mario and Sonic's joint venture was set up by Sega's acquisition of game rights for the 2008 Olympics in Beijing through a partnership with ISM, in turn designated by the International Olympic Committee (IOC) as the exclusive license holder of the 2008 games for interactive entertainment software associated with these Olympic Games.[3] Let the games begin!

It would be too easy to dismiss these enthusiastic sentiments as mere hype calculated to propel sales of a new commercial product. Yet the marketing power of associating Mario and Sonic with the Olympic Games reminds us of the lofty goals that the Olympic movement has associated for more than a century with *games*. The historical importance of the Olympics as a cultural and political phenomenon, as the "leading contemporary global mega-event," even as the "civic religion" of the 20th century, draws on the widely accepted notion that "the Games have always brought people together in peace to respect universal moral principles."[4] This chapter is motivated by the contemporary relevance of this idea, in the sense that digital games potentially provide a new playing field for it. Rather than locating the digital version of Olympism in the competition of game characters, however, my focus is on efforts to encourage international competition among *players* of digital games. My focus will be on a particular event, the annual World Cyber Games (WCG), commonly known as the "video game Olympics." The official slogan invented by the WCG's Korean organizers, "Beyond the Game," lets us in on their conception of this international tournament of digital gaming as "not just . . . a world game tournament," because in their view, it "also combines the world to create harmony and enjoyment." Founded in 2001, the organizers have insisted all along that the WCG is a "world cultural festival" that fosters respect among competitors from dozens of countries. Like the International Olympic Committee, they have expressed hope that their games might even contribute to world peace by emphasizing sportsmanship, fair competition, and other "ethical principles."[5] In the first part of this chapter I will take these claims seriously by exploring the historical background for the Olympic Ideal as a fundamental context for the World Cyber Games and other international digital game competitions.

The second and larger section of the chapter turns to the growth of competitive e-sports. The WCG has been the most ambitious international tournament in terms of expressed goals such as global harmony and a positive digital culture, while like other cyberathletic competition venues it has also been connected to efforts to build the commercial potential of these events as a form of entertainment. This tension between conciliation and commerce in the Olympic Ideal of global e-sports is hardly shocking in light of similar issues in athletic competitions of a more traditional sort. This is not the only issue with respect to the Olympic Ideal as applied to games, including mediated digital games. Another possible contradiction is built into a central principle of game theory, the notion of the "magic circle" associated with Johan Huizinga's cultural history of play. The idea that games take place in a reality separate from that of our everyday lives, with their own rules and consequences, carries significant implications for the idea of a beneficial crossover effect from the field of competition to civic affairs and international politics. Is the goal of a positive digital culture undermined when values learned from simulated game worlds can be expected to portray outcomes and consequences that vary considerably from those of "real life"? It is important to consider whether the Olympic Ideal is undercut by the magic circle and if so, whether this is sufficient reason to shake apart the claim that agonistic play in digital games can shape the values of global citizens "beyond the game."

The Olympic Ideal

In October 1993, the General Assembly of the United Nations declared 1994 to be the "International Year of Sport and the Olympic Ideal." Its resolution praised the "Olympic movement for its ideal to promote international understanding among the youth of the world through sport and culture" and endorsed efforts by the IOC to "build a peaceful and better world through sport."[6] The year marked the one hundredth anniversary of the International Athletic Congress convened by Pierre de Frédy, Baron de Coubertin (1863–1937) in Paris, at which Coubertin unveiled his proposal to establish the modern Olympics on the model of the ancient games held in Greece. He founded and became general secretary of the International Olympic Committee, which he continued to lead through the eighth modern Olympiad, held at Paris in 1924. Coubertin's influence and writings, along with the work of the IOC, define the Olympic Ideal, a set of core values that extend well beyond the Games' official motto, "citius, altius, fortius" (faster, higher, stronger) to a conception of the pedagogical, cultural, and moral value of athletic achievement and competition.

The Olympic Ideal stresses achievement alongside participation, and competition with moderation. It connects sports with cultural development and global understanding in the service of humanity. Helmut Digel, vice president of the German Olympic Committee since 1994, has summarized this conception as calling on humanity as a whole to "find a way to sports, independent of race and religion, so that sports can contribute to world peace and help to build a better world in its own way."[7] The "fundamental principles of Olympism" contained in the Olympic Charter define it as "a philosophy of life" with the goal of placing "sport at the service of the harmonious development of man, with a view to promoting a peaceful society concerned with the preservation of human dignity." These ideas put a substantial *external* burden on sports competition. They imply the desirability of universal access, thus leading to the principle that "the practice of sport is a human right."[8] There is an implicit tension, however, in this principle's connection to high-performance, competitive sports. Broad participation in a program that essentially amounts to "mens sana in corpore sano" does not seem to be directly connected to the evolution of the Olympics into an elaborate, expensive, global event featuring elite athletes in intense competition against one another. Put it another way, ideas such as "fair play" or "being there is everything" seem oddly fitted to "faster, higher, stronger." In order better to understand these tensions, it will be necessary briefly to revisit the pedagogic and philosophical underpinnings of Coubertin's Olympism.

The story of Coubertin's path to the Olympic Ideal is well known. At its core, his conception of Olympic values was born out of a sensibility that was able to juggle commitment to international understanding and a distinctly competitive national perspective.[9] There were several important influences on Coubertin's thinking. His belief in the superiority of the English educational system, with its emphasis on physical education, coupled to a Darwinist analysis of the relative superiority of national cultures, were two legs of the stool. The other was his fervent Philhellenism, fueled by the first archaeological excavations by a German team at Olympia between 1875 and 1881. Coubertin's studies and visits to educational institutions in the United States, Canada, and England, particularly the pedagogical writings of Thomas Arnold, guided his conclusion that firm commitment to athletic education had been the key factor pushing the English ahead of the French school system. He found confirmation of this notion in the ancient Olympic Games; exciting results from the Olympia dig were perfectly timed to influence the young student, whose Jesuit instructors were instilling in him a passion for ancient history. Coubertin's basic insight was in seeing the revival of international competition on a common, international playing field as fueling the creation of local and national institutions, such as in schools, for the athletic training of youth. He believed this to be a lesson learned from the ancient Games, fully expecting the "restoration of the gymnasium" as a corollary of the return of the Olympiad.[10] This was the resolution of the tension between elite competition and universal improvement, as energetic competition among top athletes would encourage attention to physical training. In turn, the improvement of education would produce better citizens, who could be expected to form a greater harmony of minds and bodies. Competition thus could be seen as a civic virtue.

The focus of this chapter is competitive e-sports based on digital games. So we must consider whether the ideals Coubertin and his followers associated with athletic competition can be transferred to any sort of agonistic competition. Coubertin, like those who have followed after him, referred to the Olympiad interchangeably as the Olympic *Games*; it is reasonable to expect that his ideals could equally be applied to athletic and ludic competition. In planning for the 1908 Olympiad in Rome, Coubertin provided a clue as to his thinking when he identified nineteen "disciplines" for the competitions, ranging from gymnastics and track and field to arts ("directly inspired by the sporting idea") and motor racing. One of these disciplines he indeed called "games," which he identified at that time as football (Rugby Union and Association), cricket, lawn tennis, and polo.[11] A few years later, in a short essay called "The Various Sports in the Olympic Program" written in 1910, Coubertin rejected the idea that any one form of competition could constitute an Olympic competition; rather, the term referred for him to "gatherings for a variety of sports." He would not narrow the focus beyond insisting that the "Olympic Games are the temple of muscular activity in *the most widely varied forms possible*, though there is no need to assign degrees within some hierarchy of beauty or nobility" (italics added). Competitions embracing the ideals of competition in a variety of disciplines and aiming for the improvement of humanity would qualify as Olympics, and Coubertin invited anyone willing to organize such events to "go ahead and use" the term.[12] Competitive digital games would seem to qualify; high-level mastery of game syntax, strategy, and tactics requires a mix of physical and mental skills, particularly in multiplayer games that create conditions for competitive player performance, even if these conditions depend on computer technology. Sports accepted by Coubertin, such as shooting, archery, fishing, and motor racing, called for comparable relationships of physical skill and mastery of equipment.

Norbert Müller argued only a few years ago that nobody would associate the Olympics with automobile racing.[13] Not so for digital games today; a proposal presented to the organizers of the 2008 Olympics could lead to the appearance of digital games as a demonstration sport in Beijing.[14] While Coubertin might have reservations about the levels of physical activity involved, there is reason to think that he would not object in principle to e-sports as part of a sufficiently broad program of competitions.

Competitive Games

What do we mean by terms such as e-sports or cyber games? Competition play has always been at the heart of computer-based gameplay as performance. The history of competitive performance in digital games began with the first popular computer game, *Spacewar!*, developed at MIT in 1962. A computer game competition of sorts was featured ten years later in a paean to hacker culture written by Stewart Brand for *Rolling Stone*; he framed a reportage about work in computer science and artificial intelligence inside a description of the "First Intergalactic *Spacewar!* Olympics" organized by him and sponsored by *Rolling Stone* in October 1972 at the Stanford Artificial Intelligence Laboratory. Brand enthusiastically described players (photographed by Annie Leibovitz) "brandishing control buttons in triumph" after winning efforts in the tournament. Ultimately, Brand's article was not about cyberathletic performance. He was more concerned with displays of technical mastery such as a superior programming trick or hack as well as a new "bonding of human and machine" presaging things to come. Yet even though the Stanford tournament was something of a staged event, Brand's article established a framework for seeing computer game performance as a tip-off about the convergence of competitive skill, programming wizardry, and player communities.[15]

Any number of head-to-head competitive and multiplayer games could be played in arcades, homes, and computer labs during the 1970s and 1980s. Public and televised competitions were infrequent, but arcade owner Walter Day's Twin Galaxies brought together top players for "high score" contests beginning in the early 1980s. Twin Galaxies organized an invitation-only "Video Game World Championship" broadcast on ABC-TV's *That's Incredible* in 1983, which Day called the "North American Video Game Olympics."[16] Networked competitive games such as *Maze War*, developed by a group at the NASA Ames Research Center in Mountain View, California, were available as early as 1973, but games of this sort were not widely available—at least not outside computer research labs—until the early 1990s.

The breakthrough titles for competitive e-sports were id Software's *DOOM* and *Quake*, both networked first-person shooter (FPS) action games. Released in December 1993, *DOOM* lead programmer and id cofounder John Carmack introduced a 3D graphics engine, fast peer-to-peer networking for multi-player gaming, and a modular design that let authors outside id create new levels. The key innovation, however, was a new mode of competitive play devised by another id cofounder, John Romero, that he dubbed "deathmatch." When Carmack told Romero of his intention to program networking *into DOOM*, so that human players could shoot each other on the various playing fields defined by the game's maps, Romero immediately realized that this technology would open up a new realm of digital play; as their biographer has put it, "Romero imagined two players shooting rockets at each other, their missiles sailing across the screen. Oh my God, he thought, no one has ever seen that in a game. Sure, it was fun to shoot monsters, but ultimately these were soulless creatures controlled by a computer. Now gamers could play against spontaneous human beings—opponents who could think and strategize and scream."[17] Other game developers and programmers experienced the same eureka! moment within a year or two of the id team, as they rapidly learned to appreciate networked multiplayer games as a competitive platform. Blizzard Entertainment's *Warcraft: Orcs and Humans* (1994) played nearly the same role in defining real-time strategy (RTS) games as a competitive game form as *DOOM* had done for FPS games. Allen Adham, one of Blizzard's cofounders, recalled that during coding and playtesting, his programming team also learned how multiplayer, networked competition would transform digital gaming as a performance space: "The feeling of sitting alone in front of a computer, looking at your screen and realizing that off in cyberspace somewhere there is another sentient being building, exploring, and plotting your destruction was exhilarating. It was a totally different feeling from the hundreds of strategy games I had played against computer AIs, or even multiplayer games where your enemy sat beside you and shared a monitor." Like Romero, Adham found that his first multiplayer game was "definitely a defining gaming moment for me." For the emerging game genres of the mid-1990s, multiplayer competition was the place to be.[18]

Romero and Adham, like many others, recognized that network technology had created a compelling basis for multiplayer competition. FPS and

RTS games mounted on local area networks (LANs) and eventually across the emerging Internet created countless arenas for human versus human competition by the mid-1990s. In 1994, during the UN's International Year of Sport and the Olympic Ideal, *DOOM* players were beginning to earn reputations for their superior skills. The immense popularity enjoyed by *DOOM* led to other games that emphasized competitive format, from FPS games such as *Quake* (1996), and *Half-Life: Counter-Strike* (1999) to RTS games such as *Warcraft*, *Warcraft II: Tides of Darkness* (1995), *Command & Conquer* (1995), and *Starcraft* (1998). Players of these games focused on high-performance play, and their time and energy established a basis for spectatorship through the making of clansites, forums, news sites, demo movies, and replays distributed by the same network technology that had made such competition possible. For example, demo movies certified the status of star players. Beginning in 1994, the Doom Honorific Title (DHT) Program, a game rating system, became "the means by which good players can objectively prove to the world that they are as good as they claim."[19] This certification process used validated replays to promote the performance of gameplay through demo movies. Players proved that they could achieve certain levels of proficiency, and lesser players watched these movies to pick up on their techniques and tricks. Establishing a basis for spectatorship by recording gameplay also encouraged growth of the player community, as individuals and regular teams of players joined together in semiofficial "clans" as another way to establish reputations based on superior play. Famous teams such as the *Quake* clan Schroet Kommando in Germany, the *Counter-Strike* team Team3D in the United States, and the multi-game 4Kings clan in Europe (with the motto "Gaming Royalty") paced the increasing status and visibility of elite players. High levels of competitive skills and the perception that computing technology served as the highly skilled player's "equipment" suggested opportunities for sponsorship. By the early 2000s, international reputations, fan clubs, Web sites, and equipment endorsements led to the expectation that corporate sponsorship validated these dream teams of cybergaming.

These factors propelled the development of competitive digital games and pulled together communities of dedicated players and fans. It was not long after the emergence of id's FPS games as a solid platform for multiplayer competition before the possibility of professional play in the form of sponsored teams and competition prizes was widely discussed. The Cyberathlete Professional League (CPL) was launched in the summer of 1997 in the United States.

It has since offered an international tournament schedule featuring primarily FPS games. A decade later, in 2007, the competitive e-sports scene comprises more than a half dozen leagues and tournament seasons, with names like the Electronic Sports League (ESL, founded in Germany in 1997), Cyberathlete Amateur League (CAL, a division of the CPL launched in 2001), the World Series of Video Games (WSVG, an international competition season launched in 2006), the Electronic Sports World Cup (ESWC, founded in France in 2003), and more. These names indicate the clear intention of the organizers to identify their events as a form of sports competition, typically called electronic sports or simply e-sports. The portrayal of the World Cyber Games as the Olympics of e-sports is echoed by the identification of these leagues and venues in terms calling up the full roster of marquee events in national and global sports, from the World Series to the World Cup. Verbally only perhaps, the ESWC has wrested the baton from the Olympics movement altogether by immodestly calling itself "the greatest sporting project of the 21st century."[20]

Spectatorship and sponsorship have been lynchpins of the competitive gaming scene since the demo movie was established as a popular replay format in *DOOM*. Today, tournaments are routinely made available via broadcast, Webcast, and Shoutcast through a variety of outlets ranging from Europe's gamer.tv and Korean cable television stations such as Ongamenet to *Half-Life* TV (HLTV) on the Web and Web sites such as GotFrag and WCReplays (for *Warcraft III* games).[21] The tight interlinkage of sponsorship, online community, and competition seasons built around tournaments is perhaps best exemplified by GameRiot, a traveling video game event from Games Media Properties (GMP) mixing demonstrations and competitive matches. It was founded in 2003 and, as an indication of the careful attention to marketing and sponsorship, toured as "SpikeTV presents GameRiot powered by Xbox." In 2004, it traveled to more than eighty cities, with sponsorship from nine media and consumer brands. For Matt Ringel, GMP's president, the connection between commerce and public events such as GameRiot was paramount and "the real lodestar . . . is to what degree does trial drive sales? People do these types of events for awareness, building buzz and preorders."[22] Competitive leagues were also heavily sponsored, such as the WSVG, "presented by Intel" and allied with CBS Sports and "official suppliers" such as Nvidia and Creative; Intel had previously been one of the lead sponsors of the CPL as well. As we are about to see, the translation of the Olympic Ideal for e-sports in the form of World Cyber Games has embraced corporate sponsorship.

The World Cyber Games

A reporter for *JIVE Magazine* observed in a story about the 2003 WCG as the "Olympics of the Video Game World" that "it only stands to reason that—with the tremendous popularity of worldwide video game competition—a best-of-the-best tournament was an inevitable prospect."[23] The World Cyber Games had been founded in Korea as the World Cyber Game Challenge in August 2000. The annual competition and festival of e-sports from the beginning was managed by the Korean company International Cyber Marketing (ICM), created only a month before the WCG. The first competition took place in October of that year at the Everland theme park located outside Yongin, Korea. The rhetoric of the WCG's publicity as well as information provided to players and referees has consistently echoed the Olympic Ideal. In 2007, for example, the official press kit for the WCG to be held in Seattle later in the year explains the WCG's "Beyond the Game" slogan as representing a "commitment to harmony and building friendships among the world's youth through e-Sports."[24] The organization, formats, and technology of the games themselves have borrowed substantially from both existing tournaments and the Olympics. Open preliminary competitions are staged in dozens of participating countries, leading to the selection of top finishers in head-to-head LAN play, as well as the selection of top teams for the *Counter-Strike* competition, arguably the marquee event at the games. At the final events at the WCG itself, teams of referees oversee the games and adjudicate disputes; players are restricted in the degree to which they are allowed to personalize the equipment (mouses, keyboards, specific software drivers, and key bindings are allowed, but the tournament venue provides computers, screens, and the networking infrastructure), and provision is made for viewing of the game by spectators at the venue in ways that prevent players from peaking at what their opponents are doing.[25] The formal rules with regard to qualifying and gameplay (such as the selection of maps for RTS games) are minimal, but provide an opportunity for emphasis on the principles of fair play and mutual respect emphasized by the WCG's goals and philosophy.

The first games held at Everland went smoothly. The finals competitions comprised four FPS, sports, and strategy titles, including *Starcraft*, by then the favorite of the rapidly growing e-sports culture in Korea; in the seventh WCG held in 2006 in Monza, Italy, the number of titles had risen to eight, with a similar mix of games tilted toward the Korean love for RTS games.[26] The four tournaments included 174 players from 17 countries, out of roughly 168,000 competitors who entered national trials in each of the nations (see table 1). It is tempting to compare these numbers to the first Olympic Games in Athens, which attracted 311 male athletes, but only 81 from outside Greece.[27] In one of his more quoted passages from his *Olympic Memoirs*, Coubertin had insisted on the value of high-performance competition, even if reached by few athletes, as a motivating force even for those with less ability: "For one hundred to go in for physical education, fifty have to go in for sport. For fifty to go in for sport, twenty will have to specialize. For twenty to specialize, five will have to show themselves capable of astounding feats."[28] In other words, the Olympic Ideal values every level of competition as a part of physical education while recognizing the multiplier effect of admiration for the athlete who pushes to extreme levels of accomplishment. Even if the number of finalists was relatively small, by the measure of participation the first WCG could be considered a success, with more than 1,000 players taking part for every 1 able to reach the finals. In 2004, more than a million players competed, with 642 qualifiers from 63 countries attending the finals in San Francisco, a ratio of more than 1,500 contestants to each finalist.[29] The 2006 WCG included 700 players from 70 countries, roughly a fourfold increase over the games held six years earlier (see table 1). While participation in the WCG has not been disappointing, these numbers still beg the question of whether the numbers are seen as a sign of success in terms of the ideals and ambitions of the games' organizers and sponsors. The spectator turnout, on the other hand, has at times been disappointing, particularly at the poorly attended 2004 WCG in the United States.

Unlike the traditional Olympic movement, the WCG has never gone through a period of adherence to the cult of the amateur athlete, upheld religiously by every president of the IOC through the end of Avery Brundage's reign in 1972. More recent scholarship has discredited Coubertin's idealization of the Greek athlete as a historical basis for this conception of the purity of the amateur.[30] Coubertin considered the "noble and chivalrous quality" of the Greek athlete as essential if sports were to continue to "play within the education of modern peoples the admirable role which the Greek masters attributed to it." He contrasted this to the "human imperfection"

World Cyber Games statistics

	2000	2001	2002	2003	2004	2005	2006
Participants	168,000	389,000	450,000	600,000	1,000,000	1,250,000	1,300,000
Finalists	174	430	462	562	642	679	700
Finalist countries	17	37	45	55	63	67	70
Number of games	4	6	6	7	8	8	8
Total prize money	$400,000	$600,000	$1,300,000	$2,000,000	$2,500,000	$2,500,000	Not available

Source: Data compiled from WCG website, history pages, e.g., https://web.archive.org/web/20070516043128/http://www.worldcybergames.com/icm/eng_history2006.html.

that transforms the "Olympian athlete into a circus gladiator." By this, he meant to criticize the tendency toward "lucre and professionalism" that already in 1894 seemed a distinct threat to amateurism, as national organizations compromised and contradicted the status of the amateur in order to keep their best athletes on the field. This was perhaps a case of overzealous national athletic organizations giving "faster, higher, stronger" precedence as a competitive principle over Coubertin's pedagogical and international goals. His clarification of the firm centrality of the amateur principle thus held sway over Olympic competitions for nearly a century.[31] Since 1986, the IOC has left the decision concerning participation of professional athletes in Olympic sports up to the cognizant international federations. When the US "dream team" won the gold medal in men's basketball at the Barcelona Olympics in 1992, the last nail had been driven into the coffin of amateurism as a key tenet of the Olympic movement; the word "amateur" is missing entirely from the text of the latest Olympic Charter (2004). The only requirements for participation are compliance with the charter, respect for the "spirit of fair play and non-violence," and adherence to the World Anti-Doping Code.[32]

The attainment of professional status and corporate sponsorship for cyberathletes as a kind of holy grail of digital game competition would seem to collide headlong with the Olympic Ideal of the amateur athlete. On this score, the WCG has followed the current practice of the Olympics and the culture of competitive e-sports, rather than Coubertin's Olympic Ideal. One of the major draws for the WCG has been the availability of substantial cash prizes for winners of the several tournaments held for each of several games. The inaugural games in 2000 offered first prizes of $25,000 for each of the four featured games: *Quake III*, *FIFA 2000*, *Age of Empires II*, and *Starcraft*. The total purse of $400,000 for the 2000 WCG had risen more than sixfold to $2,500,000 in 2005. Corporate sponsorship has played an essential role in the WCG from the very beginning. Samsung, the primary sponsor of the WCG, had since the 1970s led the growth of Korean consumer electronics, manufacturing, semiconductor, and telecommunications industries. The Samsung conglomerate of businesses had evolved into a global player competing at the level of worldwide companies such as Sony, Motorola, and Nokia by the end of the 1990s.[33] Under the leadership of Kun-hee Lee, son of the company's founder, it embarked on a program of "new management" that put new emphasis on innovation as the foundation of competitiveness in the global marketplace. Lee had been an active athlete since his youth, and he continued to promote sports in Korea from his position as a corporate leader. Samsung acquired several professional sports teams in and outside of Korea and sponsored tournaments and amateur teams in a wide variety of sports. Lee himself was recruited by Juan Antonio Samaranch to join the IOC in 1996, the year of the Olympic Games in Atlanta, Georgia; in 1997, Samsung became a Worldwide Olympic Partner for the Games.[34] Samsung's sponsorship of the Olympics thus preceded its active role in promoting the WCG. For example, it provided some fifty-eight thousand pieces of wireless telecommunications equipment to the 1998 Winter Olympics in Nagano, the 2000 Games in Sydney, and the 2002 Winter Olympics in Salt Lake City, and some fourteen thousand mobile phones to the Summer Games in Athens held in 2004; a corporate press packet justified the extent of Samsung's support as deriving from what it called the "ultimate mission" of the corporation, "fostering unity among the global community." The same document revealed the corporation's "sports and business philosophy" in similar terms, citing the "unique role" of sports "in unifying people regardless of age, race, or gender" and tracing these corporate values directly to Kun-hee Lee's personal commitment as chairman of the company. A company publication on "corporate citizenship" stressed, "Sporting initiatives have become a focus of Samsung's marketing and communications strategies in the markets where we operate. Through sports sponsorship, Samsung has built strong brand awareness, as well as a reputation for excellence in corporate citizenship."[35] Like the 1988 Olympic Games held in Seoul had done for Korea as a nation, Samsung's initiatives in the area of sports sponsorship thus testified in no small measure that the corporation had truly arrived on the international scene.

Given its corporate presence and the experience of its key executives in the international sports establishment, as well as its growing reputation as a high-technology firm, it was hardly surprising that Samsung should take a leading role in the sponsorship of international cyberathletic competition. In August 2000, the vice president of Samsung Electronics, Yun Jong Yong, assumed the chair of the WCG Organizing Committee, and Samsung has sponsored all seven of the tournaments thus far. The promotional activities of the WCG have from the beginning been closely tied to national ambitions of economic competitiveness. By the late 1990s, South Korea was well on the way to becoming the "bandwidth capital of the world."[36] Since console-based video games were largely banned due to long-standing restrictions on Japanese cultural imports, Korea's lead in broadband penetration was driven by networked games, often played in PC bangs—neighborhood cybercafes—by near-fanatic legions of players devoted to online games led by *Lineage* and *Starcraft*. By 1999, a year after the release of *Starcraft*, soon to become something of a national pastime in Korea, there were more than seven million online gamers in a nation of roughly forty-five million people. The president of NCSoft, publisher of the massively multiplayer game *Lineage*, reiterated what id's and Blizzard's developers had discovered a half decade earlier when he justified the Korean passion for competitive games as being based on the idea that "human beings want to interact with other human beings. Gaming with another person has got to be more fun than playing against a machine."[37] As ICM and Samsung were readying the WCG, the South Korean government had already begun to appreciate the significance that online games could play in promoting the country's nascent broadband industry. The Korea Game Promotion Center, later renamed the Korea Game Development and Promotion Institute (KGDI) and finally the Korea Game Industry Agency (KOGIA), was founded in 2001 under the sponsorship of the Ministry of Culture and Tourism. The ministry also joined Samsung that year as a lead organizer of the WCG. These initiatives were all part of a concerted program to promote the Korean game industry into a position of international leadership. Samsung's sponsorship of the WCG as the "Olympics of online gaming"[38] must therefore be seen as part of an established tradition of industry-government cooperation on behalf of Korea's economic competitiveness on the Asian and international playing fields.

The point here is not to berate the economic motivations of the Korean founders of the WCG by contrast to Coubertin's somehow purer form of pedagogic and cultural idealism a century earlier. Given criticism of the increasing role of marketing, licensing, professionalism, and business interests in the organization of the Olympics, it would hardly seem fair to raise a different standard for the cyber games. A more apt comparison might be from the perspective that the central idea of internationalism in both the Olympic Ideal and the WCG has been framed by strong national perspectives on international competition. For Coubertin, this was concern about the failings of the French educational system vis-à-vis nations with a more enlightened outlook about physical education. For the organizers of the WCG, the local context for global games was the drive to build South Korea's presence in the global marketplace of high-technology industries.

The Magic Circle and the Olympic Ideal

Tensions between world harmony and international competition, or between the ideal of the amateur and the rewards of professional status, naturally lead to skepticism about the contribution that sports, games, or athletics can make to intercultural communication and understanding. Ironically, it is the inevitability of contact between real-world issues—sponsorship, financing, elitism, rivalry—and the agonistic arena that seems to weaken commitment to the notion that play can make a meaningful difference of the sort the Olympic Ideal proposes for athletic games. We have seen that Coubertin's admiration for Anglo-American physical education and his idealization of the ancient Greek games intersected with patriotism and a heartfelt desire for France to catch up on the international playing field. Again, this does not strike me as so fundamentally different from the motivations of the Korean captains of industry and government who supported the World Cyber Games; like Coubertin, they also believed sincerely in the value of transcultural contact inspired by games, while at the same time rooting for the home team.

There is, however, an inherent tension in the nature of play that the Olympic Ideal and its WCG counterpart rarely confront head on. Their values clearly depend on the idea that there is something inherently valuable, from a moral point of view, in certain forms of play and competition. A key notion in this regard is that of "fair play." Coubertin considered fair play to be an essential aspect of the Olympic Ideal, meaning that competitors would essentially agree not to circumvent the rules of the game. This was the basis for inclusion, harmony, and the competitions themselves, as well as underpinning the goal of respectful transnational communication. As Kay Schaffer and Sidonie Smith have pointed out, this concept of "fair play" was another idea Coubertin had taken away from the English school system.[39] Adherence to the principle of fair play in sports and other endeavors marked the English gentleman and was witnessed by spectators at sporting events as "rituals of gentlemanly masculinity." Coubertin's deployment of this and other ideas as part of the Olympic Ideal can be seen as mobilizing athletic competition as a unifying social force, invoking moral superiority, physical culture, and rituals of competition on behalf of the state. The absolute ethos of "fair play" wrapped inside Coubertin's conception of athletic competition on sacred Olympic ground thus seems diminished when unpacked in terms of its impact in real-world terms.

This problem of the transmutability of the system inside the game to the world outside it is familiar to game studies. In *Homo Ludens*, Johan Huizinga was preoccupied by the cultural significance of play as a human activity, specifically his conviction that the "rules of a game are absolutely binding and allow no doubt." Play meant residing in a world apart, which Huizinga explicated as follows in one of the most cited passages from the book: "Just as there is no formal difference between play and ritual, so the 'consecrated spot' cannot be formally distinguished from the play-ground. The arena, the card-table, the *magic circle*, the temple, the stage, the screen, the tennis court, the court of justice, etc., are all in form and function playgrounds, i.e. forbidden spots, isolated, hedged round, hallowed, within which special rules obtain. All are temporary worlds within the ordinary world, dedicated to the performance of an act apart" (italics added). While Huizinga in fact did not too often use the term, the magic circle has since stood in most readings of *Homo Ludens* as shorthand for this idea of the game space as a "world apart." For Huizinga, violating the principle of fair play—to cheat—meant precisely that a player was merely pretending to inhabit the "magic circle while not really being there"; he denied the possibility of playing a game while trespassing against its rules.[40] Huizinga also derived characteristics of presence in the magic circle that also seem to be traced in the Olympic Ideal; for example, differences of rank are abolished there, and "whoever steps inside it is sacrosanct for the time being."[41] Huizinga's analysis of play makes it difficult to see how values such as "fair play" allow for dual citizenship both in the magic circle and the world outside the game. His insistence on the impermeability of play worlds does not so much challenge the Olympic Ideal as a charter for life in the game as it leads us to at least question the claim that it might be a basis for conscious improvement of the real world.

Conclusion: The Outside and the Inside

We have seen that the WCG's claim that these e-sport Olympics are "not just a world game tournament" faces two distinct problems. The contributions of a worldwide games competition to the desirable ends of world harmony, peace, and maximization of human potential seem to be threatened from both sides, from the outside and the inside. The outside problem is an intrusion of the "real world" on what appears to be special about the playing field. While concern about distractions such as financing and organization goes back to Coubertin, most critics would consider the current crisis as having derived from the increasing complexity and ambition of the Olympics as a global mega-event. The WCG has not reached anything near this scale in terms of participants or attention, but it has embraced close cooperation with industry and national development efforts as part of what can only be called its business plan. The inside problem gives us the other side of the coin. While too much of the outside world may contaminate the virtues of play, if the magic circle is a separate reality, how does play lead to progress beyond the field of competition? For Coubertin, answers could be found in benefits such as promoting physical culture and what participants took away from being there. The World Cyber Games also subscribes to the argument that players are a vanguard by insisting on values such as fair play and mutual respect among competitors. The WCG justifies itself in terms derived from the Olympic Ideal, but transformed for the digital age; it is a "digital cultural festival, where language and cultural festival barriers are stripped away" and promotes "harmony of humankind through e-sports" as a step toward a "healthy cyber culture."[42]

So we are left with these two questions. First, will commercialization of the World Cyber Games interfere with the values expressed in the Olympic Ideal? Does the undeniable fact of commercial sponsorship constitute an erosion or even repudiation of these values? In other terms, we might ask if the outside problem is really not a problem at all, but rather a suggestion that the WCG as the Olympics of the digital age will operate under a different set of expectations. Of course, the traditional Olympic Games face the same question. The influence of Samsung's leadership on the WGC was similar to the role played by Adidas's Horst Dassler in restructuring the sponsorship structure of the Olympics.[43] The Games held in Barcelona (1992) and Atlanta (1996) perhaps represent a high point of extravagance, gigantic spectacle, the influence of television, and runaway commercialism. This has led to concern about the future direction of the Games. Andrew Young, former mayor of Atlanta and US ambassador to the United Nations, famously responded to critics of planning for the Atlanta games by insisting that "the commercialization of sport is the democratization of sport."[44] At the time, Young was serving as chairman of the Atlanta Committee for the Olympic Games (ACOG) and was a key player in bringing the XXVIth Olympiad to Georgia, termed derisively by some as the "Coca-Cola Olympics." His declaration provided an interesting twist on a remark by Coubertin that "rowers, runners and fencers" would be the "free trade of the future," arguing that athletics presents two features conducive to this position in the modern world: "It is democratic and international."[45] Yet others saw Young's Olympics as a betrayal of the Olympic Ideal. Martin Luther King III, for example, was quoted in the *Atlanta Journal-Constitution* as declaring that "greed, exclusivity and elitism have become the symbols of Atlanta's Olympic movement."[46] These statements must, of course, also be read in terms of local politics and the organizational challenges faced by the ACOG; yet they unintentionally provide some guidance for the World Cyber Games as a product of corporate sponsorship and national economic ambition that nevertheless strives to reach "beyond the game."

Without commercial sponsorship, no organization could possibly hope to stage the Olympic Games of today. That is a reality. While far more modest in size, the World Cyber Games is perhaps even less conceivable without close ties to commercial sponsors, from hardware manufacturers to game developers. Without computer and networking technology, the competitive platforms and arenas for these competitions would not exist. When professional players such as Johnathan Wendel, a.k.a. Fatal1ty, are able to convert the perception that game performance depends closely on superior gear into numerous business partnerships and brands, it should hardly be surprising that the companies that build gaming equipment are visible at competition venues. Indeed, Wendel has become the official spokesman of the Championship Gaming Series launched in 2007, with support from an international group of satellite television companies. On its official Web site, the WCG takes higher ground when it quotes Kim Dae-jung, former president of South Korea and recipient of the 2000 Nobel Peace Prize, as a way of enunciating the value of its own games: "Cyber games are a business with high added value, based on knowledge and cultural creativity. [They] also serve as a link that interconnects young people of the world."[47] Cyberathletic competition, then, may well serve as the free trade of the future.

But what about the inside problem? If, as Huizinga argued in *Homo Ludens*, games are played for their own sake, the fundamental claims of the Olympic Ideal and the WCG's "Beyond the Game" motto become quite mysterious indeed. Huizinga drew attention to this issue as a historical problem, not so much out of concern about the relevance of play as from a sense that the contemporary world had lost its capacity to develop through the spirit and forms of play.[48] Writing in the late 1930s, Huizinga was already struck by the systematization, regimentation, and professionalization of sports, noting that "something of the pure play-quality" of sport had been lost. Play, in effect, had become serious business. This observation is hardly surprising to us, some seventy years further down the road of professionalization and commerce in sports. Huizinga then proceeds to a more provocative, tentative, even contradictory observation. Just as sport seems no longer to be a pure form of play, commerce and other serious activities have themselves adopted qualities previously associated with the magic circle. As he put it, "Certain activities whose whole *raison d'être* lies in the field of material interest, and which had nothing of play about them in their initial stages, develop what we can only call play-forms as a secondary characteristic." While sport and athletics "stiffened" into seriousness, serious business seemed to be "degenerating into play but still being called serious."[49] He traced this development to modern communication and other technologies whose impact was the creation of a kind of universal playing field for competition in every field of human activity, but especially commerce. Business, according to Huizinga, had become a continuous, multilevel, and international field of play. These observations led him to consider the play element in fields such as contemporary art and science. He was left with an apparent contradiction: on the one hand, the play element in civilization had waned in terms of its historical meaning and purpose; on the other, the contemporary world is dominated by a kind of "false play," such that "it becomes increasingly difficult to tell where play ends and non-play begins."[50]

Huizinga tempered an impulse toward pessimism about the value of play. At first, he argued, there appears to be "little opportunity for it in the field of international relationships." So much for the Olympic Ideal. And yet, he continued, it might be beneficial to look at the maintenance of a system of international law and relationships as a kind of system of rules. Like the supposed sanctity of the Olympic Games, the magic circle of international civility, even the notion of warfare as a noble game, seemed to be eroding around Huizinga as he wrote in the late 1930s: "The observance of play-rules is nowhere more imperative than in the relations between countries and states. Once they are broken, society falls into barbarism and chaos." Huizinga's argument insisted that "real civilization cannot exist in the absence of a certain play-element," and his conclusion dovetailed with the Olympic Ideal and the WCG credo in locating the critical element in the concept of "fair play."[51] True play, not a disingenuous form, and adherence to a system of rules are the core elements of this play element. *Homo Ludens* thus concludes by insisting on the importance of play as a form of commitment to an ethos of playing by the rules. This, of course, is a notion that might be considered utopian with respect to the culture of digital play, which has often given wide berth to subversion, hacking, and even forms of cheating. In a recent book on cheating in video games, Mia Consalvo has argued that the separation of games as "walled off spaces" encourages the full exercise of the imagination in these spaces.[52] This might seem to put the "cultural creativity" of cyber games at odds with Huizinga's notion of absolute adherence to a shared system of rules. Of course, cheating might be seen as an equally pervasive practice in nearly all forms of Olympic and professional athletic competition today. These issues around the permeability or impermeability of game spaces are at the heart of what ultimately remains a leap of faith regarding the Olympic Ideal's relevance for competitive e-sports. In the form of the World Cyber Games, the Olympic Ideal of global sports competition as a basis for international understanding has been reshaped for the age of cyberathleticism. At the same time, the WCG has fully embraced a notion of globalism rooted as deeply in commerce as culture and as fully committed to performance as pedagogy. Perhaps this too is simply a sign of the times.

NOTES

1. Kris Graft, "Sega Carries Torch for Beijing 2008," *Next Generation*, December 5, 2005, http://www.next-gen.biz/index.php?option=com_content&task=view&id=1791.

2. "Sega and Nintendo Join Forces for *Mario & Sonic at the Olympic Games*," press release, March 28, 2007, https://www.gamesindustry.biz/sega-and-nintendo-join-forces-for-mario-sonic-at-the-olympic-games.

3. ISM had produced official IOC-licensed video games for the 2000 and 2004 Summer Games and 2002 and 2006 Winter Games, as well as producing a variety of sports games under various licenses since the late 1990s.

4. Maurice Roche, *Mega-Events and Modernity: Olympics and Expos in the Growth of Global Culture* (London: Routledge: 2000), 99; Eilert Herms, "Die olympische Bewegung der Neuzeit. Sozialpolitisches Programm und reale Entwicklung," in *Nachdenken über Olympia. Über Sinn und Zukunft der Olympischen Spiele*, ed. Helmut Digel (Tübingen: Attempto, 2004), 74; Official Web site of the Olympic Movement, http://www.olympic.org/uk/games/index_uk.asp.

5. "Beyond the Game," World Cyber Games, http://www.worldcybergames.com/6th/2007/Overview/general.asp.

6. "International Year of Sport and the Olympic Ideal," United Nations General Assembly, A/RES/48/10, October 25, 1993, http://www.un.org/documents/ga/res/48/a48r010.htm.

7. Helmut Digel, "Idealität und Realität. Was aus der Geschichte der neueren Olympischen Spiele zu lernen ist," in *Nachdenken über Olympia: Über Sinn und Zukunft der Olympischen Spiele* (Tübingen: Attempto, 2004), 180.

8. *Olympic Charter. In Force as from 1 September 2004* (Lausanne: International Olympic Committee, 2004), 9.

9. This essay is by no means intended as a critical introduction to Coubertin's ideas. My primary guide to this subject has been *Pierre de Coubertin, 1863–1937: Olympism, Selected Writings*, ed. Norbert Müller (Lausanne: International Olympic Committee, 2000), which includes Müller's essay "Coubertin's Olympism." See also Digel, *Nachdenken über Olympia*; Louis Callebat, *Pierre de Coubertin* (Paris: Fayard, 1988); John J. MacAloon, *This Great Symbol: Pierre de Coubertin and the Origins of the Modern Olympic Games* (Chicago: University of Chicago Press, 1981).

10. Cf. "Letter to the Members of the International Olympic Committee (February, 1920)," in *Pierre de Coubertin*, ed. Müller, 673–74.

11. "Financial Planning for the IVth Olympiad in Rome," in *Pierre de Coubertin*, ed. Müller, 675–684.

12. *Pierre de Coubertin*, ed. Müller, 706–709.

13. Norbert Müller, "Coubertin's Olympism," in *Pierre de Coubertin*, ed. Müller, 42.

14. Chris Morris, "Video Games Push for Olympic Recognition," CNN Money, May 31, 2006, http://money.cnn.com/2006/05/31/commentary/game_over/column_gaming/index.htm; Tim Surette, "Olympic Torch Burns for Gaming," *Gamespot News*, May 31, 2006, http://www.gamespot.com/news/6152088.html.

15. "SPACEWAR: Fanatic Life and Symbolic Death Among the Computer Bums," *Rolling Stone,* December 7, 1972, http://www.wheels.org/spacewar/stone/rolling_stone.html; on Brand and the event, see Fred Turner, *From Counterculture to Cyberculture: Stewart Brand, the* Whole Earth *Network, and Rise of Digital Utopianism* (Chicago: University of Chicago Press, 2006), 86–88.

16. On the history of Twin Galaxies, see Walter Day, ed., *Twin Galaxies Official Video Game and Pinball Book of World Records* (Fairfield, IA: Sunstar, 1998).

17. David Kushner, *Masters of DOOM: How Two Guys Created an Empire and Transformed Pop Culture* (New York: Random House, 2003): 148–149.

18. Quoted in T. Blevins, "A Decade of Blizzard" (2001), IGN, http://pc.ign.com/articles/090/090953p1.html.

19. From "Welcome to the *DOOM* Honorific Titles!" *DOOM* Honorofic Titles, http://www-lce.eng.cam.ac.uk/~fms27/dht/dht5/#dht5.

20. Electronic Sports World Cup, http://findarticles.com/p/articles/mi_moPJQ/is_13_2/ai_n6093989/pg_1.

21. On Korean game channels, see John Anderson, "Spot On: Tuning In to What's On in South Korea, Japan," *Gamespot News*, January 20, 2006, http://www.gamespot.com/news/6141627.html.

22. Quoting Ringel, from "Q+A: GameRiot's Matt Ringel: Diversifying Market Silos," *Electronic Gaming Business*, June 30, 2004, http://findarticles.com/p/articles/mi_moPJQ/is_13_2/ai_n6093989/pg_1.

23. Mike Halekakis, "World Cyber Games 2003: The Olympics of the Video Game World," *JIVE Magazine*, November 4, 2003, http://www.jivemagazine.com/article.php?pid=858.

24. *WCG 2007, World Cyber Games: Press Kit* ([Seoul?]: WCG, 2007).

25. Despite the best efforts of the organizers, this does not always prevent cheating. For example, at the 2004 games held at the Bill Graham Civic Auditorium in San Francisco (at which I was head referee of the *Warcraft III* tournament), an entire round of the *Counter-Strike* tournament was replayed after a charge that audience members favoring one of the competing teams were flashing hand signals to its players in order to reveal opponents' strategies. A section of the grandstand was subsequently roped off to spectators.

26. The four titles represented in the 2000 games were *Age of Empires II*, *Starcraft: Brood War*, *Quake 3*, and *FIFA 2000*; in 2008, *Starcraft* and *FIFA* (2006) were the only original titles still on the list, joined by *Warcraft III*, *Warhammer 40,000: Dawn War*, *Counter-Strike 1.6*, *Need for Speed: Most Wanted*, *Dead or Alive 4*, and *Project Gotham Racing 3*.

27. John E. Findling and Kimberly D. Pelle, eds., *Historical Dictionary of the Modern Olympic Movement* (Westport, CT: Greenwood, 1996), 5.

28. Pierre de Coubertin, *Olympism: Selected Writings*, ed. Norbert Müller (Lausanne: International Olympic Committee, 2000), 663.

29. These figures are taken from the historical section of the ICM Web site, http://www.worldcybergames.com/icm/eng_history2006.html.

30. Thomas F. Scanlon, *The Olympic Myth of Greek Amateur Athletics* (Chicago: Ares, 1984).

31. Coubertin, "Circular Letter, January 15, 1894," in *Pierre de Coubertin*, ed. Müller, 301; "The Charter for Sports Reform," 1930, *Pierre de Coubertin*, 237.

32. *Olympic Charter*, 80.

33. The company was founded by Byung-Chull Lee as the retail operation "Samsung General Stores" in 1938. See *Samsung Press Information* (Seoul: Samsung, 2001).

34. Helen Jefferson Lenskyj, *Inside the Olympic Industry: Power, Politics, and Activism* (Albany: State University of New York Press, 2000), 43; Christopher R. Hill, *Olympic Politics* (Manchester, UK: Manchester University Press, 1992), 194–195; a detailed study of sports sponsorship at Samsung is provided by Hyun-Jong Park's PhD dissertation, "Sponsoring und Werbung im Sport: Eine Longitudinalstudie zur Sportsponsoringwirkung am Beispiel der koreanischen Elektronik-Konzerns Samsung" (Univ. Münster, 1995).

35. *Corporate Citizenship: Press Information* (Seoul: Samsung, n.d.), 6. On Samsung's support of the Olympics, see Lee Dongyoup, *Samsung Electronics: The Global Inc.* (Seoul: YSM, 2006), 65–66.

36. J. C. Herz, "The Bandwidth Capital of the World," *Wired* 10, no. 8 (August 2002), http://www.wired.com/wired/archive/10.08/korea.html.

37. Quoted in Moon Ihlwan, "The Champs in Online Games," *Business Week International Edition*, July 23, 2001.

38. Ihlwan, "Champs in Online Games."

39. Kay Schaffer and Sidonie Smith, "The Olympics of the Everyday," in *The Olympics at the Millennium: Power, Politics and the Games*, ed. Schaffer and Smith (New Brunswick, NJ: Rutgers University Press, 2000), esp. 214–215.

40. Johan Huizinga, *Homo Ludens*, 11.

41. Huizinga, *Homo Ludens*, 77

42. "About WCG: WCG Concept," World Cyber Games, http://www.worldcybergames.com/6th/inside/WCGC/WCGC_structure.asp.

43. Hill, *Olympic Politics*, esp. the chapter on "Financing the Games," pp. 70–89. Like Kunhee Lee, Dassler was also the son of his company's founder.

44. *Le Mouvement olympique = The Olympic Movement* (Lausanne: International Olympic Committee, 1993), 34.

45. From "Physical Exercises in the Modern World: Lecture Given at the Sorbonne (November 1892)," in *Pierre de Coubertin*, ed. Müller, 297.

46. Quoted in *Historical Dictionary of the Modern Olympic Movement*, 196.

47. "About WCG: WCG Concept."

48. In the chapter of *Homo Ludens* "Play-Element in Contemporary Civilization," 195–213. For a more detailed study of Huizinga's thoughts about this problem, see Hector Rodriguez, "The Playful and the Serious: An Approximation to Huizinga's *Homo Ludens*," *Game Studies* 6, no. 1 (December 2006), http://gamestudies.org/0601/articles/rodriges.

49. *Homo Ludens*, 197–200.

50. *Homo Ludens*, 206.

51. *Homo Ludens*, 208–211

52. Mia Consalvo, *Cheating: Gaining Advantage in Videogames* (Cambridge, MA: MIT Press, 2007). Huizinga, of course, considered cheating a form of "false" play and banned it from the magic circle; cf. Huizinga, *Homo Ludens*, 52.

IV INTERVIEW WITH HENRY LOWOOD

T. L. Taylor

T.L.: I wanted to start with your trajectory, your intellectual academic path. Can you tell me a little bit about your path into university life from your bachelor's to how you ended up in a PhD program?

Henry: My parents were both immigrants, from Germany and Austria, and I was born in St. Louis, Missouri. We left when I was less than one year old because St. Louis—my father just said, "anywhere else." I don't remember this, of course, but literally it was after looking around the Eastern Seaboard, not liking it, and they threw us all in the car. My sister was just born and we drove to California without ever having been there. No job, no nothing. So I grew up in kind of a working-class area in Central LA. I started at UCLA, then went to UC Riverside. I had started in sciences. I was a physics major. At UCLA during the first two years as a physics major (I was originally pre-engineering, then went to physics), you were allowed one elective in your first two years. So the spring quarter, I believe of my sophomore year, I took my one elective. I said, "You know, I love reading history," so "Oh, here's a class on historiography. I'll probably learn something." Little did I know that was the major's seniors' course in history taught by a guy named Peter Reill, a very good historian who later became director of a library. But anyway, I took his course, loved it, did all right even though I hadn't realized historians read so much, [laughter] and I went into his office hours. This is why I tell my kids, "Always go into your professor's office hours. Get over that hurdle, go in!" I went in, said "a physics major, I really like history. Is there such a thing as history of science?"

T.L.: What year was this?

Henry: This would have been, I think, spring of '73. I said, "Is there such thing as history of science?" And he said, "Yeah, there is actually. I recommend you go to UC Riverside." Riverside at that time was a top-ten history department. It had two historians of science on the faculty. At that time it was also the only UC I could still transfer to in the spring quarter. I got in and worked mostly with a historian named Ronald Tobey who was a historian of science. I did continue to take some physics courses and some graduate courses, but that was more like a minor and I majored in history. So that was it. It was going to be history of science. Then [for grad school] I applied to a bunch of places, but ended up going to Berkeley, which is great. It was tied for the number one history department in the country at the time, but it didn't have a history of science department. There were some history of science departments. Harvard had a history of science program. Berkeley had a couple of faculty, very good faculty, which I appreciated, but it wasn't my first choice. But my father had cancer. He ended up dying of it when I was in grad school, but I decided to stay in the Bay Area. My parents had moved up here. And so I said, "Ah, Berkeley, well, that's hardly a compromise." It just meant I didn't get a full ride and that kind of thing, but I did get enough to start. That was what got me from the sciences to history of science. It took me a while to finish because in the middle of that I started working at Stanford.

At Stanford I started very low and kept getting promoted. Eventually I hit the point where I could jump to a professional curatorial or librarian type position. They said, "You really should get a degree in library studies." Once again, when I had questions I would always go in and ask people. In this case my question was, "There's a library school at Berkeley and I'm a PhD candidate in the history department. Is there such a thing as a graduate double major? Can I do a second field while a graduate student?" I thought I might as well ask. And you know what? There was. They said, "Oh yeah, you don't have to take the GRE again. You don't have to do any of this stuff. Fill out this application." I got in. So while I'm doing my dissertation and working at Stanford I started library school at Berkeley. I take a couple of classes, maybe a cataloguing class or something. The woman who taught it, bless her heart, was somebody who retired from the Library of Congress probably in 1950 and she's talking about catalog cards. When I'm at work I'm on computers and I thought, "Oh boy, this doesn't seem very useful." So I went in again to the office and said, "You know, I have another question. I'm a doctoral candidate and I'm taking these beginning courses. Is there anything from the history department that would count toward this degree or any considerations?" And they said "Oh, in that case, yes, we'll count a few of your history courses. And also you don't have to take the beginning courses."

So I did that and it went from being this really dull, not that interesting stuff—because I already was working in a library—to really interesting courses. I took an entire course from a retired IBM engineer on non-Boolean searching. I took a course on office technology in which—this would have been the early 1980s—they were talking about expert systems and things like that. And eventually I would end up getting the archives of the people who started that whole thing at Stanford like Ed Feigenbaum. It became really interesting because it was more about information technology and what the futures were for that sort of thing. I mean, you look at my résumé and see all these graduate degrees and think "he did all this work" but actually it was all fun. I kept lucking out and getting all these breaks. All those things just kind of came together and then that's what led to my promotions at Stanford.

T.L.: But it is so interesting because it sounds like even from your earliest days as a student you were already thinking either cross field or next step. Like the history of science at that point, was that a boom period for it?

Henry: The faculty I was working with were the generation of people, kind of like you and me in game studies, where there was no field when they got their degrees. The two historians I worked with the most, one of them, for example, was Tom Kuhn's student when he was at Berkeley. And what was Tom Kuhn? Was he a physicist, a philosopher? I don't know; he was all of that. It was that generation that had the positions. There were a few like I. Bernard Cohen at Harvard who had the first American PhD in history of science, a few people like that scattered here and there, but mostly they were people who had experienced the growth of big science in the fifties and sixties. History of science was initially sparked by that. You know, the atomic bomb. All these things that raised a lot of questions, but the generation teaching still in the seventies and eighties—people who had gotten their appointments maybe in the sixties—had physics PhDs, and other kinds of degrees, other kinds of humanities degrees, maybe, other kinds of history degrees. And I was in the generation of the first group getting PhDs in history of science.

T.L.: Right. Also, and this is as an outsider, when I think of people who did early Internet studies, a lot of them are coming through library and information sciences and Berkeley was certainly a vibrant place. So it's interesting to me that you were part of that program.

Henry: Do you know the story of that program? I was still in the library and information studies school. A little after I graduated, in the early nineties, maybe mid-nineties, they closed that school and they reopened it as a school of information management systems. No library. And in the engineering school. Most of the folks who had gotten their degrees in library and information, in LIS instead of MIS, still won't give money to Berkeley. It's a very sore point. I don't feel that way about it, but a lot of the alums were very unhappy with that. They felt that it was a downgrading of librarianship as a profession to take away that degree. There are very few schools now around the country that credential librarianship as such. San Jose State University is probably the biggest in the country at the moment. Most of them have become I-schools, which tend be very interdisciplinary, research-y, not so much concerned with credentialing working librarians. So that happened and it was a very big deal at Berkeley for the reasons you mentioned. It was considered top, really, for library schools and for it to suddenly disappear and re-emerge as this engineering program upset a lot of people.

T.L.: That's really interesting. I'm curious about that other moment you described where you pivoted. It sounds like you were also thinking about a trajectory in libraries versus a traditional academic appointment?

Henry: Within five minutes of starting graduate school! [laughter] I told Roger Hahn, the professor who ended up being my first reader, I said, "If it's all right with you, I am uninterested in a wandering scholar existence and tenure track positions. I don't want that." I said, "I'm more interested in doing something—like today we would say alternative academic career—is that okay with you?" And he said, "Oh yeah, that's perfectly okay with me." I found out later, he actually didn't win the position but was second choice to become director of the Bancroft Library at Berkeley. So he was also very interested in librarianship and he was very helpful and considerate about that. And so, yes, I was very open about that and very clear with him that that's what I wanted to do. And it was fine, it was not a problem.

T.L.: How did you know so quickly and early? Some folks come to that a little bit later, but how did you know then?

Henry: This is one of these tricky ones where I'm not so sure if my memory is going to be accurate or if I'm just layering in the way things worked out. I'm in my fortieth year at Stanford and so I like stability. I didn't see that in a traditional academic career. The other thing, and this was in the work that I did back then, in 18th century science, is that I was very intent on working on something using original documentation that had not been worked on. The two things I don't like to do when I write is—and they're both faults, I'll admit that—one is, I'm not very theoretical. Number two is I don't like just commenting on other people's work. I would much rather work on something that either is directly experienced or with primary sources. I think I knew then that for me to really do something like that I had to view graduate school as more of a personal project and not worry about if I was picking the right dissertation topic for me to land the right kind of job immediately. My feeling was, look, I'm going to a really good program. I don't think I'm going to have trouble getting another kind of job with the PhD from Berkeley. I just didn't feel pulled to an academic . . . there's a German word, *Laufbahn*, a runway kind of thing. I just didn't want to enter the on-ramp to that kind of career. I just felt like I would enjoy graduate school much more if I looked at it as a personal scholarly project. It didn't buy me anything, I mean, it's not like they cut me loose from any requirements. They treated me just like anyone else. In fact, I believe there were five of us entering in that class and I was the only one that got through. So it wasn't any sort of lack of rigor.

T.L.: That makes a lot of sense. How did your family reflect on this journey at the time? I know we have sort of similar backgrounds in a lot of ways, but I forget what your folks did for work. I'm curious how your family responded.

Henry: I'm certainly the first American college person in my family. Although my sister, I guess I'm allowed to say this, we're Irish twins; my sister's less than nine months behind me. My father didn't even finish *Realschule* in Germany, which is not the highest schooling even; *Gymnasium* is the highest level of high school. He was in the trade track and didn't even finish that. My mother did. She went to the *Modeschule der Stadt Wien*, in Vienna. It's the fashion institute so she had training in fashion design. I still have her drawings. But she came to the US and that meant she basically could be a seamstress. So my mother was a seamstress and my father was a waiter. This was growing up still a little bit in the fifties and then the sixties. You could still raise a family that way. We lived in central LA, in pretty much the African American part of central LA. It was fine. Our family did all right. Today that's sort of impossible. There wasn't much college orientation or stuff like that. My mother was educated but she'd come to the US not speaking English and so she read a lot. She actually ended up speaking very good, though accented, English. Probably all of the academic inclination, both in myself and my sister, came from my mother. But there was nothing anybody could tell me about what it would be like going to college or graduate school or any of that. It was me just kind of figuring things out. There was a lot of financial aid involved. I had a full ride through undergraduate with UC. All my degrees are UC, undergrad and graduate.

I've read what you've written about your background and the one thing I did not do is attend a junior college at any point. I applied to one university. I applied to UCLA. I knew I was going to get in because I got a Regents Scholarship. I never even thought of applying anywhere else. Well, one place I thought of applying as an option was West Point. That was the only other thing I thought of because I thought I could get in. Little did I know that probably I couldn't have because who was going to recommend me at that time? I think I maybe got a brochure from them or something, but I never thought of anything else. I went to Hamilton High School in Los Angeles, which was the big academic public school at the time, and after Berkeley High School the second-highest number of UC Berkeley freshmen came from Hamilton. I remember that at that time there must've been a lot of people like me. Two students from my graduating class of 1,200 went to Stanford. Most of the white kids in my graduating class were Jewish and at the time the word on the street was that, you know, Stanford, wasn't a welcoming place for Jewish kids. The competitive place to go then, the medal of honor at that time, was UC Santa Cruz because at that time Santa Cruz still had the no grading, no majors deal. It was very, very competitive to get into. I knew one person out of that huge graduating class, out of basically one of the top public high schools, who went to an Ivy. So it just wasn't the way I thought. It's like, "Okay, UCLA, a few miles away, that's where I'll go." There really wasn't anything else to talk about until that transfer to Riverside. You've written about this in terms of your own biography. We have to try hard to remember how wonderful the higher education system was in California. There's been some chipping away at it, but for people who did not have that kind of easy way in, the way things were set up in California was really great.

T.L.: Did you play sports at all?

Henry: A little. I think by senior year, I'm in the yearbook, on the golf team, and I'd played some basketball, but I wasn't an athlete. I couldn't figure out why I was housed in the athletes' dorm. I think—at UCLA—maybe they assumed because of my name and because of where my address was, that I was African American. I never figured it out, but there were all these athletes on the dorm floor I was on. Which was fine with me. I love sports, I'm just not good enough to be quite there. I played intramurals and all that at UCLA, but that's kind of a strange thing. I always felt like I was kinda like . . . even though I had the scholarships and everything . . . that I wasn't really a blue blood in academic terms. I was still sort of lucky to be there, if you will.

T.L.: I certainly know that feeling. What did your parents make of it when you switched from being a physics major to this thing called history of science?

Henry: I don't think I even told them. My parents were up here in the Bay Area and my father . . . I had no conflict with him or anything, but when I was in school my father would look at my PE grade first. I loved him and everything but it's not like he cared that much about the academic stuff. My mother did care. I would always give her copies of my publications. She had a copy of my dissertation, all that stuff. If I was there for dinner or something, I would tell her what was going on and sometimes she'd be interested in a particular topic. Now, there's one caveat to that. My first memories are of speaking in German in the home and then when I went to school, I stopped. In college at one point I needed a foreign language. I thought, "I remember a little bit of German, let me take German." It all came back to me and I became basically fluent. When I was in grad school my outside field was German studies and I did my dissertation work in Germany. I lived in Germany. That's something that really interested my mother, in particular, because then she would ask about places I'd been; she came and visited. I wrote my dissertation on an 18th century German topic and I could actually talk to her about some of that stuff and she would be interested. She would even think of some things. That came into play a little bit more when I was working at Stanford. I was the German curator for about fifteen years. She was very interested in that side of it, that was the one thing that really sort of sparked an interest. But with the decision of moving from physics to history of science, yeah, I doubt I got on the phone and told them right away. [laughter]

T.L.: I forget if I told you my father, I think up until his dying days, thought I was a psychologist. He didn't know what sociology was. I think those of us who come from not just nonacademic families, but working-class families, the day-to-day of our academic lives just doesn't have a foothold in people's everyday experience in the same way. So there's often not follow-up questions or even much sense of what we're up to. Except for those moments where it transfers to something very direct, like your mom had access to your work about Germany.

Henry: That's very true but I think if we're writing something or doing research on something, ultimately we have to explain it to somebody. That the value is not just in doing the research and sitting in the room and writing but there's something beyond that. We have to somehow make it clear to somebody what it's about. Maybe in that sense it's a good thing to train a little bit with being relevant to your family. [chuckles]

T.L.: It's a great point! I hadn't considered that in terms of how our writing may at times be accessible or we have a different sense of the range of folks who may encounter it. Shifting gears, can we talk a bit about your work pre-gaming? I want to better situate games preservation in your broader interest in things like software preservation and archives. When I was looking at your CV, it seemed like the first games event you spoke at was in 2001, so there's this whole intellectual life and trajectory before that and I don't want to jump over that. Can you give me some highlights from that period?

Henry: A couple of things come to mind. One is that's ten years after I started working on software preservation. I use "working" lightly; let's say "talking about it," because there wasn't any work you could really do in the nineties. The first conference which I attended on preserving microcomputing software was in '91, I believe, in New York. I'd been doing that sort of thing and there were a couple of initiatives in the nineties, mostly oriented toward archives, but there was already a little bit of discussion about software too. The other thing is that there was a very strange thing that happened to a lot of other people. When I say a lot, a handful of other people of my generation and grad school, which was folks who had done work on 18th century science, and as soon as they got a job, whether it was teaching or something else, went immediately to post–World War II, either history or collecting like I did. To this day, my most cited piece of writing is something I wrote about early forestry, about 18th-century, early 19th-century forestry. I still get inquiries from people working in environmental history and so forth. So I started in 18th century, finished that, and then did this work. The first courses I taught at Stanford in the 1980s were a scientific revolution course and an enlightenment science course. So that's where I started, but then I was moving into the contemporary scene and started the Silicon Valley Archives around '83, '84.

I was a bit of a computer enthusiast. I believe I was the first to turn in a computer-printed dissertation at Berkeley and I had to go through a lot of stuff with what kind of paper to print it on. I had one of these huge Anderson Jacobson printers. They were kind of like Selectric typewriters where you physically put in the print heads. It was an impact printer, a gigantic thing with that fanfold paper. But the question was, "Where would I get paper that was of sufficient quality for the dissertation as per the requirements, but could also be fed through a printer like that?" That cost me a ton of money to get that paper. So anyway, I was doing that and I was enthusiastic about all those sorts of things and a variety of things happened that steered me toward Silicon Valley history. Being around Stanford and working on those questions led to computing history. It isn't yet software preservation, it's history of computing. My 18th- and early 19th-century work was always sort of in between history of technology and history of science. Low in science terms, maybe high in technology terms. Forestry in the late 18th in Germany was a very innovative area and the concept of sustained yield came out of that. So I'd worked on those sorts of things and I think computing appealed to me in a similar fashion because it's a very high technology, lower science compared to maybe physics or something. I just found it a comfortable area to work in as history of computing. I was doing a little bit of that and that started me thinking much more about the problems with documentation and so forth. That was the first step. And that's true of most of the major figures who wrote about software preservation. That whole thing started more about computer-generated records rather than about software. That would mean state archives and places like that, universities, where they were starting to use computer systems for record keeping, they needed to think pretty hard about how to preserve those kinds of records. In many cases, they had very important reasons to preserve those records, not historical reasons, but for keeping track of loans and laws and all manner of things like that. That's where the intellectual effort was first, around preserving documents that were in digital form.

Two important influences in that period were David Bearman, who had run the Smithsonian's computer division and then published a newsletter on documentation in computer form, digital form, and Jeffrey Rothenberg, who wrote this very important article on emulation that was published in *Scientific American* that got everybody finally thinking about software preservation in a useful way by the mid-nineties. But in that period of the nineties, nobody was doing anything sustainable. Nobody was doing anything yet that could earnestly tackle the problem. It was still a lot of thinking about what the different approaches might be, but there weren't systems yet. And there were no collections yet that needed to be preserved. The collection that we acquired at Stanford in '98, '99, 2000, the Cabrinety Collection, I've done a fair amount of research on this and I'm pretty sure it was the first historical software collection that any institution, museum, library, or archive acquired. We had no idea what we were going to do with it when we acquired it. We're still working on it. We're working on a grant-funded project now around emulation that's still using that collection as a base. I think it's the sixth project we've done over the twenty years or so involving different aspects of that collection. In the nineties that was all future, really distant, stuff. People were just beginning to wrap their heads around the problems of software preservation and historical software. What would it be used for? I remember some of the computing organizations at their annual conferences had events around history of computing and archives. Not too long ago, for something I was writing, I was looking at some of the brochures that had been handed out at one of these meetings around 1990. They never mentioned software. They only mentioned documents, ads, manuals, things like that. They never mentioned software as an object for preservation, as documentation. So it took longer for that to get on the radar and I'd say it's only in the last decade that we've been seriously imaging older software, media, creating digital objects, creating digital repositories, figuring out workflows. Now we're finally inching up to where we can deploy emulation and networks to deliver these collections to people who need them. But it's really been a long road.

T.L.: That's amazing. When you were at various professional meetings then back in the day, you would be like, "We need to talk about this!" Were you chomping at the bit at those meetings?

Henry: So I was not a rabble-rouser, that's just not my nature if that's what you're asking. [chuckles] No, but what I could do is . . . there are more people like this now than there were thirty years ago say . . . but what I could do in the nineties is I could talk to computer scientists, and I did all the time. I could talk to people in industry, at Apple, at places like that, and be one of those people who could translate across to the library and say, "Oh, there are people I know in libraries or museums, maybe we can get something going."

And there were groups that I was part of that combined all of that. There were a few other people like that. David Allison was at the Smithsonian and he was one person like that. Arthur Norberg who was at the Charles Babbage Institute, there were a few other people like that. That's the mode that I prefer to work in—configuring groups of people, getting people to talk to each other, figuring out what I could learn from computer science.

It's kind of interesting because you would think, "Well, you know, it's a computing problem, computer scientists will give you all the answers and that's that." But I learned very early, and this is part of the translation work, I learned that there's this interesting tendency among computer scientists to kind of hand-wave, to sort of say, "This is theoretically possible. That was interesting. Thanks for talking to me about it" and then just walk away from it. The thing was, it looks like there's less interest there in implementing systems for doing this because they're often talking about the really important big problems, so what you need to do is get that time to talk to them, get what's useful, and what can be translated and figure out what of that can be turned into something that maybe you can implement, or with somebody who knows more about programming to work with you, to work with somebody to do something. I've seen that, in particular, with other colleagues who did exactly that and had really good results. One example is at Stanford, we've developed this email archiving program called ePADD, which is pretty much a standard thing now in libraries that archive email. Peter Chan developed that. He is somebody I hired at Stanford, and he basically found out that there was a graduate student in the computer science department who was working on theoretical problems with different methods of access to your own personal email, different ways of categorizing and clustering recipients and correspondence and stuff like that. He sees this dissertation and thinks, "Huh, maybe that's what we can use to kind of vet email collections." I mean, he just had this idea. And they started talking and that resulted in the library developing it in consultation with this person who had done this dissertation work in the computer science department. So again, it's a kind of translation work that seemed to work. That's more what I feel comfortable doing rather than rabble-rousing. I'm not Jason Scott; he does that really well. He's very effective at that, getting himself out there and saying stuff that may be pushing some buttons, but gets people to work on things as well. It's a different approach. I like working with Jason, but we're sort of different. I don't wanna say good cop, bad cop or whatever the analogy would be where we're like two people working together but they have very different styles. I'll pass things to him that fit his style better and he'll pass things to me that fit my style better. I have to say this might be another reason that I gravitated away from the historian trajectory to the library trajectory. I really like collaborative work. I think I would miss that.

T.L.: It sounds like you're collaborating and you're actually helping make material, helping nurture things along into existence, that are then living in the world of preservation, living in the library. It's not just producing writing, it's helping produce systems and software and other kinds of things, even if you're not doing the coding yourself.

Henry: Another angle about that is I'm also very comfortable creating these collections, then telling people, "Hey, you know, there's this collection here that you can work on," and somebody else becomes the historian who works on that. I never feel like I'm creating these things for myself. Even though I do my historical work, it's like a different hat entirely. That's another thing besides collaboration, the service aspect to it. Creating this for somebody else whether they're students, whether it's for instructional purposes, or whether it's for other people to do research. You have to kind of be comfortable with putting it there and seeing who's going to come in and make something of it.

T.L.: As you were talking just now, my mind immediately went to thinking about machinima and game engines and helping lay out a space in which others are then doing something.

Henry: That's like that pyramid Will Wright used to talk about where at the very bottom you have all the players of a game, that's the broad base, and maybe at some point they do like a little blog or something and so they move up to a narrower space and eventually you get to the top, it's the people who aren't even playing the game anymore. They're just making platforms for somebody else or modding or doing something like that. There's this thing where you have a kind of career in anything you do, independent of your professional career. You might move from, in my case, being a historian to building things that help other historians do their work and being comfortable with that.

T.L.: When I think of the actors in your story, all the different components that came together to make your trajectory and story what it is, how critical is being not just at Stanford but where you are located geographically? How important is that space in the trajectory?

Henry: That's a very hard question for me to answer because the only two places I've ever been engaged in this kind of work are California and Germany. I don't really know what the rest of the US is like at all, other than visiting conferences and things like that. I suspect the Bay Area is a bit of a bubble. Stanford, very much so when I started, was really distinctive and it was so different from UC Berkeley it took me a while to get used to it. At Berkeley, and maybe it was because I was a graduate student, but I felt like there was a lot more structure to what people did or didn't do. Right from the start I was told "Stanford is a very entrepreneurial place." That doesn't just mean people creating companies, that means even within an academic context, they're entrepreneurial. I've come to think of it as both an advantage and a disadvantage because if I look at the projects I've been involved with—and that would include things like the Stanford Humanities Laboratory—it's really easy to start something at Stanford. If you have an idea and you have collaborators, getting something started, even to have an early impact, is welcomed and encouraged. You're given all the rope that you need to go out to do it. A few faculty members have ended up leaving Stanford because of this. At times, there have been cases where sustainability was an issue. Faculty members I've worked with have said, "I shouldn't have to be doing all this out of my grant funds or my research funds. This is a really good program and the library is working with me on this. It seems like it should be sustained in some way." And at that point it becomes sort of harder to do, because there's less trust, I guess, in these sort of structures than there might be at other institutions. It's a very powerful, encouraging sort of thing if you don't mind taking risks and putting yourself out there a little bit but you have to be very careful about making commitments. You really have to pay a lot of attention to the sustainability side of it. Once you get past a certain point, now I'm mixing my metaphor with the rope here, but it's like walking the plank. [chuckles] And sometimes, well, the Humanities Lab, which was a great thing, it just wouldn't work anymore and had to be shut down. That, I think, is very Stanford. I don't know if I would have had the chance elsewhere. I was pretty young when I became a librarian, and then when I became the curator for the history of science, I was still in my late twenties. I don't know if I would have been given that much opportunity, like starting the Silicon Valley Project, anywhere else. And not just because we're in Silicon Valley but just giving me that much leeway to start bringing collections in, I was very unusual. I remember I went to a few Society of American Archivists meetings at the beginning and I talked about the general approach that I had to building the archives and the archivists at those meetings were negative about it. They just sort of said, "You know, that's not the way you should do things" and so on. It's kind of interesting to me because of the projects that started in the eighties, like MIT had a Route 128 project and there were a few others like that, they've all stopped. The Silicon Valley Archives now, with the new funding that we have and the new construction that we're doing and all of that, it's sustainable. It's going to be around forever.

T.L.: Was there a certain orthodoxy around archives then that you, just by virtue of youth, were bucking but didn't know?

Henry: I think that was part of it. I remember one particular point, this one sticks out at me. There was a lot of work done in the eighties on institutional archives of a new sort. That would be like the big science centers, like the linear accelerators and things like that. How do you archive something where there's hundreds of scientists? Corporate archiving, all of that. Those kinds of things were relatively new and a lot of the work—I don't know if you call it "ideology," which is a little strong—but a lot of the guidance on that was that you basically start by making a pitch to the top of the organization with a certain rationale for the value of archives. A lot of that rationale was tied to the importance of history for decision-making. I almost never did that. I almost always at the beginning went to people like Doug Engelbart, people who had been lab directors or, in a few cases with somebody like Bruce Deal at Fairchild, who were prominent but also were pack rats. I would work more in the middle of the organization rather than going to the top, and work with those people directly. I was bringing in a lot of collections that people had just collected, but they were not institutional records in a way that archivists think of them, where you start with the core records of the institution, the "C-suite" materials which would be very central and all that stuff. I was bringing in things that look more like lab records, reports, and notebooks, things like that, lab notebooks, that sort of thing. It even sort of looked different. To me it seemed more interesting, but that's a matter of opinion. But the thing was, I remember very specifically talking about that particular twist in the way that I was bringing collections in and getting lectured to by more senior archivists in an SAA meeting who were saying, "No, you can't do that without talking to and getting buy-in from the directors of the company."

T.L.: To me that sounds like a historian of science who's, maybe not consciously, thinking about where the work of science is getting produced, is happening. It makes a lot of sense to me. Okay, let me see if we can pivot to games. Maybe a way we can start because it hits back on something you mentioned a bit ago . . . given all your earlier work, did the acquisition of the Cabrinety Collection prompt your interest in writing about games preservation in particular? When does games start coming into frame more?

Henry: Cabrinety was a pivot, but it was a little more complicated than just the acquisition of it. I was never a console gamer. I think my first console I probably bought because I was writing about games and because I had kids. Kids that became my home lab, sort of. I had done a lot with computer games and I would occasionally go into arcades, but I'm a lifelong board game player—board games and sports games, so simulation-type games. My only weakness as a collector is board games. I don't collect books that much but board games have long been a weakness. I had been collecting with the idea of eventually writing about it for a long time, probably since the late eighties at least. I worked a little bit with some companies and different things like that, but the games were just sitting there. Nothing was happening. The situation with Cabrinety was out of the blue. Stephen Cabrinety had died. He was a Stanford student who collected all this stuff. His father was an executive with Digital Equipment. Stephen died in his late twenties with some kind of invasive, bad disease, that killed him pretty quickly. He developed—even had the founding documents and everything—for a thing called CHIPS which was the Computer History Institute for the Preservation of Software. He was really thinking in a very forward way about what would happen with his stuff, had a board and all that. When he died his family was left with this huge collection.

His sister was calling around trying to figure out what to do with it and ran across a Web site or something we'd created—this would have been about '97, so a while back—about the Silicon Valley archives. Tim Lenoir, who was on the history faculty at the time, and I were pretty much partners in crime on that. We taught some courses here. She contacted both of us, Tim and myself, with this idea. As a result of that inquiry Tim and I started talking. We had worked together but we had never talked about games I don't think until then. And he says to me, "You know, I'm working on military simulation and we're starting a project on that." He had originally been mostly a historian of biology and now he was kind of doing the switch and he started telling me about it. I said, "What? You know, Tim, I'm collecting all this stuff on board games. A lot of what I played when I played competitively and all that, was wargames, board games. I know a lot about the commercial side of that." So we collaborated. We wrote some stuff together and we started getting into this much deeper conversation. We started the How They Got Game project. We started that after getting the Cabrinety Collection and from the beginning it was games and simulation. We worked with the *America's Army* folks very early on and with other simulation projects. It combined simulation and games, and digital and physical. We were pretty much interested in all of that stuff. The interesting thing for me was, "Okay, that's Tim and Henry being interested in that." But the Cabrinety Collection was a massive commitment by the library. I suppose in the back of my head it was one of those moments I thought like, "Oh, am I going to have a job in a year? Is this going to be the end of Henry's library career?" But Mike Keller, the library director, was supportive. The faculty were supportive. I'd heard around that time the UC Irvine Faculty Senate decided not to have a game studies program because it was not an academic field, so I was a little bit nervous about this commitment. Technically it was a microcomputing software collection and generally a microcomputing history collection because there's a lot of stuff that's not software. There's magazines and books and all sorts of things but it was mostly games. It's 80 percent games. And so I was worried about that, but there was nothing but support.

T.L.: What year was that?

Henry: The negotiation started in '97, I believe. The first tranche of the material came in '99 and the second big tranche in 2000.

T.L.: So pre–game studies really.

Henry: Pretty much right before. Even with the label "microcomputing history" as something maybe a little less risky, there wasn't any microcomputing history collecting going on either. There was also this thing you'd sometimes hear from computer scientists about how microcomputers were toys, like the "real computer," the "real computational issues," were something else. But everybody agreed that we should get the collection. It was funny, it went through pretty easily. I don't need to get into the details, but it was a massive amount of work with probably half a dozen colleagues to wrestle that collection into submission. To some degree, we still haven't entirely; it's such a complicated collection with so much stuff. So now we have that collection. That's when we started the How They Got Game project. It was in the first generation of the Stanford Humanities Lab as a funded project from SHL. We were doing the writing and those kinds of things out of the lab. The e-sports workshop that you and I did at Stanford was in the SHL space. I think by that time I was a codirector of SHL as well. But since then, a steady quarter to a third of my activity has been around this cluster of topics—software preservation, game preservation, writing about those things. It's been a very steady area of activity for me. And like I say, the timing was kind of right because it was just starting to be a period in which things could be implemented to deal with some of the preservation problems. Preserving Virtual Worlds and the project we did with NIST [National Institute of Standards and Technology] for example, they fit in really nicely into a sequence and we were ready. We had a collection to deal with all of that activity. Things just fit together.

T.L.: You have this deep personal history of playing board games and simulation board games, I'm probably not using the right terms, but did you ever consciously think, "There's something here intellectually."

Henry: Yeah, I just hadn't gotten around to it. I was collecting with the idea that I would do something in the future. By the way, that thing you were just thinking about, the name, there's an interesting little historical footnote. I think now more than ever that going back to the history of board games is very important because much of the writing about board game design in, let's say, the late sixties to late seventies, was extremely high level and many terms like "game design" were invented as part of that. One of the interesting things to come out of that was from Jim Dunnigan, who headed a company called SPI, Simulations Publications Inc. He wrote about this and would make a big point of saying, "You must never refer to these as wargames. They're conflict simulations" because he was always working the angle between the commercial side and a lot of work as a consultant with the military. He was involved in some of the early development of computer-based simulation out of some of these board games—wargames—that he had developed. But there was always that thing of like, "wargames" are sort of low culture, but "conflict simulations" on the other hand raise it to a level of intellectual respectability, right? There's always this nomenclature thing and there've been echoes of that in digital games studies too like, "What words do you use?"

T.L.: Yeah, it's funny though, too, because now that you say that I'm reminded of the number of times in the last year I've heard discussed on the news how a pandemic situation was previously "gamed out" in a simulation. So we now also have that idea kind of in the mainstream a little bit more, that there are these "serious" simulation games that policy wonks and governmental folks are doing. And military folks. It's kind of out there broadly now.

Henry: We have a local game store and they can't keep *Pandemic* on the shelf.

T.L.: One thing I wanted to ask about this pivot point jumps off from something you wrote in 2006 for that inaugural issue of *Games and Culture*. You wrote that Frans Mäyrä's early call for game studies reminded you of the rhetoric that launched the history of science in the sixties and seventies. I wonder what your assessment of that is now, fifteen years on? Are we still following that trajectory?

Henry: I sorta think we still are, yeah. There's two issues that I think are starting to play out just now. One is, what is the right academic preparation for doing the work? The other one is, should these be separate departments or is this a field within other departments? You know, in 2006, I don't think I could have pointed to anybody, although we know there were a few people who did get PhDs earlier, but you couldn't tell somebody who came to see you what to do if they wanted to get a PhD in game studies. It wasn't a settled thing at all. Now we're starting to see this wave of students who are getting graduate degrees focused mainly on games studies in different disciplines. That was something that also happened in history of science. That was the generation that I was kind of in—people are getting training in history of science as opposed to coming from other fields. The other thing is about departments. Well, history of science did that too. All the graduate programs I was accepted to, except Berkeley, had separate departments or programs across disciplinary programs. I think only one of those is still in existence. I think all the departments have pretty much closed down. I think maybe Indiana still has a separate department, maybe Wisconsin. What's happened is every history department has one or two historians of science. I remember this coming up a fair amount in academic summits and things like that, "Oh there's gonna be departments of games studies." I was always really skeptical about that, although I'm sure some people saw that as a badge of having arrived, to have departments. I don't think we're going to see a lot of departments of game studies. There will be people who do game studies, but they will be doing it within certain disciplinary approaches. Again, it's very similar. By the way, in 2006, when I was writing that, one of the things that was really on my mind, was ludology versus narratology. The first time I encountered it I thought, "Oh God, it's internalist versus externalist." It's exactly the same thing. It's an intellectual debate that will seem silly at some point, but it becomes kind of like a fulcrum for getting people to talk about the field in a very engaged way. You read certain things together and debate them and do all of those things as this area is emerging even though the intensity of that engagement was going to evaporate. That was when I first started thinking, "Wow, this is just like history of science" where everybody knows it's not just internal and it's not just external, but it was something in the middle. It was the same way with that.

T.L.: I completely agree, the disciplines have absorbed games, as have the traditional journals. One of the unfortunate things I see is some of that early work is lost because it exists in this strange multidisciplinary gap space that is just not in the disciplines. I see a little bit of recreating the wheel. I guess one of the things I wonder then, is this all true for games preservation and archiving as well?

Henry: I'll tell you where I see that. Like the Preserving Virtual Worlds project or with work that I've done, early work being lost is going to be part of the evolution of the field within library and archival studies, whether it's a good thing or not. It is not necessarily that this work was outside the mainstream. The Preserving Virtual Worlds project was a Library of Congress project and people like Jerry McDonough were involved. Here's the thing I think might get lost, and this will be really interesting because I think it's sort of new with games and maybe other kinds of media preservation, and that is nonacademic work. I'm not sure how this is resolved. I'll take as an example the Internet Archive. When Brewster Kahle has shown up at library meetings, I've always been very grateful that he's there because all of the negativity is directed toward him and not toward me. [laughter] Libraries have had a very hard time accepting the Internet Archive. Or I think of Jason Scott who's like an outsider too, though he wasn't always with the Internet Archive of course. I wonder because the work they've done is going to be fundamental and however that ends up surviving, whatever it looks like a hundred years from now, what Internet Archive has managed to preserve is going to be fundamental certainly in the history of the Web but also many other things as well. But I'm not sure if they're going to be viewed as part of the institutional mainstream of archival work or if projects like the Internet Archive are going to be looked at as a kind of "collector," like an antiquarian collector of the 18th century, like [James] Smithson for the Smithsonian. If the people behind those projects are going to be looked at as somebody who is outside the library mainstream, who made some great collecting decisions that fed in to institutions. Alternatively, maybe the nonacademic, nontraditional institutions will be seen as part of the mainstream. We're talking about preservation now.

We're seeing some things like the Internet Archive being first with an emulation library that is accessible with their Internet Arcade. They are very nimble, they have technical chops. They've been able to do some really good work. However, from a more mainstream institutional perspective, it doesn't look like a finished product always. It looks a little bit rough. Metadata, things like that. And I know there are some very long-term projects—Stanford's involved with one of them—on emulation. We're working on something that's very solid, sustainable, so forth, but we're probably not going to have it fully ready for another five years. So other solutions may come online five, ten, twenty years from now that end up being the mainstream that we just don't know about yet, so that may affect the ways in which the Internet Archive is viewed. But the same goes for all of these, MAME [Multiple Arcade Machine Emulator] and all these collectors of software and preservation-related projects. Many ways of archiving things that have been put together. People are just very clever, have really great ideas, but it doesn't fit the way the institutional world works very well. I'm not saying traditional institutions are necessarily going to be the winners. I just don't know how it's going to play out if we're looking at it from 2035 or something like that, what the landscape is going to be like, what the services are going to be, all of those sorts of things. My career has played out on the more traditional institutional side of it. I'm pretty confident whatever I've done is going to be in that tradition. I hope that that there's room to continue to bring in the work that's been done in other traditions, but I just don't know.

T.L.: As you're talking I'm immediately struck by how I had put up the esports workshops originally at YouTube because it was easy and quick but I wasn't happy that they were just at YouTube. Then I thought about the Internet Archive so I put them up there, but I have to admit when I put it up I was like, "What's the longevity? How long will this be around for? Is this the place that I can have these things live?" I was putting in the metadata and I was just like, "Okay, well, I'm going to try to be clear with my terms and checking that I'm using the same ones consistently." And as you're speaking, I'm thinking Stanford will probably be around in a hundred years. Is the Internet Archive going to be around?

Henry: I think the copy that you gave the Internet Archive will be around in a hundred years. I don't doubt that. I don't know if the Internet Archive will be a separate thing. It could very well be like the Apple records or the Ampex Museum; their historical collections are at Stanford now. It may be that the Library of Congress or Smithsonian or Stanford or whatever, it may be that Internet Archive collection folds in with something like that. I think they'll probably survive, but the track record for traditional libraries that were separate, outside this tradition, set up by philanthropy, that kind of thing, most of them have shut down at one point or another. The Linda Hall Library is still around in Kansas City. So some do survive. I think the important thing is that the Internet Archive dealt with a lot of really important historical documentation that, if they had not acted, would have been lost. To me, whatever format Internet Archive's collections survive in, it doesn't matter. Brewster [Kahle]'s star is solid. It's not going away. It just might be that in the future, for one reason or another, they end up doing things in a different way or other standards emerge for dealing with digital collections. The metadata that you put in around your collection may not be used in exactly that way. I think in your case, as I've offered, why not also give a copy to the Digital Repository at Stanford or somewhere else?

T.L.: Very interesting. I want to weave back to something you said earlier. You talked about not being very, I forget how you put it, "theory driven." I always joke that I'm low theory at best. [chuckles] So I smile when I hear you say that, because it resonates for me. But that said, when I read your work, I think there are really hefty questions and issues you deal with around things like materiality, users and technological determinism, how to think of capture and ephemerality and authenticity. When I look at your work there are really big-ticket items that transcend the mechanics of archiving or curating. There's something quite important and conceptual underneath it all. If there's a thousand collections and people are doing all kinds of things but aren't dealing with some of those issues when they're creating collections, if there's not that robust conceptual wrangling, does it pose a problem? Is it a challenge?

Henry: One thought that I had as I was kind of thinking about this a little bit, was what if she asks me about influences through other people? And I realized besides Ray [Raiford Guins], who I work with a lot, the two game studies people I really like to read are you and Mia Consalvo. I was thinking about that and I realized one thing the two of you have in common. I was going back looking over her bibliography and realized . . . is it called the Association of Internet Researchers? She was active in that. And you were too, right? The insight I got from that was there's something about writing about ideas with a certain kind of granularity where the idea generation is not coming from comparing theories, ideas of different writers, but it's coming bottom up. It's coming from observations of what's happening. I think both you and Mia do that a lot. Your surveys of what people are doing and all of that. And yes, that generates concepts, that generates ideas. I don't mean not doing theory is like, we hate ideas. You have to end up saying something useful. So if I'm thinking about authenticity, I'm thinking, "Who worries about authenticity?" When I was writing about that for one conference, I was looking at historical re-enactment as a way of thinking about software, authenticity. That helped me a lot more than thinking, "Well, who's a historian who has written about it?" I think about people for whom authenticity was a value. What were they doing? Or what were they saying about what they did rather than looking for the book on authenticity or the theory of authenticity that somebody had written. I couldn't even tell you who wrote the theory of authenticity. Do you think that's fair if I say that's kind of the way you approach things too?

T.L.: Yeah, and it's funny because I think probably one of the reasons I feel like you and I always connected intellectually. One of the things I've always appreciated in your work is I always see it as holding two things together at once—an eye on systems and structures and software and what I call digital materialities—the stuff of the world. And at the same time, I never see you as having that trump practice, use, and what comes up from the bottom. Going back to who you in part turned to find collections, the people who are building the stuff, then makes a lot of sense. One of the beauties of your work is you always have these two things in tandem. It's something I certainly strive to, though I probably overbalance on the user side. But we both have a foot in that science and technology studies world.

Henry: Yeah, I think that's it.

T.L.: You have a very rich conceptual framework that you approach questions through. Maybe all archivists are like this and I just don't know enough to know [chuckles], but your approach seems more than "What's the best way that people could experience this archived stuff?" It goes to very fundamental questions. I feel like we've talked about this before, maybe around the virtual world stuff. How much of culture can you actually capture, notate? Should we, even? Do you remember when Raph Koster had that talk at a GDC on trying to create a fully notable system for virtual worlds? Of course, the qualitative sociologist to me is like, "Oh my God, at some point not everything in the world maybe can and should be captured." I love that you've always offered a really rich way of trying to think through those issues. When we think about game preservation and archiving, have some of those early conversations been lost or are they in the field now?

Henry: I think there's a lot of rediscovery of issues that maybe could have been found in the earlier discussions if somebody had looked. A lot of the debate around what you've just been talking about had to do with a tendency . . . maybe I do see this less now . . . when so much of the energy around preservation of software and digital worlds was coming from people with backgrounds in the engineering side of it. There's a tendency because by necessity their professional practice involves them sweeping through the code, through the data, generating information about behaviors and everything that's going on in these worlds. There's a tendency to think that data captures everything.

When I talked about hand-waving earlier with computer scientists, definitely twenty years ago a lot of computer scientists would tell you, "You don't need to worry about digital because everything is captured in virtual worlds." That's something that I've been concerned about for a long time and have written about this idea that because the platform is a digital platform that everything that's important is recorded in some way. It seems very strange to me. It's probably a transition of mentality that somebody should sit down and really think deeply about because I don't think anybody ever would've dared say that historical reality is captured entirely by any kind of documentation we've ever had available to us. We always know that things are lost and we know that some digital stuff is lost. But even just the idea that what happened was even *captured*. What is it about the digital that there's this conceit that since there's encoding involved at a fundamental level therefore everything is caught? Sometimes people are surprised by this, that I don't believe that there's any possibility of a recreation, any idea about authenticity that gets you to a perfect recreation of history. I'm a historian but I find historical re-enactment very futile. All you can talk about is how we experience what we experience of the past. The only way we can find out about how somebody experienced something in the past is to look at the evidence we have of that. We can't construct an environment in which we experience the past reality in such a way that we can talk about it as authentic.

I don't really need to go to that too much, but I do find that interesting and worthy of investigation, how did we get to that conceit? What drove that? I think that would be a very important thing to look at. What it is about the impact of digital technologies that encouraged a lot of people to think that way. That's very important and it does have an impact on practical preservation strategies. Like how far do you go? That's something I have written about, how far are you going to go to say it's an authentic thing? Does somebody need to be in a space that resembles 1980s space to get the full impression? Does the arcade cabinet have to be exactly the same? Do you have to be listening to the same music? How far are you going to take that sort of thing? I'm not a believer in that sort of transportation of authenticity to the present. I just don't see any way to do it.

T.L.: I often say to folks one of the values of watching gaming is that issues are showcased, are rehearsed, years or decades before they hit the mainstream. All kinds of things bubble up in gaming before they become mainstream. As we're talking I'm thinking about how much folks over the last handful of years have been re-debating this issue of capturing everything. It's about capturing all of Twitter or all of Facebook . . . the same quest for completion and "full understanding" by having the total collection. Which is to me a very strange idea.

Henry: For example, how would you use the firehose of Twitter, of tweets around Trump? How would you turn that into an understanding of the 2016 and 2020 elections without all kinds of other contexts? And once you have all that other context, how much of that firehose do you really need? We're not even getting into the privacy issues and all those sorts of things but just as a practical matter of the benefit of having everything. If you're really focused on dealing with that entire archive, even if you are using it in a way that's using machine learning and using all kinds of things to do the distant reading of all of that stuff. . . . Usually when I read that sort of thing, they usually end up going back then to explain their results by trying to understand what happened by looking at more general contextual things. I'm just very skeptical about that. There's never been a phenomenon that has been captured in its entirety on the documentation side.

This is why, by the way, when you always say to me, "I'm not a historian," I say to you, "Yes, you are," or at least you're an ally to historians because I can read the books that you've written about Twitch and about online gaming and you make arguments in there but there's so much there that is about the experiences of people at the time. And you've done the work of a certain filtering of that, of selectivity. I know you don't think you're producing a work of historical documentation, but you are, because it's a way of getting at authentic experiences that people had in that time and are documented. That's what I meant about your work and Mia's work, that it's grounded in what people at that time were experiencing or saying about what they were experiencing. It's not grounded in having access to all the code of *World of Warcraft* or Twitch or something like that to try to explain it through. I wish maybe more people did this sort of work, where they paid very close attention at that level of granularity, whatever their argument is going to end up being. Tell us about what people were doing at a particular point, in a particular context. Use some degree of episodic or anecdotal reporting or oral history or just different things like our personal experiences, which you've done. To me, there's an authenticity there that you just will not get from reenactment of any kind.

T.L.: As we start wrapping up, I had this other cluster of questions that are more about domain areas you've worked in where I wanted to see what your thoughts were these days. One, of course, is thinking about e-sports stuff. One of the fun things about knowing you as long as I have is we've both also seen a trajectory in game studies, a trajectory of topics that have come and flowed and changed. I don't know if I've ever quite expressed this to you personally, but I was so influenced and it really just did something to my thinking when I read your piece on Grubby* and your argument that new things can occur even in these computational spaces. That things like virtuosity and the revealing of a move were happening in digital games. I remember when I first encountered that idea in your work being really blown away by it. I wanted to take a moment to reflect on where you think that e-sports conversation has gone over the last fifteen, nearly twenty years. Are there

*See Henry Lowood, "'It's Not Easy Being Green': Real-Time Game Performance in *Warcraft*," in *Videogame, Player, Text*, ed. Barry Atkins and Tanya Krzywinska (Manchester, UK: Manchester University Press, 2007), 83–100.

remaining big issues or questions that you want to see that space tackle? Have you seen the conversation change in particular ways that's worth noting?

Henry: There's a few things. First thing I'll just say about my own trajectory is that I'm in a really sad e-sports space right now, which is e-sports around sports games. [laughter]

T.L.: Oh, right, you are working on the book *EA Sports FIFA: Feeling the Game*.

Henry: Yeah, I'm working on *FIFA*. The other kind of simulation game I've always been interested in, besides historical simulations and wargames, is sports games. Lately for digital games mostly that's what I'm playing, sports games. If I look at *FIFA*'s efforts to promote e-sports through Electronic Arts, boy, what a disaster it's been so far. It's been terrible. I remember talking to a former student of mine who is now the main community guy at Riot. He came to speak at Stanford, I want to say about four years ago, and I'm talking to him and said, "Well, what do you guys think about sports games?" for e-sports. It just seems so obvious, sports and e-sports. And of course, sports teams at that time were just starting to promote e-sports teams and all of that sort of thing. He looked at me and said, "Oh, that's like the least interesting area in e-sports because sports games will never work. The problem being that you don't know if what just happened is because of you or because of the player who's simulated in the game." Is it just because you have a Messi on your team and the Messi in the game is better, or is it because your skills are better?

I've just written a piece about this that's going to be in the book that has to do with how players as users figure out how the game works, because there's been a lot of discussion in the *FIFA* community about whether there's scripting, where there's different things to account for the fact that people, for example, might have different qualities of teams. It's not going to be any fun if I have worse players in my ultimate team playing *FIFA* than you, because you're always going to beat me. If I'm a better player I could win, or maybe EA will do something to make sure that the better team is handicapped. There's a huge discussion around handicapping, but nobody knows if that's actually in the game or not. (I doubt it.) There are these huge discussions among players as users of how the software works without being able, of course, to see any of the code or any of that sort of stuff. There are issues of, "Am I doing it or is the player that I happened to have on my team doing it? Or is the code doing it because EA's manipulating the results of the game?" A light went on when he told me that. Yeah, sports are a problem for e-sports.

If you're playing *League of Legends* you would hope that the maps and the characters were balanced sufficiently that there weren't these issues. There's nothing about any of those characters that depends on a real-life simulation that that character has to be better than the other characters. That's one thing that's interested me, looking in the sports space. That's mostly what I've looked at the last few years, keeping track of that. The other thing is, in this book that Raiford's editing, there is an unpublished essay that I wrote fifteen years ago about e-sports. It's about the Olympics. I showed it to a writing group that we have here in the library, a bunch of people who are actively writing things. I showed it to them actually before this book project and I said, "It's always bugged me that I never published this, but now that it's fifteen years old I'm thinking how do I update this?" How do you do that? And this group said to me, "Don't update it. Find a way you can publish it and introduce it because it's really much more interesting to read it as something that was written fifteen years ago," partly because there's this concern in the article about how this Olympic Ideal was reshaped to serve a thing that's more oriented toward sponsorship and a world of games that is more controlled by the publishers, the commercial context to it. That was 2006 and, T.L., you've tracked this much more closely than I have, but I think it's fair to say that that's what happened. It's one of the problems with growing e-sports.

T.L.: Yeah, the Olympics has certainly wrangled with it for all kinds of reasons, and it's never taken off in quite the way some folks thought it might.

Henry: The other thing I wanted to acknowledge that I think is a really important area to look at is the whole area of inclusivity in the e-sports scene. It is one thing about e-sports that could have an impact on all the rest of digital gaming. Because that's an area that is so not, was not, inclusive. If you show a way that you can make inroads into that and talk about it and develop ways of dealing with that, I think that's going to be something that will resonate in other communities, even people who aren't interested in e-sports will be looking at how e-sports changes. So I think that's really important.

T.L.: I agree. I always say e-sports is where traditional sports was pre–Title IX. It's so behind the curve. So the other topic area I was thinking a lot about in relation to your work was machinima. I recall when I was out at Stanford for my sabbatical years ago, when we did our e-sports workshop, during that period I also went to a machinima conference you were holding.

Henry: The law conference. Yeah, you were there. Actually, I remember a comment you made there, where I could see people reacting to you. You were using the term "co-creation" at the conference and I could tell there were people who had never heard "co-creation" as a word before. You could kind of see the light bulb. They were sort of like, "Oh, wow. What an interesting thing." I'll never forget that.

T.L.: I was thinking about machinima and how discussions around it were so vibrant at one period. It was at every turn and animating so much conversation. I don't know if it's the world I'm in now, but I feel like it has dropped off in terms of concern. Now, when I talk about intellectual property or co-creation, it's not resonant in the way it used to be. Is whatever animated that interest . . . have we moved from that moment? I know you still do a lot of stuff on the DMCA [Digital Millennium Copyright Act] and working with the EFF [Electronic Frontier Foundation] trying to push to get the ability to share things. But are we just in a different moment with regard to those issues?

Henry: This is just my personal take, I haven't really documented it. There's some machinima being made and I jury on festivals. It is very much a category within the art world. It's mostly artists who are doing it and it's not as distributed and as widespread as it used to be. I think what kind of killed all of that was the game industry building it into their games. For example, a lot of games now are being run as online services. If you want to capture a clip, you do something in the game interface, it gets sent to a server somewhere, it's shared with all your friends. You can share this clip with all your friends pretty easily through social media, but it isn't encouraged, really that easy, to get to the raw platform or if you're interested in generating things or get to the video to start editing. In a way it's been so co-opted within the game software, so that there's really not a lot of need or encouragement to be particularly creative about it. It's just there. And so for the creative work that's being done, very few games are being used. A lot of the work is done using platforms that are designed for capturing video. *Second Life* is still being used for that sort of thing, where you create an environment, but it is also a platform for that kind of creativity, rather than being a game. It's no longer, "Oh, I'll see how the game engine works. And I'll capture this and do some simple editing."

The mid-nineties to mid-2000s was kind of a sweet spot for that where the complexity of the systems and the way the systems were managed was very conducive to people jumping in and just redirecting the games that they were given. Today, it's not just that the systems are more complicated, it's also that you can do so much using the system as is, that some of the gateway stuff that you would do initially, like taking a simple video and putting a skin on something to be funny, you just do in the game now. Or you get a skin by playing the game, so "Why do it myself?" I feel that the energy was almost a one-time thing. I accept it, but I think it's kind of unfortunate, kind of sad that that's the case. I can say that Matteo Bittanti still does a machinima festival in Milan. I was just jurying for that a couple of months ago. The work is great, but making it is not widely accessible like that kind of machinima was. As a result of not using games as a platform quite so much, there isn't the built-in spectator community of game players. *World of Warcraft* was the center of machinima for a while because there were sites and forums and you would post something and immediately thousands, tens of thousands, of people would see those videos. Now that's just not really built into the ecosystem.

T.L.: I'm mindful that I've kept you over two hours, so maybe one thing I could ask just as a way of gesturing toward a closure [laughter] . . . When I think about those earlier moments in game studies, of machinima, conversations around virtual worlds, there was a lot happening, a lot of vibrancy. When you look out at the landscape now of games studies, game preservation, curation, where do you think the most energetic hotspots or animating critical questions are? What either you're most excited by or you see others most excited by?

Henry: Certainly for me, and I believe judging by the submissions we're getting at *ROMChip* and other things that I see, board games is one area. Of course, they were leading computer games historically, but they're lagging in terms of scholarship. Right now there's a pretty amazing second golden age of board game design that's occurring. And there is a fair amount of writing around that. So, board games, and also, if you bring board games into the fold, I think more generalized writing about games that's not so focused on digital games. I'm interested to see what comes out of that.

On the preservation side, the uncharted territory right now is all around access. I think a lot of the technical problems with long-term preservation are seeing a solution coming. A lot of work has been done with digital repositories and so forth, but there are few convincing, if any, really convincing access models. Part of that problem is legal. It's not technical. As you mentioned, I've been talking to lawyers in this area particularly around the DMCA exemptions and things like that. I just straight up asked them if there's scholarship that exists in the area of copyright across regimes. For example, if Stanford puts up a game, a digital disc image, and we put it up according to American copyright law and we're buttoned down as far as that's concerned, but then somebody from Germany or England or Canada or China wants to look at that: What is the relationship? How does the US copyright regime interact with the Chinese copyright regime? Or the English copyright regime? That's going to be something really interesting.

I also think there are going to be some very interesting new models. We're starting to see some. One thing that's kind of interesting that was talked about right from the beginning with the Internet, but it's really not developed so much, is the whole idea of applying micropayments and things like that to access these kinds of materials. I'm starting to see some work, particularly in Germany around the applicability of something like the regime around music, which is very much built around micropayments, applying to other kinds of digital cultural objects so that you might pay a few pennies, or a library might pay a few pennies, for use. How would that all play out? Like taking an ASCAP [American Society of Composers, Authors, and Publishers] model and applying it to cultural objects. There's some serious work being done on that. And then emulation and we get into the authenticity aspects of it. What does it deliver? What does the delivery as part of the access solution have to do? What are the problems with that? What's the best form of the object? If I get a request to see a game from the 1980s, is it the disc image? Is it the files so the person can install it themselves? Is the thing running in an emulator? There's a lot of that kind of stuff. That's on the library side. I think there's still a lot of work to be done on access. Metadata and capture and things like that, we're much further along on, but the access side is really still in its infancy. Those are a couple of things.

T.L.: This was such an interesting conversation. Thank you so much. I'm so excited that this book is coming out. When I went back and was rereading all your stuff, I was like, "Yes, this is going to be a fantastic collection!" I think it's so cool.

Henry: Great. Thank you. Thank you for participating. That means a lot to me really.

T.L.: I'm honored. Thanks, Henry!

July 12, 2021

Bibliography

Aarseth, Espen. "Computer Game Studies, Year One." *Game Studies* 1, no. 1 (2001). http://www.gamestudies.org/0101/editorial.html.

"About WCG: WCG Concept." World Cyber Games. http://www.worldcybergames.com/6th/inside/WCGC/WCGC_structure.asp.

Adham, Allan, quoted in T. Blevins. "A Decade of Blizzard." IGN, 2005. http://pc.ign.com/articles/090/090953p1.html.

Adrian, George. *The Curator's Handbook: Museums, Commercial Galleries, Independent Spaces*. New York: Thames and Hudson, 2015.

AHIKS. "Message from the AHIKS." *Avalon Hill General* 3, no. 3 (September 1966).

AHIKS. *AHIKS Member's Guide*. 1981.

Alcorn, Al. "Simplified Block Diagram Pong." PowerPoint presentation and lecture, Stanford University, January 13, 2005.

Allen, Thomas B. *War Games: The Secret World of the Creators, Players, and Policy Makers Rehearsing World War III Today*. New York: McGraw-Hill, 1987.

Allison, David K. "Preserving Software in History Museums: A Material Culture Approach." In *History of Computing: Software Issues*, edited by Ulf Hashagen, Reinhard Keil-Slawik, and Arthur L. Norberg, 263–272. Berlin: Springer, 2002.

American Federation of Information Processing Societies. "Preserving Computer-Related Source Materials." 1979. Software History Center, http://www.softwarehistory.org.

Anderson, John. "Spot On: Tuning In to What's On in South Korea, Japan." *Gamespot News*, January 20, 2006. http://www.gamespot.com/news/6141627.html.

APX/Atari Program Exchange Product Catalog: *Fall Edition 1983*. Internet Archive scan, 1983. https://archive.org/details/Atari_Program_Exchange_catalog_Fall_1983.

Arthur, Brian. *The Nature of Technology: What It Is and How It Evolves*. New York: Free Press, 2009.

Arvold, Alan R. "A Comprehensive Index to Panzerblitz." n.d. http://grognard.com/info1/pbartrev.html.

Atari Inc. "Atari Expands Worldwide!" flyer, Arcade Flyer Database, 1972, http://www.arcadeflyers.com/?page=thumbs&db=videodb&id=3303.

Au, Wagner James. "Making 'Molotov': How the Man behind the HBO/Cinemax Special Created His Avatar-Based Documentary, and Why." *New World Notes*, May 15, 2008. http://nwn.blogs.com/nwn/2008/05/making-molotov.html.

Au, Wagner James. "HBO Buys U.S. TV Rights to *Second Life* Machinima Series, Promotes It as Oscar Nominee Contender." *New World Notes*, September 24, 2007. http://nwn.blogs.com/nwn/2007/09/second-life-mac.html.

Avalon Hill Game Company. *Play-by-Mail Instruction*. 1964.

Avalon Hill General. *Index and Company History, 1952–1980. Volume 1–Volume 16*. Baltimore: Avalon Hill, 1980.

"The Avalon Hill Philosophy—Part 3." *Avalon Hill General* 4, no.1 (1964).
"Avalon Hill Philosophy—Part 24, "Why Panzerblitz?" *Avalon Hill General* 7, no. 4 (1970).
"Avalon Hill Philosophy—Part 53." *Avalon Hill General* 12, no. 5 (1976).
Avedon, Elliot, and Brian Sutton-Smith. *The Study of Games*. New York: Wiley, 1971.
Baer, Ralph H. *Videogames in the Beginning*. Springfield, NJ: Rolenta Press, 2005.
Banks, Paul N. "Preservation and Format Issues." Unpublished paper. Henry Lowood Papers, Stanford University Libraries, Box 3, folder 35.
Bartle, Richard. *Richard Bartle Papers, 1979–1997*. Stanford University Libraries, 1992.
Bauer, Richard. "Thoughts on *Strategy I*—Part 1." *S&T Supplement* 3 (1970): 7–11.
Bauer, Richard. "More Thoughts on *Strategy I*." *S&T Supplement* 4 (1970): 18–22.
Bearman, David. *Collecting Software: A New Challenge for Archives and Museums*. Archival Informatics Technical Report 2 (Spring 1987).
Bearman, David. "What Are/Is Informatics? And Especially, What/Who Is Archives and Museum Informatics?" *Archival Informatics Newsletter* 1, no. 8 (Spring 1987).
Bearman, David, and Margaret Hedstrom. "Reinventing Archives for Electronic Records: Alternative Service Delivery Options." In *Electronic Records Management Program Strategies*, edited by Margaret Hedstrom, 82–98. Archives and Museum Informatics Technical Report 18. Pittsburgh: Archives and Museum Informatics, 1993.
Becker, D. "The Return of King Pong." Interview. CNET News.com, March 15, 2005. http://news.com.com/The+return+of+King+Pong/2008-1043_3-5616047.html.
Becker, Hans, Arthur Chapman, Andrew Daviel, Karen Kaye, Mary Larsgaard, Paul Miller, Doug Nebert, Andrew Prout, and Misha Wolf. "Dublin Core Element: Coverage." September 30, 1997. http://www.alexandria.ucsb.edu/historical/www.alexandria.ucsb.edu/docs/metadata/dc_coverage.html.
Bell, Gordon. "Towards a History of (Personal) Computer Workstations (Draft)." *Proc. ACM Conf. History of Personal Workstations* (ACM Press, 1986).
Bennett, Jim. "Scientific Instruments." Department of History and Philosophy of Science, University of Cambridge. 1998. Archived August 16, 2002. https://web.archive.org/web/20040406120525/http://www.hps.cam.ac.uk/research/si.html.
Berry, D. B. *Game Design Memoir*. 1992. https://web.archive.org/web/20110725030024/http://www.anticlockwise.com/dani/personal/biz/memoir.htm.
"Beyond the Game." World Cyber Games, http://www.worldcybergames.com/6th/2007/Overview/general.asp.
"Biding Your Time with Computerized Chess." *PC Magazine*, September 1983, 449–458.
Blizzard Entertainment. "*World of Warcraft* Movie Contest: Rise to Power." March 12, 2010. www.wow-europe.com/en/contests/alienware-2010/winners.html.
Bloom, Steve. "The First Golden Age." In *Digital Deli: The Comprehensive, User-Lovable Menu of Computer Lore, Culture, Lifestyles, and Fancy*, edited by Steve Ditlea, 327–332. New York: Workman Publishing, 1984.
Bogost, Ian. *How to Talk about Videogames*. Minneapolis: University of Minnesota Press, 2015.
Bogost, Ian. *Things to Do with Videogames*. Minneapolis: University of Minnesota Press, 2011.
Bolter, Jay David, and Richard Grusin. *Remediation: Understanding New Media*. Cambridge, MA: MIT Press, 1999.

Brand, Stewart. "SPACEWAR: Fanatic Life and Symbolic Death among the Computer Bums. *Rolling Stone*, December 7, 1972. http://www.wheels.org/spacewar/stone/rolling_stone.html.

Brand, Stewart. "Escaping the Digital Dark Age." *Library Journal* 124, no. 2 (February 1999): 46–49.

Brewer, Garry D., and Martin Shubik. *The War Game: A Critique of Military Problem Solving*. Cambridge, MA: Harvard University Press, 1979.

Burnham, Van. *Supercade: A Visual History of the Videogame Age, 1971-1984*. Cambridge, MA: MIT Press, 2001.

Butterfield, Herbert. *The Origins of Modern Science*. London: G. Bell and Sons, 1958.

Cabrinety, Stephen M., Collection in the History of Microcomputing, circa 1975–1995. Department of Special Collections and University Archives, Stanford University Libraries, Stanford, CA.

Callebat, Louis. *Pierre de Coubertin*. Paris: Fayard, 1988.

Campbell-Kelly, Martin. "Development and Structure of the International Software Industry, 1950–1990." Conference abstract. Conference on "History of Software Engineering." Schloß Dagstuhl, August 26–30, 1996. http://www.dagstuhl.de/DATA/Reports/9635/campbell-kelly.html.

Card, Orson Scott. "Gameplay." *Compute (Greensboro)* 104 (January 1989).

Carmack, John. "*DOOM 3*: The Legacy." Transcript of video retrieved June 2004 from the New *DOOM* website, http://www.newdoom.com/interviews.php?i=d3video.

Carmack, John. "Re: Definitions of Terms." Discussion post to Slashdot, January 2, 2002. http://slashdot.org/comments.pl?sid=25551&cid=2775698.

Carlson, Marvin. *Performance: A Critical Introduction*. London: Routledge, 1996.

Cassell, Justine, and Henry Jenkins, eds. *From Barbie to Mortal Kombat: Gender and Computer Games*. Cambridge, MA: MIT Press, 1998.

Ceruzzi, Paul. *A History of Modern Computing*. Cambridge, MA: MIT Press, 1998.

Chang, Alenda Y. *Playing Nature: The Ecology of Games*. Minneapolis: University of Minnesota Press, 2019.

Chaplin, Heather. "Is That Just Some Game? No, It's a Cultural Artifact." *New York Times*, March 12, 2007.

Charles Babbage Foundation Software Task Force. *Final Report*. November 7, 1998.

Cohen, Scott. *Zap! The Rise and Fall of Atari*. New York: McGraw-Hill, 1984.

Commission on Preservation and Access and the Research Libraries Group. *Preserving Digital Information: Report of the Task Force on Archiving of Digital Information*. Washington, DC: Commission on Preservation and Access, 1996.

Consalvo, Mia. *Cheating: Gaining Advantage in Videogames*. Cambridge, MA: MIT Press, 2007.

"Contest No. 1." *Avalon Hill General* 1, no. 1 (May 1, 1964).

Cortada, James W. *Archives of Data-processing History: A Guide to Major U.S. Collections*. New York: Greenwood, 1990.

Costikyan, Greg. "A Farewell to Hexes." Internet Archive copy, 1996. http://web.archive.org/web/20040212100739/http:/www.costik.com/spisins.html.

Costikyan, Greg. "Board Game Aesthetics." In *Tabletop: Analog Game Design*, edited by Greg Costiky and Drew Davidson, 179–184. Pittsburgh, PA: ETC, 2011.

Courbertin, de Pierre. "Physical Exercises in the Modern World: Lecture Given at the Sorbonne." In *Pierre de Coubertin, 1863–1937: Olympism, Selected Writings*, edited by Norbert Müller. Lausanne: International Olympic Committee, 2000.

Coughlan, P. J., and D. Freier. "Competitive Dynamics in Home Video Games: The Age of Atari." *Harvard Business School Industry and Competitive Strategy Cases*, 9-701-091, June 12, 2001.

"Cover Story." *Avalon Hill General* 4, no. 1 (1967).

Crawford, Chris. *The Two Cultures, Maybe Three*. 2004. http://www.igda.org/columns/ivorytower/ivory_May04.php.

Crawford, Chris. "Why Is Interactivity So Hard?" *Interactive Entertainment Design* 9 (1995–1996). http://www.erasmatazz.com/library/JCGD_Volume_9/Why_so_Hard.html.

Crawford, Chris. "The Future of Computer Wargaming." *Computer Gaming World* 1, no. 1 (1981): 3–7.

Crawford, Chris. *The Art of Computer Game Design*. Berkeley, CA: McGraw-Hill, 1982.

Crawford, Chris. "*Easter Front*: A Narrative History." *Creating Computing* 8, no. 8 (1982): 100–107.

Crogan, Patrick. *Gameplay Mode: War, Simulation, and Technoculture*. Minneapolis: University of Minnesota Press, 2011.

Cross, Scott. "The Art of Spooning." Atlantic Guard Soldiers' Aid Society. July 13, 2016. http://www.agsas.org/howto/outdoor/art_of_spooning.shtml. Originally published in *Company Wag* 2, no. 1 (April 1989).

Day, Walter, ed. *Twin Galaxies Official Video Game and Pinball Book of World Records*. Fairfield, IA: Sunstar, 1998.

De Koven, Bernard, with Holly Gramazio. *The Infinite Playground: A Player's Guide to Imagination*, edited by Celia Pearce and Eric Zimmerman. Cambridge, MA: MIT Press, 2020.

DeMaria, Rusel, and Johnny L. Wilson. *High Score: The Illustrated History of Electronic Games*. New York: McGraw Hill/Osborne, 2002.

Digel, Helmut. "Idealität und Realität. Was aus der Geschichte der neueren Olympischen Spiele zu lernen ist." In *Nachdenken über Olympia: Über Sinn und Zukunft der Olympischen Spiele*, edited by Helmut Digel, 179–200. Tübingen: Attempto, 2004.

Dongyoup, Lee. *Samsung Electronics: The Global Inc.* Seoul: YSM, 2006.

Dorosh, Michael. "Tactical Game 3." *Tactical Wargamer*, 2008. http://www.tacticalwargamer.com/boardgames/panzerblitz/tacgame3.htm.

Douglas, Susan J. "Some Thoughts on the Question 'How Do New Things Happen?'" *Technology and Culture* 51, no. 2 (April 2010): 293–304.

Douglas, Susan J. "The Turn Within: The Irony of Technology in a Globalized World." *American Quarterly* 58, no. 3 (2006): 619–638.

Dunnigan, James F. "Transition: S&T Change Publishers." *Strategy and Tactics. Book IV: Nrs. 16–18*. N.p: Simulations Publications, Inc., n.d.

Dunnigan, James F. "Designers Notes: The Game Is a Game." *Strategy and Tactics* 22 (1970).

Dunnigan, James F. *The Complete Wargames Handbook*. New York: Morrow, 1980.

Dunnigan, James F. *Wargames Handbook, Third Edition: How to Play and Design Commercial and Professional Wargames*. Lincoln, NE: Writers Club Press, 2000.

Dunnigan, James F., and Redmond Simonsen. "The Blitzkrieg Module System." *Strategy and Tactics* 19 (1969): 17–24.

EaaSI: Emulation-as-a-Service Infrastructure. https://www.softwarepreservationnetwork.org/emulation-as-a-service-infrastructure/resources/.

Edwards, Dan J., and J. M. Graetz. "PDP-1 Plays at *Spacewar*." *Decuscope* 1, no. 1 (April 1962): 2–4.

Edwards, John. "Stalingrad: Australian Style." *S&T Supplement* 3 (1970): 12–18.

Edwards, John. "Interview: John Edwards." *Avalon Hill General* 15, no. 1 (1978): 16–17.

Edwards, Paul. *The Closed World: Computers and the Politics of Discourse in Cold War America*. Cambridge, MA: MIT, 1996.

Electronic Sports World Cup website, retrieved June 2007. http://findarticles.com/p/articles/mi_moPJQ/is_13_2/ai_n6093989/pg_1.

Ernst, Wolfgang. *Digital Memory and the Archive.* Minneapolis: University of Minnesota Press, 2012.

Farrow, Daniel W., IV. "Avalon Hill Games, 1952–1998." Game Ludography, 2005. http://users.rcn.com/dwfiv/games/avalonhillgames.html.

"Feminist Game History." *Feminist Media History* 6, no. 1 (2020).

Findling, John E., and Kimberly D. Pelle, eds., *Historical Dictionary of the Modern Olympic Movement*. Westport, CT: Greenwood, 1996.

Finn, Bernard. "Collectors and Museums." In *Exposing Electronics*, edited by Bernard S. Finn, Robert Bud, and Helmuth Trischler, 175–191. Amsterdam: Harwood, 2000.

Fourmilab. "The Analytical Engine: Is the Emulator Authentic?" Fourmilab. March 21, 2016. http://www.fourmilab.ch/babbage/authentic.html.

Friedl, Markus. *Online Game Interactivity Theory*. Boston: Charles River Media, 2003.

Galloway, Alexander. *Gaming: Essays in Algorithmic Culture*. Minneapolis: University of Minnesota Press, 2006.

Galloway, Alexander. *The Internet Effect*. Cambridge, UK: Polity, 2012.

GAMECIP: The Game Metadata and Citation Project. https://gamecip.soe.ucsc.edu/.

"Game History: A Special Issue." *Game Studies* 13, no. 2 (2013).

Gapps, Stephen. "Mobile Monuments: A View of Historical Reenactment and Authenticity from inside the Costume Cupboard of History." *Rethinking History: The Journal of Theory and Practice* 13, no. 3 (2009): 395–409.

Gardner, Martin. "The Fantastic Combinations of John Conway's New Solitaire Game 'Life.'" *Scientific American* 223, no. 4 (October 1970): 120–123.

Gardner, Martin. Papers. SC647, Stanford University Libraries.

Gates, Bill. *The Road Ahead*. New York: Viking, 1995.

Gazzard, Alison. *Now the Chips Are Down: The BBC Micro*. Cambridge, MA: MIT Press, 2016.

"General McAuliffe Added to Advisory Staff." *Avalon Hill General* 1, no. 6 (1965): 1–2.

Goffman, Erving. *The Presentation of Self in Everyday Life*. Garden City, NY: Doubleday: 1959.

Gordon, John Steele. "Review of Seth Shulman, *The Telephone Gambit*." *Wall Street Journal*, January 16, 2008, D10.

Graetz, J. M. "The Origin of Spacewar!" *Creative Computing Video and Arcade Games* 1, no. 1 (Spring 1983): 78–85.

Graft, Kris. "Sega Carries Torch for Beijing 2008." *Next Generation*, December 5, 2005, http://www.nextgen.biz/index.php?option=com_content&task=view&id=1791.

Great Britain. *Rules for the Conduct of the War-Game. 1884.* London: Stationery Office, 1884.

Gregory, Jason. *Game Engine Architecture.* Wellesley, MA: A. K. Peters, 2009.

Grenadier: Company Level Combat 1700–1850. Designed by James Dunnigan. Simulations Publications Inc., 1971.

Grossberg, Lawrence. *Bringing It Back Home: Essays on Cultural Studies.* Durham, NC: Duke University Press, 1997.

Guins, Raiford. *Game After: A Cultural Study of Video Game Afterlife.* Cambridge, MA: MIT Press, 2014.

Haigh, Thomas, ed. *Histories of Computing.* Cambridge, MA: Harvard University Press, 2011.

Halekakis, Mike. "World Cyber Games 2003: The Olympics of the Video Game World." *Jive Magazine*, November 4, 2003.

"Harvard Sportsmanship." *Princeton Alumni Weekly* 35, May 17, 1935.

Harvey, Ross. *Digital Curation: A How-To-Do-It Manual.* New York: Neil Schuman, 2004.

Harvey, Ross. "From Digital Artefact to Digital Object." In *Multimedia Preservation: Capturing the Rainbow: Proceedings of the Second National Conference of the National Preservation Office, 28–30 November 1995*, 202–216. Canberra: National Library of Australia, 1996.

Hedstrom, Margaret L., Christopher A. Lee, Judith S. Olson, and Clifford A. Lampe. "'The Old Version Flickers More': Digital Preservation from the User's Perspective." *American Archivist* 69, no. 1 (Spring–Summer 2006): 159–187.

Hedstrom, Margaret L., and David Bearman. "Preservation of Microcomputer Software: A Symposium." *Archives and Museum Informatics* 4, no. 1 (Spring 1990).

Hedstrom, Margaret L. *Archives and Manuscripts: Machine-Readable Records.* SAA Basic Manual Series. Chicago: Society of American Archivists, 1984.

Herman, Leonard. "The Untold Atari Story." *Edge* 200 (April 2009): 94–99.

Herman, Leonard. *Phoenix: The Fall and Rise of Videogames.* 3rd ed. Springfield, NJ: Rolenta, 2001.

Herrmann, Laura. Email from Laura "BahdKo" Herrmann to Henry Lowood, January 28, 2004.

Herms, Eilert. "Die olympische Bewegung der Neuzeit. Sozialpolitisches Programm und reale Entwicklung." In *Nachdenken über Olympia. Über Sinn und Zukunft der Olympischen Spiele*, edited by Helmut Digel, 65–82. Tübingen: Attempto, 2004.

Herz, J. C. "The Bandwidth Capital of the World." *Wired* 10, no. 8 (August 2002).

Hill, Christopher R. *Olympic Politics.* Manchester, UK: Manchester University Press, 1992.

Hoefler, Don C. "Silicon Valley USA." *Electronic News*, January 11, 1971.

Hofer, Margaret. *The Games We Played: The Golden Age of Board and Table Games.* Princeton, NJ: Princeton Architectural Press, 2003.

Horwitz, Tony. *Confederates in the Attic: Dispatches from the Unfinished Civil War.* New York: Pantheon Books, 1998.

Howgego, Tim. "Exploration Is Dead. Long Live Exploration!" August 17, 2008. http://timhowgego.com/exploration-is-dead-long-live-exploration.html.

Huhtamo, Erkki. "Slots of Fun, Slots of Trouble: An Archaeology of Arcade Gaming." In *Computer Game Studies Handbook*, edited by Joost Raessens and Jeffrey Goldstein, 3–22. Cambridge, MA: MIT Press, 2005.

Huizinga, Johan. *Homo Ludens: A Study of the Play Element in Culture.* Boston: Beacon, 1950.

id Software. "id Software Backgrounder." id Software, retrieved February 2004. http://www.idsoftware.com/business/home/history/.

Ihlwan, Moon. "The Champs in Online Games." *Business Week International Edition*, July 23, 2001.

Interactive Digital Software Association. *Essential Facts about the Computer and Video Game Industry*. Washington, DC: IDSA, 2000.

"International Year of Sport and the Olympic Ideal." United Nations General Assembly, A/RES/48/10 (25 Oct. 1993). http://www.un.org/documents/ga/res/48/a48r010.htm.

JCB III [pseud.]. "The Curious Saga of the D10 "Postal" Combat Results Table." Map and Counters blog, November 17, 2011.

Jenkins, Henry. "Art Form for the Digital Age." *Technology Review* (September–October 2000). http://www.techreview.com/articles/oct00/viewpoint.htm.

Jenkins, Henry. *Fans, Bloggers, Gamers: Exploring Participatory Culture*. New York: New York University Press, 2006.

Jerz, D. 'Somewhere Nearby Is Colossal Cave: Examining Will Crowther's Original "Adventure" in Code and in Kentucky." *Digital Humanities Quarterly* 1, no. 2 (2007), http://www.digitalhumanities.org/dhq/vol/001/2/000009/000009.html.

Johnson, Luanne. "A View from the Sixties: How the Software Industry Began." *IEEE Annals of the History of Computing* 20, no. 1 (1998): 36–42.

Johnson, Margaret, et al. *Computer Files and the Research Library*. Mountain View, CA: Research Libraries Group, 1990.

Joint Chiefs of Staff. *Publication 1, Department of Defense Dictionary of Military and Associated Term*. Washington, DC: Government Printing Office, 1987.

Jones, Caitlin. "Seeing Double: Emulation in Theory and Practice. The Erl King Study." Paper presented to the Electronic Media Group. June 14, 2004. Electronic Media Group. http://cool.conservation-us.org/coolaic/sg/emg/library/pdf/jones/Jones-EMG2004.pdf.

Juul, Jesper. *Half-Life: Video Games between Real Rules and Fictional Worlds*. Cambridge, MA: MIT Press, 2005.

Kahn, Herman. *Thinking about the Unthinkable*. New York: Horizon Press, 1964.

Kaltman, Eric. "Current Game Preservation Is Not Enough." *How They Got Game* (blog), June 6, 2016. http://web.stanford.edu/group/htgg/cgibin/drupal/?q=node/1211.

Kaltman, Eric, Noah Wardrip-Fruin, Mitch Mastroni, Henry Lowood, Glynn Edwards, Marcia Barrett, Greta de Groat, and Christine Caldwell. "Implementing Controlled Vocabularies for Computer Game Platforms and Media Formats in SKOS." *Journal of Library Metadata* 16, no. 1 (2016): 1–22.

Karampelas, Gabrielle. "Stanford Libraries' Transformative Gift Creates Hub Highlighting Silicon Valley History." *Stanford News*, January 31, 2019. https://news.stanford.edu/2019/01/31/intendo-libraries-transformative-gift-creates-hub-highlighting-silicon-valley-history/.

Kent, Steven L. *The Ultimate History of Video Games*. New York: Three Rivers Press, 2001.

Kerecman, L. "Computer Space." Arcade History Database. http://www.arcade-history.com/?n=computer-space&page=detail&id=3388.

Kinder, Marsha. *Playing with Power in Movies, Television, and Video Games: From Muppet Babies to Teenage Mutant Ninja Turtles*. Berkeley: University of California Press, 1991.

Kirschenbaum, Matthew. “War Stories: Boardgames and (Vast) Procedural Narratives.” In *Third Person: Authoring and Exploring Vast Narratives*, edited by Pat Harrigan and Noah Wardrip-Fruon, 357–371. Cambridge, MA: MIT Press, 2009.

Kirschenbaum, Matthew. “Materiality and Matter and Stuff: What Electronic Texts Are Made Of.” *ebr* 12 (2002). http://www.altx.com/ebr/riposte/rip12/rip12kir.htm.

Kittler, Friedrich. “There Is No Software.” *C-Theory: Theory, Technical, Culture* 32 (October 18, 1995). http://www.ctheory.com/article/a032.html.

Kline, Ronald, and Trevor Pinch. “Users as Agents of Technological Change: The Social Construction of the Automobile in Rural America.” *Technology and Culture* 37, no. 4 (1996): 763–795.

Kramer, Jerry. *Instant Replay: The Green Bay Diary of Jerry Kramer*. New York: World, 1968.

Kranzberg, Melvin. “Kranzberg’s Laws.” *Technology and Culture* 27, no. 3 (July 1986): 544–560.

Kuhn, Thomas. S. *The Structure of Scientific Revolutions*. 2nd ed. Chicago: University of Chicago Press, 1970.

Kushner, David. *Masters of DOOM: How Two Guys Created an Empire and Transformed Pop Culture*. New York: Random House, 2003.

Le Mouvement intendo = The Olympic Movement. Lausanne: International Olympic Committee, 1993.

Lenoir, Timothy, and Henry Lowood. “Theaters of War: The Military-Entertainment Complex.” In *Collection, Laboratory, Theater: Scenes of Knowledge in the 17th Century*, edited by Helmar Schramm, Ludger Schwarte, and Jan Lazardzig, 427- 56. Berlin: de Gruyter, 2005.

Lenoir, Tim. “All but War Is Simulation: The Military-Entertainment Complex.” *Configurations* 8, no. 3 (September 2000): 289–335.

Lenskyj, Helen Jefferson. *Inside the Olympic Industry: Power, Politics and Activism*. Albany: State University of New York Press, 2000.

Library of Congress. *Annual Report of the Librarian of Congress for the Fiscal Year Ending September 30, 1988*. Washington, DC: Library of Congress, 1989.

Library of Congress. “Collections Overview: Computer Files.” http://www.loc.gov/acq/devpol/colloverviews/computer.pdf.

Library of Congress. “Moving Image Materials.” Library of Congress Collections Policy Statements. November 2008. http://www.loc.gov/acq/devpol/motion.pdf.

Linzmayer, O. W. *Apple Confidential: The Real Story of Apple Computer, Inc.* San Francisco: No Starch Press, 1999.

List, Steve. “Game Design: Down Highway 61, through State Farm 69, around Tactical Game 3, and into *Panzerblitz*.” *Strategy and Tactics* 22 (1970).

Loftus, Geoffrey R., and Elizabeth F. Loftus. *Mind at Play: The Psychology of Video Games*. New York: Basic Books, 1983.

Lowood, Henry. “Animation Technology and Computer Graphics in the United States.” In *Oxford Encyclopedia of the History of Science, Medicine, and Technology in America*, edited by Hugh Slotten, 63–65. Oxford: Oxford University Press, 2014.

Lowood, Henry. “A Brief Biography of Computer Games.” In *Playing Video Games: Motives, Responses, and Consequences*, edited by Peter Vorderer and Jennings Bryant, 25–41. New York: Lawrence Erlbaum Associates, 2006.

Lowood, Henry. "The Calculating Forester: Quantification, Cameral Science, and the Emergence of Scientific Forestry Management in Germany." In *The Quantifying Spirit in the Eighteenth Century*, edited by Tore Frängsmyr, J. L. Heilbron, and Robin E. Rider, 315–342. Berkeley: University of California Press, 1990.

Lowood, Henry. "Computer and Video Games." In *Encyclopedia of 20th-Century Technology*, edited by Colin Hempstead, 180–181. Milton Park, England: Routledge, 2004.

Lowood, Henry. "A 'Different Technical Approach'? Introduction to the Special Issue on Machinima." *Journal of Visual Culture* 10, no. 1 (2011): 3–5.

Lowood, Henry. Email message to author, August 25, 2021.

Lowood, Henry. "Forbidden Areas: The Hidden Archive of a Virtual World." In *Rough Cuts: Media and Design in Process. A MediaCommons Project*, edited by Kari Kraus (2012) http://mediacommons.org/tne/pieces/forbidden-areas-hidden-archive-virtual-world.

Lowood, Henry. "Found Technology: Players as Innovators in the Making of Machinima." In *Digital Youth, Innovation, and the Unexpected*, edited by Tara McPherson, 165–196. Cambridge, MA: MIT Press, 2007.

Lowood, Henry. "Game Counter." In *The Object Reader*, edited by Fiona Candlin and Raiford Guins, 466–469. Abingdon, UK: Routledge, 2009.

Lowood, Henry. "Game Engine." In *Debugging Game History: A Critical Lexicon*, edited by Henry Lowood and Raiford Guins, 203–209. Cambridge, MA: MIT Press, 2016.

Lowood, Henry. "Game Engines and Game History." *Kinephanos*. History of Games International Conference Proceedings, January 2014. http://www.kinephanos.ca/2014/game-engines-and-game-history/.

Lowood, Henry. "Game History." In *The Johns Hopkins Guide to the Digital Media*, edited by Marie-Laure Ryan, Lori Emerson, and Benjamin J. Robertson, 206–212. Baltimore: Johns Hopkins University Press, 2014.

Lowood, Henry. "Game Studies Now, History of Science Then." *Games and Culture* 1 (January 2006): 78–82.

Lowood, Henry. "The Hard Work of Software History." *RBM: A Journal of Rare Books, Manuscripts, and Cultural Heritage* 2 (Fall 2001): 141–161.

Lowood, Henry. "High-Performance Play: The Making of Machinima." In *Videogames and Art: Intersections and Interactions*, edited by Andy Clarke and Grethe Mitchell, 59–79. Bristol: Intellect Books, 2007.

Lowood, Henry. "Impotence and Agency: Computer Games as a Post-9/11 Battlefield." In *Games Without Frontiers—War Without Tears: Computer Games as a Sociocultural Phenomenon*, edited by Andreas Jahn-Sudmann and Ralf Stockmann, 78–86. London: Palgrave Macmillan, 2008.

Lowood, Henry. "It Is What It Is, Not What It Was." *Refractory: A Journal of Entertainment Media* 27 (2016).

Lowood, Henry. "'It's Not Easy Being Green': Real-Time Game Performance in *Warcraft*." In *Videogame, Player, Text*, edited by Barry Atkins and Tanya Krzywinska, 83–100. Manchester, UK: Manchester University Press, 2007.

Lowood, Henry. "Joga Bonito: Beautiful Play, Sports and Digital Games." In *Sports Videogames*, edited by Mia Consalvo, Konstantin Mitgutsch, and Abe Stein, 67–86. London: Routledge, 2013.

Lowood, Henry. "Jon Haddock, *Screenshots*: Isometric Memories." In *GameScenes: Art in the Age of Videogames*, edited by Matteo Bittanti and Domenico Quaranta, 15–39. Milan: Johan and Levi editore, 2006.

Lowood, Henry. "The Lures of Software Preservation." *Preserving.exe: Toward a National Strategy for Software Preservation* (October 2013): 4–11. http://www.digitalpreservation.gov/multimedia/documents/PreservingEXE_report_final101813.pdf.

Lowood, Henry. "Memento Mundi: Are Virtual Worlds History?" In *Digital Media: Technological and Social Challenges of the Interactive* World, edited by Megan Winget and William Aspray, 3–25. Lanham, MD: Scarecrow Press, 2011.

Lowood, Henry. "The New World and the European Catalog of Nature." In *America in European Consciousness, 1493–1750*, edited by Karen O. Kupperman, 295–323. Raleigh: University of North Carolina Press, 1995.

Lowood, Henry. "The Obstacle Course: Documenting the History of Military Simulation." In *America's Army PC Game: Vision and Realization*, edited by Margaret Davis. Monterey, CA: US Army and MOVES Institute, 2004.

Lowood, Henry. "Oral History of Al Alcorn. Interviewed by Henry Lowood." Computer History Museum, X4596.2008, transcript 2, April 2008.

Lowood, Henry. Papers. Department of Special Collections and University Archives, Stanford University Libraries, Stanford, CA.

Lowood, Henry. *Patriotism, Profit, and the Promotion of Science in the German Enlightenment: The Economic and Scientific Societies, 1760–1815*. New York: Garland, 1991.

Lowood, Henry. Personal note. March 23, 1990.

Lowood, Henry. "Perspectives on the History of Computer Games." *IEEE Annals of the History of Computing* 31, no. 3 (2009).

Lowood, Henry. "Putting a Stamp On Games: Wargames, Players, and PBM." *Journey Planet* 26 (2015).

Lowood, Henry. "*Quake*: Movies." In *How To Play Video Games*, edited by Matthew Thomas Paine and Nina B. Huntemann, 285–292. New York: New York University Press, 2019.

Lowood, Henry. "Shall We Play a Game: Thoughts on the Computer Game Archive of the Future." Bits of Culture: New Projects Linking the Preservation and Study of Interactive Media Conference. Stanford University, 2002.

Lowood, Henry. "Software Archives and Software Libraries." In *Challenging Collections: Approaches to the Heritage of Recent Science and Technology*, edited by Alison Boyle and Johanes-Geert Hagmann, 68–86. Washington, DC: Smithsonian Institution Scholarly Press, 2017.

Lowood, Henry. "Storyline, Dance/Music, or PvP? Game Movies and Community Players in *World of Warcraft*." *Games and Culture* 1 (October 2006): 362–382.

Lowood, Henry. "Video Capture: Machinima, Documentation, and the History of Virtual Worlds." In *The Machinima Reader*, edited by Henry Lowood and Michael Nitsche, 3–22. Cambridge, MA: MIT Press, 2011.

Lowood, Henry. "Video Games in Computer Space: The Complex History of *Pong*." *IEEE Annals in the History of Computing* 31 (2009): 5–19.

Lowood, Henry. "*Warcraft* Adventures: Texts, Replay, and Machinima in a Game-Based Story World." In *Third Person: Authoring and Exploring Vast Narratives*, edited by Pat Harrigan and Noah Wardrip-Fruin, 407–427. Cambridge, MA: MIT Press, 2009.

Lowood, Henry. "War Engines: Wargames as Systems from the Tabletop to the Computer." In *Zones of Control: Perspectives on Wargaming*, edited by Matthew Kirschenbaum and Patrick Harrigan, 83–105. Cambridge, MA: MIT Press, 2016.

Lowood, Henry. "'Where There Is Smoke, There Is Fire . . .': The FIFA Engine and Its Discontents." In *EA Sports FIFA: Feeling the Game*, ed. Raiford Guins, Henry Lowood, and Carlin Wing. New York: Bloomsbury Academic Press, 2022.

Lowood, Henry, and Raiford Guins, eds. *Debugging Game History: A Critical Lexicon*. Cambridge, MA: MIT Press, 2016.

Lowood, Henry, Eric Kaltman, and Joseph Osborn. "Screen Capture and Replay: Documenting Gameplay as Performance." In *Histories of Performance Documentation: Museum, Artistic, and Scholarly Practices*, edited by Gabriella Giannacchi and Jonah Westerman, 149–164. New York: Routledge, 2017.

Lowood, Henry, and Tim Lenoir. "Theaters of War: The Military-Entertainment Complex." In *Collection, Laboratory, Theater: Scenes of Knowledge in the 17th Century*, edited by Helmar Schramm, Ludger Schwarte, and Jan Lazardzig, 427–456. Berlin: Walter de Gruyter, 2005.

Lowood, Henry, Devin Monnens, Zach Vowell, Judd Ethan Ruggill, Ken S. McAllister, and Andrew Armstrong. "Before It's Too Late: A Digital Game Preservation White Paper." *American Journal of Play* 2, no. 2 (2009): 139–166.

MacAloon, John J. *This Great Symbol: Pierre de Coubertin and the Origins of the Modern Olympic Games*. Chicago: University of Chicago Press, 1981.

MacCombe, Leonard. "Valuable Batch of Brains: An Odd Little Company Called RAND Plays Big Role in U.S. Defense." *Life*, May 11, 1959, 101–107.

MacGowan, Rodger B. "20 Years Later and 10 Years after Squad Leader." *Fire and Movement* 53 (1987): 34–37.

Maclean, A. "Computer Space Restoration." http://www.ionpool.net/arcade/archuk/computer_space_restoration.html.

Madeja, Victor. "Midway, D-Day, Tactics II, Stalingrad Re-Worked." *Avalon Hill General* 1, no. 5 (1965).

Mahoney, Michael. "The History of Computing in the History of Technology." *IEEE Annales of the History of Computing* 10, no. 2 (1988): 113–125.

Mahoney, Michael. *The Mathematical Career of Pierre De Fermat, 1601–1665*. Princeton, NJ: Princeton University Press, 1994.

Malone, Michael S. *The Big Score: The Billion-Dollar Story of Silicon Valley*. New York: Doubleday, 1985.

Manovich, Lev. *The Language of New Media*. Cambridge, MA: MIT Press, 2001.

Martin, Rex A. "Cardboard Warriors: The Rise and Fall of an American Wargaming Subculture, 1958–1998." PhD diss. Pennsylvania State University, 2001.

Mäyrä, Frans. "The Quiet Revolution: Three Theses for the Future of Game Studies." DiGRA. December 13, 2005. http://www.digra.org/hardcore/hc4.

McAneny, Larry. "*Panzerblitz*: Hex by Hex." *Avalon Hill General* 12, no. 5 (1976): 3–13, 34.

McGonigal, Jane. *Reality Is Broken: Why Games Make Us Better and How They Can Change the World*. New York: Penguin, 2011.

McGonigal, Jane. *Superbetter: The Power of Living Gamefully*. New York: Penguin, 2016.

McDonough, Jerome P., Robert Olendorf, Matthew Kirschenbaum, Kari Kraus, Doug Reside, Rachel Donahue, Andrew Phelps, Christopher Egert, Henry Lowood, and Susan Rojo. *Preserving Virtual Worlds Final Report*. August 31, 2010. https://www.ideals.illinois.edu/handle/2142/17097.

McLuhan, Marshall. "'It Will Probably End the Motor Car': An Interview with Marshall McLuhan." *Pay-TV*. August 1976, 26–29.

Media Archaeological Fundus. January 21, 2016. http://www.medienwissenschaft.hu-berlin.de/medientheorien/fundus/media-archaeological-fundus.

"Midway—Newest Battle Game!" *Avalon Hill General* 1, no. 3 (1964): 1–2.

Miller, Greg. "Software Trove Is Testament to Its Collector." *Los Angeles Times*, August 12, 1996.

Milner, R., and S. Mayer. "Stella at 20: An Atari 2600 Retrospective." Videotaped interview with Atari engineers filmed in August 1997. CyberPuNKS, 2000.

Monnens, Devin, and Martin Goldberg. "Space Odyssey: The Long Journey of *Spacewar!*, From MIT to Computer Labs around the World." *Kinephanos* (June special issue, 2015): 124–146.

Montfort, Nick. *Twisty Little Passages: An Approach to Interactive Fiction*. Cambridge, MA: MIT Press, 2003.

Morris, Chris. "Video Games Push for Olympic Recognition." CNN Money. May 13, 2006. http://money.cnn.com/2006/05/31/commentary/game_over/column_gaming/index.htm.

Müller, Norbert, ed. *Pierre de Coubertin, 1863–1937: Olympism, Selected Writings*. Lausanne: International Olympic Committee, 2000.

Mulligan, J. "Talkin' 'bout My . . . Generation." Skotos. January 22, 2002. http://www.skotos.net/articles/BTH_17.shtml.

Myers, David. "Chris Crawford and Computer Game Aesthetics." *Journal of Popular Culture* 24, no. 2 (1990): 17–32.

Nelson, Theodor H. *Computer Lib: You Can and Must Understand Computers Now.* N.p.: Hugo's Book Service, 1983.

"New Game History." *American Journal of Play* 10, no. 1 (2017).

Newman, James. *Best Before: Videogames, Supersession, and Obsolescence*. London: Routledge, 2012.

Nintendo v. Magnavox, US District Court, Southern District of New York, documents 81, 100, 112. NARA Central Plains Region, duplicate photocopies of selected records at Stanford University.

Nitsche, Michael. "Demo." In *Debugging Game History: A Critical Lexicon*, edited by Henry Lowood and Raiford Guins, 103–108. Cambridge, MA: MIT Press, 2016.

Nitsche, Michael. Claiming Its Space: Machinima. *Dichtung Digital: Journal für digitale Ästhetik* 37 (2007). http://www.brown.edu/Research/dichtung-digital/2007/Nitsche/nitsche.htm.

Norberg, Arthur L., and Judy E. O'Neill. *Transforming Computer Technology: Information Processing for the Pentagon, 1962-1986*. Baltimore: Johns Hopkins University Press, 1986.

NoSkill Memorial Site. 2004. http://www.doom2.net/noskill/index.htm.

Nutting Associates. "How Computer Space Works and Produces." *Computer Space* flyer, November 1971.

Obrist, Hans Ulrich. *Ways of Curating*. New York: Farrar, Straus and Giroux, 2014.

Olsen, Ken H. to Peter Elias. "The Story of . . . PDP-1." Internal corporate document, Digital Equipment, September 15, 1961. http://research.microsoft.com/~gbell/Digital/timeline/pdp-1story.htm.

Olympic Charter. In Force as from 1 September 2004. Lausanne: International Olympic Committee, 2004.

Olympic Movement. Lausanne: International Olympic Committee, 1993.

Oudshoorn, Kelly, and Trevor Pinch, eds. *How Users Matter: The Co-Construction of Users and Technology*. Cambridge, MA: MIT Press, 2005.

Owen, D. "The Second Coming of Nolan Bushnell." *Playboy*. June 1983.

Owens, Trevor. "Archives, Materiality, and the 'Agency of the Machine': An Interview with Wolfgang Ernst." *The Signal* (blog). February 8, 2013. http://blogs.loc.gov/digital preservation/2013/02/archives-materiality-and-agency-of-the- machine-an-interview-with-wolfgang-ernst/.

Owens, Trevor. "Preserving.exe: Toward a National Strategy for Software Preservation." In *Preserving.exe: Toward a National Strategy for Software Preservation*, 2–3. Washington, DC: Library of Congress, 2013. http://www.digitalpreservation.gov/multimedia/documents/PreservingEXE_report_final101813.pdf?loclr=blogsig.

Owens, Trevor. "Yes, the Library of Congress Has Video Games: An Interview with David Gibson." *The Signal* (blog). September 26, 2012. http://blogs.loc.gov/digitalpreservation/2012/09/yes-the-library-of-congress-has-video-games-an-interview-with-david-gibson/.

Packer, Randall. "Net Art as Theater of the Senses: A HyperTour of Jodi and Grammatron." *Beyond Interface: Net Art and the Art of the Net*. Online Exhibition, Walker Art Center, 1998.

Park, Hyun-Jong. "Sponsoring und Werbung im Sport: Eine Longitudinalstudie zur Sportsponsoringwirkung am Beispiel der koreanischen Elektronik-Konzerns Samsung." Ph.D. diss. University Münster, 1995.

Patent Arcade website. "The Video Game Lawsuits." http://www.patentarcade.com/2005/05/feature-video-game-lawsuits.html.

Patrick, Stephen B. "The History of Wargaming." In *Strategy and Tactics Staff Study Nr. 2: Wargame Design*, 1–29. New York: Hippocrene, 1977.

Parker Pearson, Jamie, ed. *Digital at Work: Snapshots from the First Thirty-Five Years*. N.p.: Digital Equipment, 1992.

Perica, Jon. "Putting More Realism into Tactics II." *Avalon Hill General* 1, no. 4 (1964): 7–12.

Perla, Peter P. *The Art of Wargaming: A Guide for Professionals and Hobbyists*. Annapolis, MD: Naval Institute Press, 1990.

Peterson, Jon. *Playing at the World: A History of Simulating Wars, People, and Fantastic Adventure, from Chess to Role-Playing Games*. San Diego, CA: Unreason Press, 2012.

Peterson, Stacey. "Open to the Public." *Computer System News*, January 2, 1989.

Pitts, Bill. "The Galaxy Game." Computer History Exhibits, Stanford University. October 29, 1997. http://www-db.stanford.edu/pub/voy/museum/ galaxy.html.

Perry T. E., and P. Wallich. "Design Case History: The Atari Video Computer System." *IEEE Spectrum* 20, no. 3 (March 1983): 45–51.

Pope, Tristan. Crafting Worlds Web site. 2005. http://www.craftingworlds.com.

"Preserving Computer-Related Source Materials." Brochure. Reproduced in the *IEEE Annals for the History of Computing* 2 (January-March 1980, orig. 1979): 4–6.

"Preserving Virtual World II: Methods for Evaluating and Preserving Significant Properties of Educational Games (2010–2013)." Stanford Libraries. https://library .stanford.edu/projects/preserving-virtual-worlds.

Pursell, Carroll. *From Playgrounds to PlayStation: The Interaction of Technology and Play*. Baltimore: Johns Hopkins University Press, 2015.

Raessens, Joost, and Jeffrey Goldstein, eds. *Computer Game Studies Handbook*. Cambridge, MA: MIT Press, 2005.

Rasmussen, Frederick N. "Charles S. Roberts, Train Line Expert, Dies at 80." *Baltimore Sun*, August 28, 2010.

Raymond, W. Boyd. "Electronic Information and the Functional Integration of Libraries, Museums, and Archives." In *Electronic Information Resources and Historians: European Perspectives*, edited by Seamus Ross and Edward Higgs, 227–233. St. Katharinen, Germany: Scripta Mercaturae, 1993.

Reddoch, Russell. "Comments on Module *Blitzkrieg*." *S&T Supplement* 2 (1970): 9–11.

Ringel, Matt. "Q + A: GameRiot's Matt Ringel: Diversifying Market Silos." *Electronic Gaming Business*, June 30, 2004. http://findarticles.com/p/articles/mi_moPJQ/is_13 _2/ai_n6093989/pg_1.

Rivlin, Robert. *The Algorithmic Image: Graphic Visions of the Computer Age*. Redmond, WA: Microsoft Press, 1986.

Roberts, Charles S. "Charles S. Roberts: In His Own Words." 1983. http://www.alanemrich .com/CSR_pages/Articles/CSRspeaks.htm.

Roche, Maurice. *Mega-Events and Modernity: Olympics and Expos in the Growth of Global Culture*. London: Routledge: 2000.

Rodriguez, Hector. "The Playful and the Serious: An Approximation to Huizinga's *Homo Ludens*." *Game Studies* 6, no. 1 (December 2006). http://gamestudies.org/0601/articles /rodriges.

Romero, John. *Oral History*. Interviews conducted and edited by Henry Lowood. Publication forthcoming.

Rothenberg, Jeff. *Avoiding Technological Quicksand: Finding a Viable Technical Foundation for Digital Preservation*. Washington, DC: Council on Library and Information Resources, 1999.

Rothenberg, Jeff. "Ensuring the Longevity of Digital Documents." *Scientific American* 272 (January 1995): 42–47.

Ruberg, Bonnie. *The Queer Games Avant-Garde*. Durham, NC: Duke University Press, 2020.

Russell, Steve, quoted in Stewart Brand. "SPACEWAR: Fanatic Life and Symbolic Death Among the Computer Bums." *Rolling Stone*, December 7, 1972; http://www.wheels.org /spacewar/stone/ rolling_stone.html.

Rütimann, Hans. letter of invitation to Henry Lowood, November 19, 1989. Henry Lowood Papers, Stanford University Libraries, Box 3, folder 35.

Salen, Katie, and Eric Zimmerman. *Rules of Play: Game Design Fundamental*. Cambridge, MA: MIT Press, 2004.

Scanlon, Thomas F. *The Olympic Myth of Greek Amateur Athletics*. Chicago: Ares, 1984.

Schaffer, Kay, and Sidonie Smith. "The Olympics of the Everyday." In *The Olympics at the Millennium: Power, Politics, and the Games*, edited by Kay Schaffer and Sidonie Smith, 213–223. New Brunswick, NJ: Rutgers University Press, 2000.

Schechner, Richard. *Performance Studies: An Introduction*. London: Routledge, 2002.

Schweikert, Annie, and Ethan Gates. "EaaSI Case Study #1: The Would-Be Gentleman." Software Preservation Network. January 2021. https://www.softwarepreservationnetwork.org/eaasi-case-study-1-the-would-be-gentleman/.

"Sega and Nintendo Join Forces for Mario and Sonic at the Olympic Games." Press release, March 28, 2007, http://www.nintendo.com/newsarticle?articleid=t5eT-QaVMFMIfXI3Z_abfVviRBskksJr.

Sheff, David. *Game Over: How Nintendo Conquered the World*. Wilton, CT: GamePress, 1999.

Shimer, Eric R. "Meanwhile—Back at Tactics II" *Avalon Hill General* 1, no. 5 (1965).

Shustek, Len. "What Should We Collect to Preserve the History of Software?" *IEEE Annals of the History of Computing* 28 (October-December 2006): 110–112.

Simonsen, Redmond A. "Image and System: Graphics and Physical Systems Design." In *Wargame Design: The History, Production and Use of Conflict Simulation Games*, edited by Staff of *Strategy and Tactics Magazine*. New York: Hippocrene, 1977.

Simonsen, Redmond A. "Physical Systems Design in Conflict Simulations." *Moves* 7 (1973): 22–24.

"Slot Machines and Pinball Games." *Annals of the American Academy of Political and Social Science* (May 1950): 62–70. (Anonymously written.)

Smith, Merritt R., and Leo Marx, eds. *Does Technology Drive History?: The Dilemma of Technological Determinism*. Cambridge, MA: MIT Press, 1994.

Snoman [Nicholas Fetcko]. *Wandering Dreamscape*. Warcraft Movies. June 29, 2007. http://www.warcraftmovies.com/movieview.php?id=42529.

Snoman [Nicholas Fetcko]. *Ephemeral Dreamscape*. Warcraft Movies. August 28, 2007. http://www.warcraftmovies.com/ru/movieview.php?id=45896.

Specht, Robert D. Specht. "War Games." RAND Report P-1041, March 18, 1957.

Stahl, Roger. *Militainment, Inc.: War, Media, and Popular Culture*. New York: Routledge, 2010.

Sudnow, David. *Pilgrim in the Microworld*. New York: Warner Books, 1983.

Suominen, Jaakko. "How to Present the History of Digital Games: Enthusiast, Emancipatory, Genealogical, and Pathological Approaches." *Games and Culture* 12, no. 6 (2016): 544–562.

Surette, Tim. "Olympic Torch Burns for Gaming." *Gamespot News*, May 31, 2006. http://www.gamespot.com/news/6152088.html.

Sutton-Smith, Brian. *Toys as Culture*. New York: Gardner Press, 1986.

Swade, Doron. "Collecting Software: Preserving Information in an Object-Centred Culture." In *Electronic Information Resources and Historians: European Perspectives*, edited by Seamus Ross and Edward Higgs. St. Katharinen, Germany: Scripta Mercaturae, 1993.

Swade, Doron. "Preserving Software in an Object-Centred Culture." In *History and Electronic Artefacts*, edited by Edward Higgs, 195–206. Oxford: Oxford University Press, 1998.

Swalwell, Melanie, ed. *Game History and the Local*. London: Palgrave Macmillan, 2021.

Swalwell, Melanie. *Homebrew Gaming and the Beginnings of Vernacular Digitality*. Cambridge, MA: MIT Press, 2021.

Tactics II. Designed by Charles Roberts. Avalon Hill Company, 1958.

Taylor, T. L. *Raising the Stakes: E-Sports and the Professionalization of Computer Gaming*. Cambridge, MA: MIT Press, 2012.

Taylor, T. L. *Watch Me Play: Twitch and the Rise of Game Live Streaming*. Princeton, NJ: Princeton University Press, 2018.

"Test Series Games." *S&T Supplement* (December 1969-January 1970).

"The Ultimate Wargame." *Avalon Hill General Special Issue* 59 (1988).

Tobin, Samuel. "Save." In *Debugging Game History: A Critical Lexicon*, edited by Henry Lowood and Raiford Guins, 385–391. Cambridge, MA: MIT Press, 2016.

Tran, Khanh T. L. "U.S. Videogame Industry Posts Record Sales." *Wall Street Journal*, February 7, 2002.

Turner, Fred. *From Counterculture to Cyberculture: Stewart Brand, the* Whole Earth *Network, and Rise of Digital Utopianism*. Chicago: University of Chicago Press, 2006.

"TV's Hot New Star: The Electronic Game." *Business Week—Industrial Edition*, December 29, 1975.

Urban, R. J. "Gambling Today via the 'Free Replay' Pinball Machine." *Marquette Law Review* 98 (Summer 1958). http://scholarship.law.marquette.edu/mulr/vol42/iss1/12.

Valdespino, Anne. "The Big Screen Keeps Pulling Us In." *Los Angeles Times*, July 1, 2002.

Valve LLC. "Valve Unveils Steam At 2002 Game Developer's Conference." Press release. March 21, 2002.

Vanore, John J. "Interview: Charles S. Roberts—Founder of the Avalon Hill Game Company and Founding Father of Board Wargaming." *Fire and Movement* 56 (1988).

Vintage Gaming Network. "Al Alcorn Interview." Vintage Gaming. http://atari.vg-network .com/ aainterview.html.

Von Hippel, Eric. *Democratizing Innovation*. Cambridge, MA: MIT Press, 2005.

Wagner, Christopher. "Background on S&T Nrs. 16 & 17." *Strategy & Tactics. Book IV: Nrs. 16–18*. N.p: Simulations Publications, n.d.

Walker, John. Introduction to *The Analytical Engine: The First Computer*. Fourmilab. March 21, 2016. http://www.fourmilab.ch/babbage/.

Waxman, Sharon. "Hollywood's Great Escapism: 2001 Box Office Receipts Set a Record." *Washington Post*, January 4, 2002.

Weiser, Mark. "Computer for the 21st Century." *Scientific American* (September 1991): 94–104.

Weiser, Mark. "Ubiquitous Computing." http://www.ubiq.com/hypertext/weiser /UbiHome.html.

"Welcome to the *DOOM* Honorific Titles!" *DOOM* Honorific Titles. http://www-lce.eng .cam.ac.uk/~fms27/dht/dht5/#dht5.

Whannel, Garry. "Television and the Transformation of Sport." *Annals of the American Academy of Political and Social Science* (September 2009): 205–218.

Wheatley, Paul. "Digital Preservation and BBC Domesday." Paper presented at the Electronic Media Group, Annual Meeting of the American Institute for Conservation of Historic and Artistic Works, Portland, Oregon, June 14, 2004.

White, Douglas. "NIST National Software Reference Library (NSRL) Efforts in Preserving Software in the Stanford University Libraries (SUL) Cabrinety Collection." Video. September 11, 2013. https://www.loc.gov/preservation/outreach/tops/white/white.html.

White, Hayden. *The Content of the Form: Narrative Discourse and Historical Representation*. Baltimore: Johns Hopkins University Press, 1987.

White, Hayden. *Figural Realism: Studies in the Mimesis Effect*. Baltimore: Johns Hopkins University Press, 2000.

White, Hayden. "The Historical Text as Literary Artifact." *Clio* 3, no. 3 (1974): 277–303.

White, Hayden. "The Question of Narrative in Contemporary Historical Theory." *History and Theory* 23, no. 1 (February 1984).

Wing, Carlin, Raiford Guins, and Henry Lowood. *EA Sports FIFA: Feeling the Game*. New York: Bloomsbury Press, 2022.

Wolf, Mark J. P., ed. *The Medium of the Video Game*. Austin: University of Texas Press, 2001.

Woods, Stewart. *Eurogames: The Design, Culture and Play of Modern European Board Games*. Jefferson, NC: McFarland, 2012.

Yost, Jeffery R. "From The Editor's Desk." *IEEE Annals of the History of Computing* 31, no. 3 (2009).

Zocchi, Lou. "Gettysburg." In *Hobby Games: The 100 Best*, edited by James Lowder. Seattle: Green Ronin, 2007.

Index